REGIONAL OPPORTUNITIES FOR SUSTAINABLE DEVELOPMENT

STUDIES IN ECOLOGICAL ECONOMICS

VOLUME 1

Regional Opportunities for Sustainable Development

Theory, Methods and Applications

by

H.J. DE GRAAF, C.J.M. MUSTERS and W.J. TER KEURS

Environmental Biology,
Institute of Evolutionary and
Ecological Sciences,
Leiden University,
Leiden, The Netherlands

KLUWER ACADEMIC PUBLISHERS
DORDRECHT / BOSTON / LONDON

A C.I.P. Catalogue record for this book is available from the Library of Congress.

ISBN 978-90-481-5213-1

Published by Kluwer Academic Publishers,
P.O. Box 17, 3300 AA Dordrecht, The Netherlands.

Sold and distributed in North, Central and South America
by Kluwer Academic Publishers,
101 Philip Drive, Norwell, MA 02061, U.S.A.

In all other countries, sold and distributed
by Kluwer Academic Publishers,
P.O. Box 322, 3300 AH Dordrecht, The Netherlands.

Printed on acid-free paper

CONTENTS

Acknowledgements

In this book we have attempted to develop theoretical concepts and design methods aimed at identifying regional opportunities for sustainable development, and to apply those concepts and methods to real-life decision-making. The support that we received during this study focused on the theoretical aspects, the applications, or the work as a whole.

We are grateful to Gart Zweers and Wim Hafkamp for their encouragement and insightful commentary, and Wouter de Groot, Jan Schoonenboom and Wiero Beek for their comments on earlier versions of this book.

Our theoretical and methodological elaboration of the idea of 'sustainable development' would not have been possible without the foundations laid in the early eighties by the work of Wouter van der Weijden, Harm van der Wal, Nico van Brussel and Theo Bakker within the project 'Integrated Agriculture'. Nor would it have been possible without the help of the Ministry of Housing, Physical Planning and Environment. This entailed much more than financial support alone. The discussions with Lelia Oosterhuis, Rob Brinkman, Sylvie Warmerdam and Ruben Huele were extremely fruitful. The views put forward within the Netherlands Scientific Council for Government Policy were inspiring. We are grateful to Oskar de Kuijer and Henk van Latensteijn for their critical comments. And, of course, there were numerous discussions with our colleagues and students at Leiden University. Marcel Dix, Erik Gertenaar, Jeroen Hendriks, Tom van der Meij, Marcel Noordervliet, Ton Orleans, Gert-Jan Oudshoorn, Ronald de Schrijver, Jory Sjardijn and Paul Vos made an indispensable contribution to the realization of this volume.

The analyses of the proposed methods were made possible by the Ministry of Education, Culture and Sciences and the Zuid-Hollandse Milieu Federatie. We would especially like to thank Hanneke op den Brouw, Bert Bakker, René Walenberg and Arno Steekelenburg for their cooperation. Harm van der Wal, Don Jansen, Emmy Bolsius, Jan Vreke and David Verhoogh commented on earlier versions of the descriptions of these analyses. Koos Poot and Cees van Diepen were prepared to furnish additional information at a moment's notice.

The Interdepartmental Program for Sustainable Technological Development enabled us to apply our methods in a real-life decision-making situation. This was done in close collaboration with the project team 'Sustainable Multiple Land-use' and the members of the working group 'Development and Monitoring'.

And last but not least, we would like to thank those who helped us to present our findings in readable English: Barbara Fasting and Max Brecher.

Research for this book was carried out with the support of the Ministry of Housing, Physical Planning and Environment (Chapter II-V). The analysis of proposed methods and procedures in Chapter IX and X was supported by the Ministry of Education, Culture and Sciences. The Interdepartmental Program for Sustainable Technological Development made it possible to apply the proposed procedures for sustainable development (Chapter XI-XIII).

CHAPTER I

Introduction and Survey

CONTENTS

I.1 Introduction

In adopting Agenda 21 (UNEP, 1992), most of the world's governments have acknowledged that "*the ecosystems on which we depend to sustain life on earth*" are in danger, and that national policies should be directed towards sustainable development. What is needed now is a policy change aimed at "*a development that meets the needs of the present without compromising the ability of future generations to meet their own needs*" (WCED, 1987). This will mean creating the necessary conditions to meet the needs and wants of all the people involved, here and elsewhere, in the present and in the future.

Decision making aimed at such a development faces great uncertainties, not least with regard to the natural resources assumed to be available. The predictions made in the last twenty to thirty years concerning resource depletion and food shortages have proved to be wrong (Anonymous, 1997). There are fundamental uncertainties about the capacity of ecosystems to assimilate environmental impacts (WRR, 1995; Tjallingii, 1996). Moreover, it is impossible to say what people will need and want in the future. In the striving to meet all the needs and wants of present and future generations, sustainable development will always involve environmental and social risks which must be taken into account (WRR, 1995; Lélé & Norgaard, 1996). Thus decision making on sustainable development will inevitably be accompanied by great "*uncertainties about guiding values*", "*uncertainties about the working environment*" and "*uncertainties about related decisions*" (Friend & Hickling, 1987). According to Friend & Hickling (1987), this will require "*clearer objectives*", "*more information*" and "*more coordination*", although obviously it may also lead to *ad hoc* decisions. However, if policy is not to be reduced to 'trial-and-error' decision making, the concept of sustainable development will require not only new policies, but also information-gathering procedures that are tailored to these policies (Pezzey, 1992).

The call for new policies marks a shift in the way the environment is viewed (Leroy, 1994). Awareness of environmental problems began to dawn around the second half of the twentieth century, with dire warnings from scientists (e.g., Carson, 1962 ; Meadows *et al.*, 1972). In the early seventies, the focus shifted to problem solving, resulting in countless government reports and memorandums (Leroy, 1994). The WCED (1987) initiated another change by shaping new policies aimed at economic, social, cultural, and ecological goals, while at the same time solving existing environmental problems and preventing future ones.

The need to realize this broad spectrum of goals is pressing, in view of the millions of people whose socioeconomic prospects are poor, and who are endangered by environmental hazards, while the human, financial and biophysical resources of our earth are neither unlimited nor equally distributed (WRI, 1996; Brown *et al.*, 1998). The scarcity of resources demands not only a change in the way the present generation utilizes the resources presently known, but also responsibility for the resources that have not yet been valued, or that may be lost due to side effects of the present use of the environment. Therefore, sustainable development encompasses a value-oriented approach, which addresses the present needs and wants of the people involved, and a 'resources-and-use' approach for the future. Decisions on sustainable development should take into account the possibilities to realize certain values, as well as the way resources can be used to do so. Thus in effect, the information on which sustainable development is based is about opportunities, i.e., about which values are attainable and how best to proceed.

I.2 Information tailored to decision making

I.2.1 Conventional information-gathering

Assuming that decision makers need information on opportunities for development, the question is whether conventional ways of collecting and disseminating information on values and resources are sufficient for this purpose. Do publications devoted to the state of the environment provide that information (e.g., Weinreich & Musters, 1989; RIVM, 1991; Heywood, 1995; WRI, 1996; Costanza *et al.*, 1997; Brown *et al.*, 1998)? Admittedly, even general information has the power to awaken public awareness that "*something has to be done*" (Ester & Mandemaker, 1993). However, the shift from 'awakening awareness' to 'creating policy opportunities' for decision makers entails a change in the type of information needed. Often publications such as those referred to above are unable to make it clear to the decision makers how their decisions on the national or regional level affect the values described. The information collected does not "*fit the perceived needs of those making decisions*", because it is not written from the perspective of the decision makers (Underwood, 1995; Risser, 1996); it cannot be regarded as 'decisive information' which makes it possible to utilize opportunities (Bingham *et al.*, 1995). The uncertainty on this point increases the call for information (Friend & Hickling, 1987).

The possible effects of decisions have been often studied by forecasting (Van Doorn & Van Vught, 1978; Anonymous, 1997). Does conventional forecasting provide the required information on opportunities for development? Applications of this technique are also aimed at 'awakening awareness', because they are based on extrapolations of past developments and existing practices (e.g.,

Meadows *et al.*, 1972; Van der Meij *et al.*, 1995; Rotmans & De Vries, 1997). Although the technique may present different scenarios for the future based on different constraints, it tends to ignore the opportunities that may arise in the future (Schoonenboom, 1995; Anonymous, 1997).

To find new opportunities which take as their starting point the goals to be reached, a 'backcasting' approach would be more appropriate. This technique aims at sketching a potential future on the basis of the required values - regardless of whether these can be realized by means of the present technology - and exploring the opportunities for reaching such a future. A strategy is then designed to make use of these opportunities (WRR, 1983; Schoonenboom, 1995). The focus is squarely on identifying the changes needed to meet the goals which have been set, rather than projecting the past, incorporating assumed developments and measures, and calculating the impacts. In this sense, backcasting makes possible a value-oriented approach. It can be used to focus on what needs to be changed in the available resources and their use, and how to bring about those changes.

I.2.2 The shift required in information-gathering

In the interest of sustainable development, decision makers must be given information that is immediately applicable to the decisions they are able to make regarding the opportunities for development. Providing them with information on values and resources that falls outside the range of their 'options for steering' is useless in that context and possibly even counter-productive, confusing both the decision makers and the issues. At present, conventional information-gathering does not seem capable of providing the desired type of information.

The selection of relevant information can be facilitated by designing procedures for assessment (Bingham *et al.*, 1995). For purposes of sustainable development, such a procedure should be capable of defining economic, social, cultural and ecological information pertaining to the future as well as the present (WCED, 1987; Giampietro, 1997). This information must cover not only values, but also the available resources and their use, and must be suitable both for the system whose development is to be steered and for its context (Giampietro, 1994). Ideally such a procedure will be embedded in protocols that allow clear, efficient and well-supported decision making (Bingham *et al.*, 1995).

The aim of this volume is to improve existing procedures by giving them a sound and appropriate methodological foundation. To that end, concepts will be developed, methods designed, procedures for information-gathering elaborated, and specific case studies analyzed, in order to evaluate proposed methods and procedures.

and the measures to be taken can be derived. Once measures have been implemented, a new information-gathering phase starts: monitoring the effects of the measures, in effect, evaluating the development (Vos *et al.*, in prep.).

A procedure such as the above lends structure to decision making. Many different decisions must be made during each stage of the procedure, each with its own consequences for the subsequent stages. Ultimately, all decisions have political consequences, and should be made by, or in consultation with, representatives of the people involved. However, the manner in which the participation of the people involved is embedded in the information-gathering will depend largely on the specific application of the procedure. In order to gain insight into the various possibilities, three different ways of applying the proposed information-gathering procedure are distinguished, saying from applications with less direct participation of the people involved to those with quite a lot of participation:

- *Support for scientific research.* The methodology to be developed must be usable in the search for new opportunities for development. Furthermore, it should be suitable to serve as a model for analysing existing plans for sustainable development. In this research application, the people involved or their representatives can be consulted or can even participate in information-gathering and selection. Consultation can also be restricted to studying documents pertaining to the needs and wants of the people involved.
- *Support for parties involved in the debate on sustainable development.* The procedure should be suitable as an instrument to support parties developing their own strategy, for example for purposes of a debate on the development of a sector or region. Since parties, especially non-governmental organizations, may have limited time and resources for gathering information, the method should be applicable at low resolutions without losing its accuracy.
- *Support for sectoral, regional and local planning.* The procedure should be a model for the planning of developments. In this kind of application, participation can play an important role. Representatives of the people involved may not only be consulted, they may also become part of the group that makes the decisions on information-gathering and the subsequent stages in the procedure.

As already noted, uncertainties about goals, the environment and the impact of decisions may lead to a call for clearer objectives, more information, and more coordination. Satisfying the need for more information requires a scientifically sound research method in which step-by-step detailed information is gathered about goals, available resources, and the impact of coordinated steering. This may be difficult to realize in real-life applications. As Friend & Hickling (1987) observe, the process of planning developments tends to be non-linear. In this volume, the design of the procedure starts with the development of concepts and methods of scientifically sound information-gathering. These are embedded in a procedure that makes possible iterative planning processes. These are then analyzed in the light of practical applications.

A study like this is by it very nature interdisciplinary (W.T. de Groot, 1992), i.e., it is based on concepts and methods from different disciplines and techniques, such as ecosystem theory, planning theory, economic valuation, land evaluation, future research, etc. As a consequence, the focus is on connecting these diverse concepts and methods to form a new and coherent scheme for searching for sustainable development. The information from different disciplines and application is selected at a detail level that is appropriate for the development of integrated methodology and procedures.

I.4 Survey

Part One starts with a discussion on the major issues involved in a search for *sustainable development* (Chapter II). The Chapters III-VI are devoted to theoretical and conceptual issues pertaining to the collection of information for purposes of sustainable development (Box XIV.1 in Chapter XIV gives an overview of the definitions and concepts elaborated in this part).

Part Two focuses on examining the strengths and weaknesses of the procedure for information-gathering (Chapter VII-X). Can the conceptual frame, methodology, and the procedure proposed in Part One provide the information they are supposed to provide and can they do so within the temporal and financial constraints of real-life applications?

Part Three describes a real-life application in which the information-gathering procedure is embedded in a decision making process in the search for new multiple sustainable land use activities. Can the procedure developed in Part One and modified on the basis of the analyses in Part Two retain its coherence, i.e., does it have the flexibility to be modified in response to the actual decisions taken, without losing its consistency (Faludi, 1987; Tjallingii, 1996)? Although the application was not aimed at regional planning or design, there are certain similarities, since the application was embedded in decision making on the development of new land-use activities (Musters & De Graaf, 1996; De Graaf & Musters, 1997; Kuijer *et al.*, 1997a). The application is a logical intermediate step between the simulated applications in Part Two and a real-life planning application on a regional or sectoral level. The study is introduced in Chapter XI, and the applications of the procedure are described in Chapter XII and XIII.

And finally, in Part Four (Chapter XIV), the results and conclusions of the three parts are summarized. It ends with a discussion of some fundamental issues encountered during this study.

PART ONE

Theory

Part One is devoted to theoretical and conceptual issues pertaining to the collection of information for purposes of sustainable development. It starts with a discussion about the major issues involved in a search for sustainable development (Chapter II). In Chapter III, the problems and possibilities of defining the object to be steered as a socio-environmental system are investigated. Chapter IV aims at the development of a framework that relates the various types of needs and wants of the people involved to a defined system. This means the needs and wants of the people involved both directly and indirectly, and those of both present and future generations. In Chapter V a method is developed for measuring available resources: the resource state of the system. The focus is also on the relation between the resources measured and their use. Chapter VI integrates the elements from the preceding chapters into the outlines of an information-gathering procedure embedded in a procedure for steering developments. It incorporates the methods which have been developed into a logical sequence of phases.

CHAPTER II

Sustainable Development: Looking for New Strategies[1]

CONTENTS

[1] *Reprinted from Ecological Economics 16: Graaf, H.J. de, C.J.M. Musters & W.J. ter Keurs (1996), Sustainable development: looking for new strategies, pp: 205-216, copyright 1996, with permission from Elsevier Science.*

II.1 Introduction

Until recently, national and international policy makers have appeared to be capable of recognizing an environmental problem only when it was already almost impossible to solve. While acid rain was discovered in Manchester, England in 1850, and the first book on the subject appeared in 1872, it took until 1985 for a treaty designed to reduce international air pollution in Europe to be ratified. By that time, one quarter of the lakes in Scandinavia were 'dead' (Ponting, 1991).

Can the careful planning, monitoring and steering of human activities prevent such scenarios in the future? In 1987, the World Commission on Environment and Development (WCED) suggested 'sustainable development' as a possible solution. The WCED realized that environmental problems cannot be isolated from other problems, such as poverty and social disintegration. It saw sustainable development as an option that minimizes the risk of creating new problems or exacerbating existing ones, in other words, "*a development that meets the needs of the present without compromising the ability of future generations to meet their own needs*".

Since then, sustainable development has evolved into an umbrella concept, encompassing all issues pertaining to the interrelationship between the environment and human development (Dovers & Handmer, 1993; Frazier, 1997). However, this raises a number of questions. What exactly is sustainable development? What kind of development are people talking about? What must be developed in a sustainable way? Is it the earth, mankind, a country, a society, or sectors such as agriculture or transport? How long must a development be sustainable? How can today's decision makers take the needs of future generations into account?

Many politicians and scientists have tried to answer these questions (e.g., Archibugi & Nijkamp, 1989; Goodland *et al.*, 1991; Kuik & Verbruggen, 1991; Musters & De Graaf, 1992; Pezzey, 1992; Dutch Committee for Long-Term Environmental Policy, 1994; Beckerman, 1994; WRR, 1995). The strategies proposed for sustainable development range from 'business as usual' (Beckerman, 1994) to policies to control population growth (e.g., Myers, 1993), from changing attitudes (Mohr, 1994) to turning economic growth into increased prosperity (e.g., Parker, 1993). But do these strategies really reduce the risk of future problems?

In order to select a strategy, it is important to identify the type of problem to be solved. Economists often regard environmental problems as simply the unwanted side effects of economic activities, i.e., as negative externalities (e.g., Pezzey, 1992). But not only production activities have negative side effects. Any activity, whether its goals are economic, social, cultural, or ecological, can have negative side effects, leading to economic, social, cultural, or ecological problems. How can problems linked to the use of our environment be prevented, given the complexity of the relationship between activities and problems? Measures to steer isolated activities are not appropriate, since that would rule out solutions based on optimal combinations of activities. When the focus is on solving existing problems, these optimal combinations can be taken into account, but such an approach does not always prevent new problems from arising.

This chapter is in effect a search for the broad outlines of a comprehensive strategy for sustainable development. This strategy should cover all important activities, together with their relevant side effects. It should also be effective, i.e., not only acceptable to, and supported by, as many of the people or parties involved as possible, but also capable of dealing with the often substantial differences between the power and influence of the various groups and institutions. Are any such strategies currently available? If not, can they be developed?

The analysis starts with a discussion of traditional sets of instruments. Three types can be distinguished: legal instruments, economic instruments, and instruments aimed at voluntary participation (Van Ierland, 1993). It is obvious that when economic and legal instruments prove ineffective, voluntary participation becomes more important. An analysis of the effectiveness of the first two will make clear to what extent voluntary participation is crucial.

We will examine the legal and economic instruments available to deal with environmental problems (Section II.2); both of these have previously also been discussed by Rennings & Wiggering (1997). This will provide insight into the problems and questions encountered during the search for strategies aimed at sustainable development (Section II.2.3). The discussion centres on possible ways of solving these problems, and the specific knowledge and skills required (Section II.3). The chapter ends with a number of conclusions about the research needed to develop methodological concepts for the collection of information (Section II.4).

II.2 Analyzing traditional instruments for solving environmental problems

As noted above, several strategies have been proposed to meet the demands of sustainable development, most of which acknowledge that the limits of our environment have been reached. Obviously, ways must be found to remain within those limits. Two possible approaches will be discussed here.

In the first approach, it is assumed that any human society is part of, and dependent on, an ecosystem (e.g. Rees, 1990; Goodland, 1991; Meadows *et al.*, 1991). Since the ecosystem places its own constraints on the development of that society, the carrying capacity of the ecosystem must be established. This carrying capacity can then be used to set limits to known activities, with a view to attaining sustainability.

In the second approach, it is assumed that if environmental decline is regarded solely as a cost factor, then Western economics is quite capable of dealing with environmental problems (Goodland *et al.*, 1993; Common *et al.*, 1993). The fact that human society is capable of self-regulation is considered the driving force behind sustainable development. If economics incorporates the environment into cost-benefit calculations, then economic development will become sustainable.

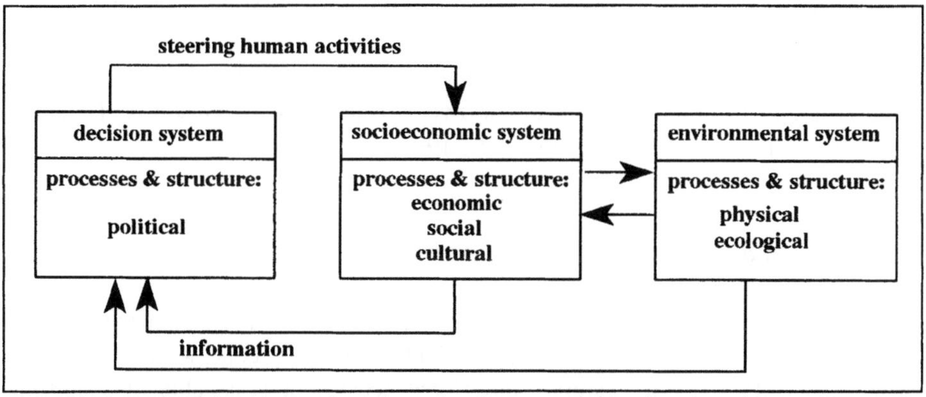

Figure II.1: Traditional strategies of steering human activities based on information on the socioeconomic and environmental systems (adapted after R.S. de Groot, 1992)

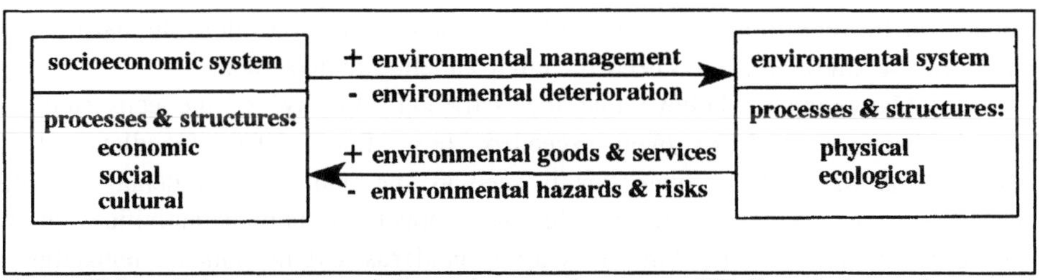

Figure II.2: Relations between the socioeconomic system (SEcS) and the environmental system (EnS) (adapted after R.S. de Groot, 1992).

Both approaches are based on information about the environment and are intended to bring about a change in human activities, with a view to influencing interactions between the environment and the socioeconomic system (Figure II.1). In Figure II.2, the relations between the two systems are specified as the impact of the socioeconomic system on the environment in terms of management (+) and deterioration (-) or, conversely, as the impact of the environment on the socioeconomic system in terms of the provision of goods and services (+) and hazards and risks (-) (adapted after R.S. de Groot, 1992). The two approaches require different types of instruments to influence these relations. The first approach calls for legal limits to be imposed on human activities, in order to reduce the unlimited use of environmental goods and services, and thus the negative impact on the environment. In the second approach, the emphasis is on pricing both the use of environmental goods and services and the negative environmental impact.

These two approaches, which in the past have often been effective in solving existing problems (WRR, 1992a; Von Weizsäcker & Schmidt-Bleek, 1994), have more in common than is generally recognized. For the sake of brevity, only the contrasting instruments - imposing legal limits and pricing - will be discussed here, although in practice a mix of the two is often seen (Atkinson & Tietenberg, 1991; Van Ierland, 1993; Shogren *et al.*, 1993).

II.2.1 Maintaining carrying capacity by imposing legal limits

In the 'legal limits' approach, it is assumed that ecosystems and populations of organisms have only a limited capacity to cope with environmental stress (Rees, 1990). For example, the global ecosystem is capable of absorbing a certain determinable amount of pollution, and this amount cannot be exceeded without catastrophic consequences. In terms of Figure II.2, this means that if it is possible to keep the use of environmental goods and services and the negative impact on the environment within certain limits, then environmental hazards and risks are low and there is no need for environmental management.

Many experts believe that the 'carrying capacity' of the ecosystem has already been reached, or even exceeded, (e.g. Rees, 1990; Goodland, 1991; Meadows *et al.*, 1991). Since this is seen as the main environmental problem, measures must be taken to reduce the impact of human activities on the environment. This can be done by setting standards and introducing legislation to reduce the material consumption per capita, enforce the use of less wasteful and polluting technologies, and reduce population growth (Goodland *et al.*, 1993).

The success of this strategy depends, first of all, on convincing the public, policy makers and all other parties involved that the limits of our environment have been reached. The carrying capacity of the environment for all possible human activities must be established. Although many people have emphasized how crucial this is, to our knowledge the necessary instruments and methods have not yet been found or have proved to be inaccurate (Anonymous, 1997). This means that the limits are still unknown. They cannot be analyzed by means of science alone, because they also depend on human needs and wants, and thus on political choices (Haila & Levins, 1992; Musters *et al.*, 1994b; Cohen, 1997).

Furthermore, even if the limits were known, it may be difficult to convince the parties involved to accept and respect them. For example, a population confronted with food shortages will try to increase production, in spite of the exhaustion of the soil. In this kind of situation, environmental limits are not the paramount issue. The people involved will play down the problem of soil exhaustion, saying that it is 'not as bad as it seems'. Therefore, setting legal limits to human activities usually has to be accompanied by a costly system designed to uphold the law (WRR, 1992a).

There are several other reasons why this approach does not work. In practice, it is difficult to achieve the desired changes in population growth, technology, and material and energy consumption through legislation alone. First of all, population policy in itself is insufficient. Both "*developing and developed countries are experiencing major difficulties in sustaining their population on their environmental resources*" (Myers, 1993). Despite all their efforts, some of them quite extreme, the Chinese authorities are still confronted with population growth (WRI, 1994). Moreover, material and energy consumption is still on the increase, even in rich countries that have the know-how and resources to do something about environmental problems. And yet, to date, there has been no move to alter 'resource-wasteful life-styles' (Myers, 1993). And, finally, the availability of less polluting technologies does not automatically guarantee that they will be used. Arguments other than 'technical availability' may be decisive here (Opschoor & Van der Straaten, 1993). For example, the implementation of new technologies may be too expensive or 'non-competitive'.

To sum up, it is difficult to demonstrate that environmental limits exist and, if they do, what they are. It is perhaps even more difficult to convince people that those limits must be respected, and to provide the necessary instruments to do so. The ' legal limits' approach overestimates the present level of knowledge on human carrying capacity, and underestimates the importance of socioeconomic factors.

II.2.2 Pricing the environment

Economic activities can cause environmental problems. Using the goods and services of the environment can result in depletion and other negative environmental impacts. As a rule, this does not entail any immediate cost, since large parts of the environment are either not 'owned' by anyone, or have disputed ownership. Moreover, the value of the environment cannot easily be translated into financial terms (Pezzey, 1992). According to the laws of economics, unpriced production factors can, and therefore will, be consumed without restriction

(Kozlowski, 1993; Dietz, 1994). Thus pollution and the wastage will go on as long as they are - technically speaking - 'unpriced'. However, as soon as they cease to be free and 'freely available', activities will change. Politicians will make other decisions, new consumption patterns will arise, and producers will search for new technologies. According to this view, pricing the environment is the obvious solution (Common *et al.*, 1993; Goodland *et al.*, 1993). This approach assumes that if environmental goods and services and the impact on the environment are priced, environmental hazards and risks will decline, and funds will be raised to pay for environmental management (Figure II.2). However, it has been argued that if the 'discount rate' is higher than the 'growth rate' of a resource, that resource will be exploited to extinction even if ownership exists and the resource is properly priced (Clark, 1973; May, 1994).

Pricing the environment is not the same in macro- and micro-economics. In macro-economics measures are often proposed to price the loss of natural resources and incorporate those losses into the Gross National Product (GNP) (Hueting, 1990; Goodland *et al.*, 1993; Cobb *et al.*, 1995). If this can be done, the GNP will then become an indicator of economic growth and prosperity (Hueting, 1990). 'Greening' the GNP can help to inform politicians about the overall changes in man-made and natural capital.

The main problem in correcting the GNP is that the price of many natural resources is not set by the market. It has to be determined indirectly, and this is only possible where there are priced substitutes for those resources (Booth, 1994). Unique resources - such as rare species and landscapes - are difficult to price. The Contingent Valuation Method (CVM), which is aimed at measuring the price of such natural values (Harris & Brown, 1992), has not been able to tackle this problem satisfactorily (Hampicke, 1994). Although it may be possible to determine what people are prepared to pay for the conservation of individual species, the CVM cannot determine whether the financial generosity exists to save all species. The price of separate species cannot simply be added up. The CVM can be used to ask what people are prepared to pay for all species together, but then a - political - decision will be required to answer the question What is meant by 'all species'? Quite apart from these problems, it must be acknowledged that some authors regard it as unacceptable to price nature, because this implies a denial of its inherent and unique value (Hourcade *et al.*, 1992; Redclift, 1993).

Another possible pricing strategy is based on the cost of restoration. However, this is often impossible, as in the case of extinct species or the casualties of environmental disasters. Hueting (1990) has proposed a system of pricing based on the cost of the measures required to meet biophysical standards for environmental functions. Applying this approach means that these standards must be assessed, for example, by determining the 'environmental utilisation space'

(EUS) (Opschoor & Weterings, 1994). But can the EUS be established? The same problems arise as in the assessment of the limits discussed in Section II.2.1. It is impossible to determine an EUS without political decisions (Musters *et al.*, 1994b).

In micro-economics pricing results in a form of taxation - either positive or negative - which is imposed on resources, production processes, and products, in order to change production and consumption patterns (Baumol and Oates, 1988; WRR, 1992a). The tax need only be high enough to steer the economic actors in the appropriate direction. The details of taxing and the appropriate tax level can be estimated and adjusted in a trial-and-error process.

The level of taxation is likewise a political issue, which means that it will be necessary to get the support of the people involved. It is important that the money collected from this kind of tax is either put back into the market or used to restore abused environments. Otherwise, the government would gain from the transgressions of others, and might be perceived as a transgressor itself (Shogren *et al.*, 1993).

Taxes can be highly effective in changing behaviour (WRR, 1992a). But this kind of financial instrument relies heavily on a strong economy. What happens when the economic system is under pressure? What happens when important social issues, such as unemployment and social security, are at stake? In such cases, the imposition of taxes as sanctions becomes a political issue.

The conclusion is that, although pricing may be an important instrument for the internalization of environmental issues, it is not sufficient. This approach underestimates the importance of 'priceless' values such as culture, nature, or landscape, while overestimating the possibilities of pricing under difficult political and social circumstances.

II.2.3 Problems encountered

The above discussion illustrates the shortcomings of the two types of instruments. They are aimed at one goal, in that the instruments are designed almost exclusively to solve existing environmental problems. This disregard for other goals can make sustainable development impossible. If tight ecological standards result in poverty, or if the pricing of natural resources causes unemployment, sustainable development will not be realistic.

Many authors recognize that the instruments proposed for sustainable development are only partial solutions (e.g., Von Droste & Dogsé, 1991; Goodland *et al.*, 1993; Peterson & Peterson, 1993; Robinson, 1993; Balstad Miller, 1994; Vatn & Bromley, 1994; Upreti, 1994). Other propose integrated models or multiple-goal techniques (Munda *et al.*, 1994; Ojima *et al.*, 1994; Riebsame *et al.*,

1994). What is more important is the fact that many researchers seem to consider sustainable development a purely technical problem (e.g., Kozlowski & Hill, 1993; Goodland & Daly, 1996; Robèrt *et al.*, 1997). And yet, as we have seen, political decisions are crucial in assessing carrying capacity, prescribing legal limits, adjusting the GNP, and imposing taxes. To our knowledge, no approaches have been published that also cover the decision making process (also noted by Hourcade *et al.*, 1992, and Redclift, 1993), although the relationship between environmental planning and policy-making is occasionally mentioned (e.g., Pearce, 1993; Dutch Committee for Long-Term Environmental Policy, 1994; WRR, 1992a & 1995).

Once the decision making has become part of a strategy for sustainable development, the major questions are as follows. How are goals for development to be set that are supported by all parties involved in decision making? Who will manage the process in order to ensure that the necessary measures will be accepted? And how is the information needed for decision making to be gathered? These considerations - which will be taken up in the following section - must be part of a new strategy for sustainable development, not only because decisions on sustainable development are political decisions, but also because they affect the more technical aspects of a strategy. Workable instruments will only be produced if it is made explicit who is to work with them, and how they are to be used.

II.3 Towards a new strategy

Environmental problems can only be prevented if economic, social, cultural and ecological aspects are taken into account (WCED, 1987, and many subsequent authors). And this approach will only be successful if economic, social, cultural, and ecological processes and structures are regarded as interrelated parts of the same system: the *SocioEnvironmental System* (*SEnS*) (Figure II.3). Consequently, decisions pertaining to sustainable development must be made on the basis of such unified systems. This decision making involves information-gathering, consensus-building, goal-setting, steering and evaluation. The object to be steered will be defined as a *SEnS*, and the goals set will be concerned with the characteristics of this system. To avoid any possible misunderstanding, decision making is placed outside the system, in order to be able to define the object to be steered: the *SEnS*. The *decision making system* encompasses the processes and structures needed for decision making and steering. Not all decision makers will be actors *in* the *SEnS*, and not all actors *in* the *SEnS* will necessarily take part in decision making.

As a political item, goal-setting should be the result of a serious debate

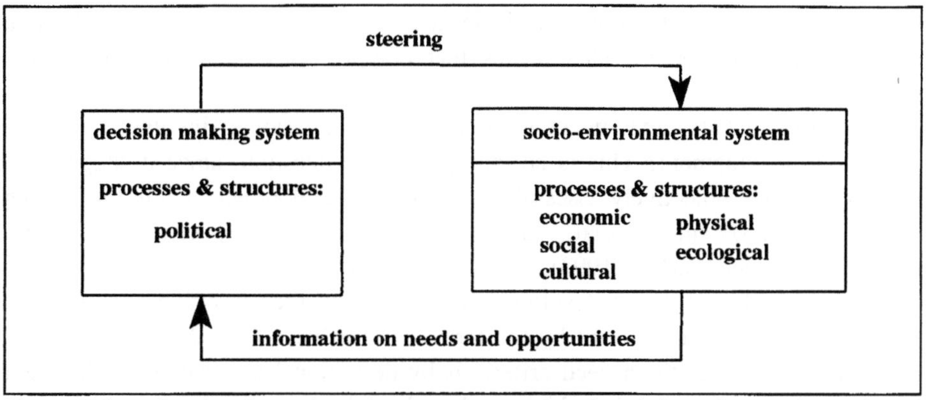

Figure II.3: Steering the *socio-environmental system* based on information on the needs and wants of the people involved and on the resources of the SEnS

involving people concerned about the satisfaction of their needs and the values at stake, now and in the future, within and outside the system. Of course, this will be successful only if the people involved feel responsible, in other words, if they too see the system as an entity. Therefore, the ideal strategy is one designed to develop a recognizable system.

The notion of a 'complete system approach' is not new. Meadows *et al.* (1972), among others, analyzed the relationship between the human population and the complete biosphere. More recently, Common and Perrings (1992) and Grizzle (1994) have advocated a system approach, and this is supported by the findings of John *et al.* (1994). Here this system approach will be used to determine what procedures are needed, and what information on the object to be steered should be available in order to describe it as a *SEnS* and to discuss the development of such a system.

We will first look at various aspects of the debate (Section II.3.1). The following sections are concerned with managing the debate (Section II.3.2), and analyzing the type of information required (Section II.3.3).

II.3.1 Debate and negotiation

The first question which presents itself is whether there is in fact room for debate? At first glance, this seems problematic, since human, financial and biophysical constraints will limit the possible solutions for sustainable development. These constraints do exist and can be analyzed in retrospect (e.g., Ponting, 1991). However, decisions on the action to be taken today must be based on prognoses for the future. And due to the inherent uncertainties of those prognoses, there will always be differing views (WRR, 1995). Hourcade *et al.* (1992) have articulated this, saying that it is often not a 'decision in uncertainty' that has to be made, but a 'decision in controversy'. This means that compromises must be arrived at by negotiation.

Will debate and negotiation between the people involved guarantee sustainable development? This is not an appropriate question, since it suggests that there is only one objectively assessable sustainable development, while the concept, with all its economic, social, cultural, and ecological aspects, is actually a normative concept (Parker, 1993). If negotiations lead to agreement on which goals must be reached, and which development is desirable, then in effect agreement has been reached on what is meant by the term 'sustainable'. Thus development which is in accordance with the agreed criteria is by definition 'sustainable development'. But does agreement on goals, which are based on debate and negotiation, guarantee a development that prevents future problems? Unfortunately, it does not. The risk of future problems can only be reduced. This can be done by including *representatives* of the people who will be involved in the future and people from elsewhere in the debate, and by taking their needs and wants into account during the negotiations (Hourcade *et al.*, 1992).

In short, one can opt to describe sustainable development as a result of debate and negotiation on the part of the people involved in a *SEnS*. Such an option must be based not only on theoretical and ideological arguments, but also on practical experience. Debate and negotiation have worked in a wide range of difficult and politically charged situations, as documented by Susskind & Cruikshank (1987). Moreover, on the basis of the results of the Dutch government's environmental policy, several authors have advocated a more 'horizontal' form of policy making (e.g., Van der Weijden *et al.*, 1984b; WRR, 1992a; Glasbergen, 1993; Hidding, 1993; De Bruijn & Ten Heuvelhof, 1994; Crul & Schelleman, 1995; WRR, 1995). There have been experiments with consensus-building in environmental policy (Hidding, 1993; Richard, 1993). In fact, according to Beck (1992), Western societies are increasingly ruled by 'sub-politics', that is, arrangements between the groups of people involved. It is logical to seek agreement on sustainable development through negotiation, because all parties may expect to gain something from a development that lowers the risk of future problems.

However, a major complication in open debate and negotiation is the fact that the parties involved usually differ widely when it comes to their ability to influence the outcome. For this reason, some form of professional management of the negotiations must be available to ensure that representatives of the people involved are given an equal say (Richard, 1993). In this way, their needs and wants can be taken into account (Susskind & Cruikshank, 1987). This means that managing the negotiation is an important aspect of the strategy for sustainable development (Section II.3.2).

These negotiations can only be brought about if all the parties have the same information at their disposal, and are able to interpret that information. For example, all parties should know what is meant by the *socio-environmental system* defined, and be aware of the needs and wants of the other people involved (Grizzle, 1994). They must also know what possibilities the system has to fulfil those needs and wants, and the consequences of the different options proposed (Pezzey, 1992).

The negotiations should result in measures to steer the system. While these measures may be unpleasant for some of the people involved, if the negotiations are conducted well, they will broaden the basis for appropriate formal measures. Moreover, discussions between the people involved may generate new ways of handling old problems. Such negotiations can fall back on one important resource: the creativity of people who are genuinely motivated to find solutions (Van der Weijden *et al.*, 1984b).

The strategy discussed here is particularly suitable for the local and regional level, since it is here that the commitment of the people involved will be highest. They will also be more likely to feel they are part of the *SEnS* and connected to the other people involved. Development can build on the knowledge and skills of those people, and on cultural and social traditions (Zubrow *et al.*, 1995). But how to deal with the higher-scale interests, i.e., the risk that a sustainable development of a *SEnS* will be accompanied by an unsustainable development on a higher scale? It has been argued that in developing countries local environmental problems should have a higher priority than global problems (Parikh & Painuly, 1994). But others take the view that where sustainable development is concerned, all the important questions must be answered on a global level (Pezzey, 1992; Giampietro, 1994a). These answers should be reflected in the steering on lower levels. Since the national government is always one of the parties involved in developments at lower as well as higher levels, it should represent the needs and wants of the higher level of scale where the parties who could play this role are lacking, or are not capable of negotiating in an adequate way. This will help to ensure that those directly involved are not exporting their problems to their neighbours or to posterity.

On the higher levels, from the national to the global, political structures already exist which are designed to facilitate negotiations. The issue here is not managing the negotiations, but rather placing sustainable development on the agenda, a step which has already been initiated by the WCED. On these levels, the main concern is how to provide the parties involved with the necessary information (Section II.3.3).

To operationalize the strategy proposed here, further study is needed in at least two main areas: the management of consultation, participation and negotiation, and the supply of information. How should this management problem be addressed (Section II.3.2), and what information is relevant (Section II.3.3)?

II.3.2 Managing the debate and negotiations

It is not enough to simply proclaim that negotiation on sustainable development is necessary. There are many questions that must be addressed, among them:

- *Who will initiate the negotiation process?*
  Any party involved should have the right to start the search for sustainable development. On the local and regional levels, the most obvious parties are the people directly involved: those who are concerned about their surroundings. However, in most cases governmental institutions will start negotiations by designating a region or a sector that is to be made sustainable. Moreover, the national government is always a party in the search for sustainability on the local and regional level of scale, in order to ensure the representation of those who cannot represent themselves.
- *Who will manage or assist in the negotiations?*
  Government institutions could be entrusted with this task, but since the government is also a party, it is preferable for this task to be taken over by an independent body, responsible for the quality, effectiveness and efficiency of the process.
- *Why should parties negotiate?*
  Why should a company that can decide for itself where to build a factory want to enter into negotiations on social structure, the landscape or nature conservation? Obviously the parties involved will only be prepared to negotiate if they can better their own position. Assuming that sustainable development entails lowered risk of future problems, one could say that everybody involved wins. For example, a company may save time and trouble by signing a covenant which reduces the likelihood of legal proceedings. It could also save money by lowering the risk of insurance claims after the factory has been built. But it is clear that not all parties involved have the same expectations with respect to future problems. Commuters may not be concerned about environmental deterioration, because they may think that it does not affect them. In that case, it must be made clear that they may later be faced with environmental taxes. What happens when not all parties are convinced of the necessity of

bringing about a sustainable development? If it is not possible to persuade all parties to participate, the question arises of whether the search for sustainable development is the right strategy under the circumstances. One important reason for parties to negotiate could be the knowledge that they actually have a say in what is going to happen. Therefore, this should somehow be guaranteed.

The task of those managing the negotiations will be to:

1. Specify what system is to be employed in order to attain a sustainable development
2. Bring the people involved - or their representatives - together around the negotiation table
3. Convince the parties that negotiations are needed
4. Guarantee that all the parties have the same information pertaining to the possibilities for sustainability, and that they understand the consequences of the proposed measures
5. Provide all parties with suitable instruments to search for optimal solutions to their problems
6. Guarantee that the parties have a voice in the proceedings
7. Ensure that the results of the negotiations are monitored, i.e., that the people involved, or their representatives, can hold decision makers responsible for the results
8. Monitor the development of the system, in order to assess the attainment reaching of goals, and to take action where necessary.

On the basis of these tasks, a formal procedure for the negotiations must be developed, which is itself beyond the scope of the present study. Such a procedure could be based on the literature of consensus-building (e.g., Susskind & Cruikshank, 1987; Richard, 1993) and on past experience with Environmental Impact Assessment (EIA) in various countries (e.g., Van Straaten, 1993; Wood, 1995). However, it is important to recognize that EIA assesses the environmental impact of a single project, while the formal procedure which is involved here must be aimed at the economic, social, cultural, and ecological aspects of a complete system.

Although further research is needed to work out the best way to manage negotiations on sustainable development, the focus here will be on the information required; aspects of information-gathering related to negotiation will only be discussed where pertinent.

II.3.3 Information required

What information is needed? First, the essence of the relationship between needs and wants and the potential of the system to contribute to those needs and wants must be discussed. We will use the example of the occurrence of environmental problems to gain insight into this relationship and, on the basis of that insight, to elaborate the type of information required.

An environmental problem is defined as a deviation from the observed state of the biophysical environment and the state in which 'people want it to be'. Or, as W.T. de Groot (1992) puts it, as a 'fact/value discrepancy'. But how does such a discrepancy arise?

In the first place, the fact/value discrepancy may be a result of changes in the biophysical environment: the facts have changed or, to be more precise, the facts as one perceives them have changed. Suppose there is a shortage of quality surface water. This may be due to non-human causes, such as volcanic eruptions causing a decline in rainfall, or human causes, such as pollution. However, the difference between 'non-human-caused' and 'human-caused' is not always clear, nor is it always of crucial importance when it is a matter of preventing problems in the future. The risk of a non-human caused change in facts can be reduced, for example, by building dikes to prevent flooding. The risk of human-caused changes in facts, however, may be difficult to control. For example, the problems of drug abuse sometimes seem insurmountable. The most relevant distinction is that between problems that can be prevented and problems that cannot, and between developments that can be steered and developments that cannot.

Second, the fact/value discrepancy may be a result of changing values, or what one perceives to be values. Changes in needs and wants, for example, can be stimulated by specific human activities, such as scientific research, technological development, or commercial advertising. It may also be the result of autonomous or uncontrolled processes. Here problems are created by sociocultural processes that result in new values and standards, and not by the state of the environment itself. Not the facts, but the values have changed.

As stated above, the *SEnS* as a whole is crucial when it comes to meeting the various needs and wants with respect to sustainable development. Therefore, the entire system should be planned and steered in such a way that the discrepancy between the way things are and the way people want them to be is minimized. And, as noted above, both the state of the environment and the state of people's minds can change. Thus the information required falls into two main categories: values and facts. In other words, (i) What human needs and wants are to be satisfied by using a particular system? (ii) What are the possibilities for satisfying those needs and wants through that system? However, these questions require information about the boundaries of the system and about the availability of resources within the system. This leads to four categories of required information.

1. Defining the system in space and time.
The system must be defined in such a way that it can be evaluated with respect to its ability to contribute to various human needs and wants. Such a system has been referred to as a *SEnS*. The definition of the system will be elaborated in Chapter III.

2. Selecting the values and related system characteristics
A strategy for sustainable development should strive for solutions that have no unwanted side effects. This means that it should be capable of considering all the system characteristics that are relevant when it comes to meeting human needs and wants. Those needs and wants must be identified and related to system characteristics. This will be discussed in Chapter IV.

3. Measuring the resource state
An important step is the quantification of the resources available in a *SEnS*. How much space is available for large mammals? How much water is available for the collection of drinking water? How much skills are available in the human population? Just as goals should not be analyzed in isolation, these questions about available resources should not be answered in isolation. Chapter V deals with the development of a method to measure the resource state.

4. Exploring the opportunities for development
It should be possible to use model studies to explore the modifications to the available resources and their use that will be needed to meet the needs and wants selected. For example, if drinking water, wood and food are required, what is the potential to provide these goods and services? Which technology must be employed to make use of the resource potential, and which technology and education programme is best able to change that potential? These issues are discussed in Chapter V.

II.4 Conclusions

Most approaches aimed at sustainability focus on one set of goals, while underplaying others. They ignore or even deny the existence of certain constraints. They do not explain how the proposed measures are to be embedded in a decision making process, and therefore do not rise above the level at which the solution of environmental problems is seen as a technical problem. The conclusion is that a new strategy is needed.

The above analysis leads to two definitions of sustainable development. One is formal, but not operational: a *Sustainable Development* is any development of a *SocioEnvironmental System* with a high potential for continuity, because it fulfils economic, social, cultural, and ecological goals. The other definition is procedural, but does not guarantee continuity: a *Sustainable Development* is any development of a *SEnS* on which representatives of the people involved have reached agreement, following a discussion of the opportunities for fulfilling economic, social, cultural, and ecological goals.

A strategy for sustainable development calls for a procedure which takes into consideration the goals and opportunities mentioned above. However, because of the uncertainties of these issues, a procedure for debate and negotiation is needed.

In order to follow such a strategy aimed at determining the goals and the opportunities, information will be needed on:

- the system that is to be developed in a sustainable way
- the needs and wants related to characteristics of the system
- the availability of interrelated resources
- the opportunities to modify the available resources and their use.

This should provide the necessary information about feasible strategies aimed at a sustainable development. Can this information actually be obtained, and can it be presented in such a way that it can be used by all parties? These questions will be answered in the following chapters. Concepts will be worked out, with a view to developing instruments for gathering information and making it available to the parties. The management of the negotiations is beyond the scope of the present work and will not be dealt with in detail.

CHAPTER III

Defining Socio-Environmental Systems for Sustainable Development[1]

CONTENTS

---

[1] *Reprinted from Ecological Economics 26, Musters, C.J.M., H.J. de Graaf & W.J. ter Keurs (1998), Defining socio-environmental systems for sustainable development, pp: 243-258, copyright 1998, with permission from Elsevier Science.*

III.1 Introduction

The World Commission on Environment and Development (WCED, 1987) defined sustainable development in terms of "*a global agenda for change*". Although decisions on steering developments may be prepared on a global level (UNEP, 1992; Masood, 1997), the actual steering takes place at lower levels, i.e., nations, regions, and even cities. Many countries have declared sustainable development to be one of their policy goals, including the United States, Canada, the United Kingdom, and the Netherlands (VROM, 1989; Anonymous, 1994; Lemons & Brown, 1995). In fact many people all over the world are searching for solutions that will contribute to global, national or local sustainable development. Urban planners strive to design the 'sustainable city' (Haughton & Hunter, 1994), while agriculturists search for 'sustainable agriculture' (Yunlong & Smit, 1994).

Not all parts of the world, however, are equally suited to serve as units of sustainable development. Suitability will depend to some extent on the possibilities for human steering. Here 'steering' refers to any deliberate action that is carried out in order to influence the course of developments. To be truly effective, such steering should be hindered as little as possible by uncontrollable processes, either within the system or outside it. Furthermore, such steering should not trigger uncontrollable processes in other systems, frustrating the steering of those systems. In Chapter II it was suggested that in order to ensure the effective steering of developments, the object to be steered should be defined as a *SocioEnvironmental System* (*SEnS*). This forms the basis for the assessment of the needs and wants to be satisfied via this *SEnS*, for the measurement of the available resources, and for the exploration of the opportunities for development. The goals of development can then be derived and measures selected.

In this chapter several critical aspects of defining a *SEnS* are analyzed, such as, why such a definition is needed (Section III.2), the fundamental issues in describing systems (Section III.3), and the essential parts of an actual definition (Section III.4). Since the research and conclusions are based primarily on experience with defining a *SEnS* at the regional level, most of the examples will be drawn from that level. This may result in a degree of bias. However, this chapter does not aim to provide a final solution, but rather to help to elucidate a topic that is regarded as "*crucial*" (e.g., Costanza & Patten, 1995).

III.2 The need for an adequate system definition

The strategy proposed in Chapter II entails five steps: (i) define the object to be steered or planned, (ii) assess the needs and wants to be satisfied, (iii) measure the resources that can be used to satisfy those needs and wants, (iv) explore the opportunities to meet these needs and wants, and (v) formulate the goals of the development and the measures required. To tailor the process to the striving for a sustainable development and the need to prevent side effects, the first of these steps was reformulated in Chapter II as the need to describe any object to be steered as a *SEnS*, in order to take into account the economic, social, cultural, and ecological aspects of development.

The definition of the object to be steered is in itself value-loaded (Wallace & Norton, 1992), and agreement must be reached on this point, as part of the overall decision making process aimed at sustainable development. It is self-evident that the object must be described adequately if it is to be steered. Inadequate description may be the result of carelessness, but since a definition has political consequences, it is often part of a political power play. There may be strong forces at work which discourage attempts to make the boundaries and content of a system explicit (Van de Meij *et al.*, 1997a). For example, if building areas are needed, it may be advantageous to maintain a low profile where the interests of those using potential areas are concerned, or to exclude these areas from the system. The parties participating in a debate have a greater stake in the explicit definition of the system, because it determines their possibilities to put forward their needs and wants in subsequent phases. After all, it would be difficult to emphasize the importance of nature areas, if the definition of the system led to the exclusion of such areas. Examples of such power play will be discussed in Chapters IX and X. Thus the need to take care in defining the system is at once self-evident and crucial, in order to prevent confusion and uncertainties about sustainable development.

This chapter is to consolidate our knowledge of system definitions, and to clarify how this knowledge can be applied in defining a steerable object such as a *SEnS*, and determining the problems which remain to be solved. The following discussion is patterned on the suggestion of Von Bertalanffy (1968) that system theory should be regarded as a "*guiding idea*" for structuring the discussion, rather than a mathematical modelling tool, although if modelling is desired, it may be very useful. Therefore, in the present chapter less attention will be devoted to a system definition with specific mathematical applications, on which a large body of literatuur exists (e.g. Bennett & Chorley, 1978).

III.3 Aspects of the definition of a socio-environmental system

As shown in Figure III.1, the *SEnS* is assumed to consist of a set of related subsystems (including actors) and processes which use inputs and deliver

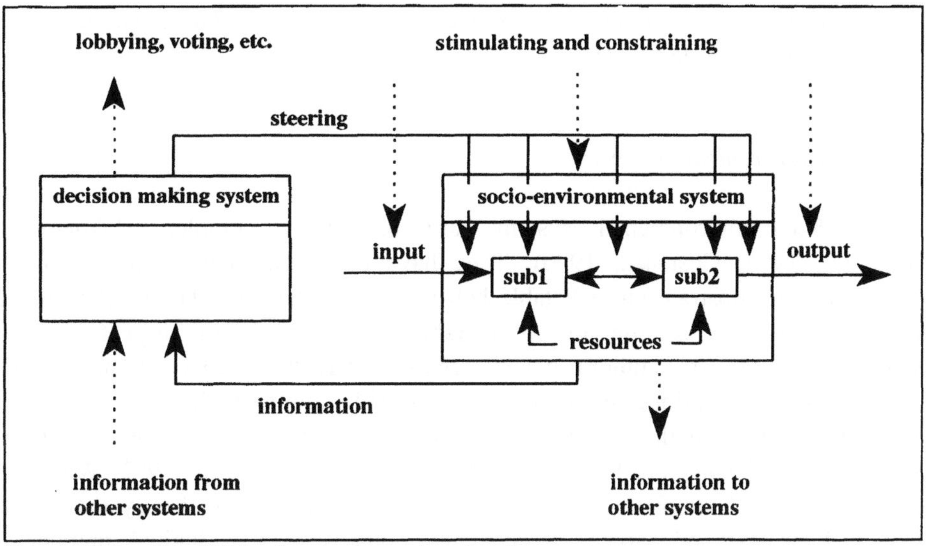

Figure III.1: Possibilities for steering. sub: subsystem.

outputs to other subsystems and to the outside. The functioning of the system as a whole, with its subsystems, is influenced by internal and external constraints. A system can be steered not only by manipulating its input and output, but also by modifying its subsystems or the relationship between them. Manipulation of the input and output of the system means that the boundaries of the system are opened or closed for specific flows of energy, material, or information. Import taxes are one example of such a manipulation. They act upon the legal input of 'foreign' goods into the system. However, as Bailey (1990) observes, 'boundary control' is usually in the hands of higher-level decision makers. Specific inputs and outputs of the system are stimulated or constrained by overall measures taken at a higher level than that of the system to be steered. For example, the amount of oil that a country exports will depend on oil prices, which are influenced by other oil-producing countries. While decision makers in any given system may try to influence decision making on a higher level - by lobbying, for example - they often encounter difficulties because the higher-level system itself is made up of many other systems. For this reason, steering a system is largely a matter of modifying the number of subsystems, their processes, or the relationship between them, while respecting the internal constraints, within the external constraints imposed upon the system. Thus steering requires a thorough knowledge of the subsystems, their relationships, their contribution to the functioning of the system, and the internal and external constraints on the functioning of the system. This leads to the following three core aspects of system description, which will be used to formulate specific rules for the definition of an object as a *SEnS*: *hierarchical*, *functional and structural*, and *steering* aspects.

III.3.1 Hierarchical systems

Socio-environmental systems are undeniably complex (Tikunov & Trofimov, 1995; Costanza, 1996). They have many different - but unknown - states, may demonstrate unpredictable behaviour, and may be adaptive (Clayton & Radcliffe, 1996), resulting in considerable uncertainty with respect to their development. Several authors have suggested that the theory of hierarchical ecosystems - as described by Allen & Starr (1982), and O'Neill *et al.* (1986) - is sufficient to describe the interconnectedness of systems and levels of scale (e.g., Norton, 1991; Jennings & Renagold, 1991; Norton & Ulanowicz, 1992; Warren & Cheney, 1993; Giampietro, 1994a & b). This idea allows the system to be described as part of another system and composed of a limited number of subsystems. But does this theory provide practical principles for describing a *SEnS* in such a way that information required for decision making can be collected? This is the first fundamental question.

A steered development of a *SEnS* should not frustrate the development of other systems, or endanger the sustainable development of the earth in general (Wallace & Norton, 1992; Giampietro, 1994a). And the development of a system can be influenced by developments in other systems. Thus when discussing the development of the system, the development of other systems, i.e., the '*environment*' of the system, must be taken into account (Hall & Fagen, 1956; cited in Bailey, 1990). Since '*environment*' is also used to designate the biophysical resources within a *SEnS*, in this volume the system's surroundings will be called its '*context*'.

A *SEnS* is an open system, with numerous relations to other systems. It is not possible to fully describe all the possible relations between the various elements of the system and the system's context. And even if it were, the amount of data would be so large that it could not be discussed or dealt with without information reduction. At first glance, it may seem easy to analyze the system as part of a network of systems, describing it only in terms of its relations with other systems, with no concern for the different levels of scale. This approach might work if one were considering only a small number of well-defined problems. For example, environmental Life Cycle Assessments follow this strategy (Heijungs, 1992). However, in the search for sustainable development, the main interest is in finding opportunities for development. In the light of opportunities, many more relations - both actual and potential - will be relevant. And by disregarding levels of scale, one disregards the fact that changes in one system may indirectly affect other systems, via the common higher-scale system (Norton & Ulanowicz, 1992). Thus any method for describing a *SEnS* must recognize different levels of scale. To the best of our knowledge, hierarchical theory is the only theory that deals expressly with this problem, and which may provide elements with which to elaborate the definition of a *SEnS*, while recognizing that it is part of a higher-scale system.

Describing a *SEnS* hierarchically means that it will be regarded as a system which consists of a relatively small number of subsystems, and is part of one higher-scale system (Simon, 1962). The subsystems are regarded as black boxes, of which only the position and the processing of input and output are described. Similarly, the relationship of the system to its context is described only as the input and output of the system that may be relevant for the sustainable development of the higher-scale system to which it belongs.

Thus a hierarchical approach is needed to arrive at an adequate definition of the object to be steered as a *SEnS*. Many types of hierarchies, including ecological, geographical, social, and governmental, can be used as a starting point. In any case, the description must provide information on the possibilities for steering, and therefore the description should be based on the organization of human activities. Such a description may not reveal all the relevant aspects of a *SEnS*: the importance of natural processes, for example, may easily be overlooked. Because human activities are often seen as inextricable from such processes, for the present, the description of human activities entails describing the natural processes that are affected by these activities. This assumption will be discussed in detail in Section III.4. As far as the context is concerned, in principle all higher scales should be taken into account. But in order to keep the description of the context manageable, only one higher-scale system will be chosen. The description of this context should likewise be based on the organization of human activities.

III.3.2 Structural or functional description

The subsystems and their position provide the system with a certain structure, and enable the system to function in a specific way, i.e., to transform inputs into outputs. The question at issue is whether a *SEnS* should be defined in structural or functional terms or both, in other words, as a set of structure elements or as a set of processes. This is a familiar issue in the literature on systems (e.g., O'Neill *et al.*, 1986; Bailey, 1990; Rosen, 1991). It is the second fundamental aspect of system definition.

People value a *SEnS* not only for the availability of structure elements, such as houses and landscapes, but also for its processes, such as the possibility to earn a salary or to produce corn. This view on valuable characteristics demands a description of the system in terms of both processes and structure elements. Both O'Neill *et al.* (1986) and Rosen (1991) state that a structural description of a system cannot be equated by a functional, and vice versa. In other words, describing the value of landscape in terms of flowing variables like money or energy, is insufficient for an overall representation of a system, as is a description of the economy in terms of the amount of cash that people have in their pockets. Saying that a structural and a functional description cannot be equated implies that neither one can be regarded as a full isomorphic description of the system. Consequently, the system description must comprise *both* structural and functional approaches, if it is to cover all the valuable characteristics. This need for a structural as well as a functional description makes the construction of overall computer models for sustainable development complex, and possibly beyond present reach. But that is no reason to abandon either set of descriptions. Computer calculation is not the only way to obtain information for steering developments: the debate can also be based on qualitative information.

O'Neill *et al.* (1986) argue that there is also a need for structural and functional descriptions of ecosystems, and they suggest that the two sets of descriptions always define a different level of scale. Does this also hold true for a definition of a *SEnS*? Obviously, when the system's inputs and outputs are described, they are being defined functionally, with the system as a black box. When the system is described as a network of subsystems, it is being defined on a lower level, producing in effect a description of the inside of the black box. However, the network description should be interpreted as a structural description of the system as a whole. Consequently, the variables describing the input/output of the system and those describing the internal organization of the system are variables of the system as a whole, and thus describe the same level of scale.

Thus a basic rule for defining a *SEnS* is that it must make it possible to describe the characteristics of the system in terms of both structural and functional variables.

III.3.3 System boundaries and steering

Steering a system is aimed at altering certain characteristics of that system. But which characteristics can be altered? Certainly not all of them. Steering may be impossible due to the boundaries chosen. The problem of steering a *SEnS*, the third and final core aspect, will be discussed below.

It has been pointed out that the term 'sustainable development' contains a contradiction: 'sustainable' refers to 'stability' and 'steady state', while 'development' presumes 'change' (Dovers & Handmer, 1993). Must a *SEnS* be defined as a system that can achieve and maintain stability? What can be learned from stability aspects of other system approaches? It is a popular misconception that ecosystems are steady-state systems, i.e., systems that are either stable or developing in the direction of a stable 'climax' state. However, it has long been recognized by ecologists that this representation of ecosystems is incorrect, or at best imprecise. Many systems never reach a climax, and systems that are thought to be in climax go on changing in important respects (Golley, 1993; Budiansky, 1995), or even change radically (Holling, 1986). Similarly, in sociology the idea that social systems can be in a state of 'equilibrium' has been widely criticised (Bailey, 1990). Thus from both a biophysical and a human stand point, it cannot be assumed that a *SEnS* must reach stability.

But there is more to this issue than the obvious statement that any system is constantly changing. The point is that in any defined *SEnS* some characteristics may be regarded as constant and others as in flux. The system as a whole may change, but certain characteristics of the system, within a specific temporal and spatial horizon, may be regarded as stable. What does this mean for the definition of a *SEnS*? The concept of '*phase space*', which often appears in the literature, may provide rules that can be useful here (e.g., Chadwick, 1971; Rosen, 1991; Clayton & Radcliffe, 1996).

The *phase space* is defined as the set of possible states that the system can have. A system can be described by means of a number of variables, the 'variable set', and the state of the system at a given moment is defined by the values of the variables at that moment. This state, also known as the 'observation set' (O'Neill *et al.*, 1986), is seen as one point in a multi-dimensional space, where the variables are the dimensions and the values the coordinates. Each variable of the system has a certain, limited range of values. For example, if a rural area can produce corn up to a theoretical maximum of x tons, than the variable 'corn production' will have values ranging from zero to x tons. The variables may be constant (area), stable but with a limited fluctuation (solar energy), evolving (land productivity), or unpredictable within a limited range (casualties of a disaster). Since all variables have a limited range of values, the system has only a limited range of combinations of values for all its variables, i.e., a limited set of states. This set of possible states can be seen as a set of points within the above mentioned multi-dimensional space, the *actual phase space*.

For the purpose of steering, it is important to know what characteristics can be steered and on what level. The decision makers on the level of the *SEnS* cannot steer all the characteristics of the system as a whole, due to external and internal constraints. Thus the number of states that can be attained by steering, form a subset of the set of all possible states, the '*feasible phase space*'. Steering can be done at a higher or lower level of scale. For example, it is difficult to steer the price of fuel at the local level, but to some extent this is possible on the national scale. This means that steering on other levels may alter the *feasible phase space* available to the decision makers. Furthermore, both external and internal constraints may change over time, autonomously or intentionally, and this will also alter the feasible space. Since decision makers have only limited possibilities to change the system, the decision on whether the available feasible space is acceptable, and consequently whether the chosen system is acceptable as the object to be steered, will be a political one. The basic rule that can be derived from this is that in defining a system, it must be clear beforehand how much room for decision making is required, and that this will determine which system should be chosen.

III.3.4 The basic rules for defining a socio-environmental system

The search for sustainable development involves an ongoing societal debate about *socio-environmental systems* that are always in flux. The debate should result in steering that is based on the present and future needs and wants of the people involved in the system as well as those outside the system but within its *context*. Steering is limited to the *feasible phase space* of the system. However, all *socio-environmental systems* display unpredictable behaviour and a *feasible phase space* that may change in unpredictable ways. As a result, the debate will be surrounded by uncertainties. Thus in order to ensure that measures do not lead to uncontrollable results, the description of a *SEnS* must be as explicit as possible. Such a description involves a hierarchical structuring and embedding of the object to be steered, on the basis of the organization of human activities on at least three levels of scale: the system level itself, one scale higher, and one scale lower. The description has to comprise functional as well as structural variables, and it must be clear what can be steered and to what extent steering is required. These basic rules are applied in the following discussion on the definition of *socio-environmental systems*.

III.4 Defining a socio-environmental system

A group of people or a governmental organization may be concerned about developments in certain aspects of their economy, society, or environment, i.e., the system they are involved in. Such concern may lead to a striving for the sustainable development of that particular system. Thus in its present state, this system has certain values that are under threat or certain undesirable properties that should be changed. In this respect, it is seen as a value-producing system.

Section III.2 ended with the conclusion that the description of the system is value-loaded and that decisions must be made about the description. For this reason, the people involved must be identified, if their issues are to be taken into account. But, they cannot be identified before the system has been delineated in terms of space, time and organization. In this sense, there is something paradoxical about defining the object to be steered as a *SEnS*. The question which presents itself is 'At what point does the process begin?'

This process can best be initiated by the people or government institutions that feel the need for sustainability. The first step is to delineate the system in space and time. This will entail the identification of the system (1), its hierarchical organization (2), and the people involved (3). They - or their representatives - are asked how they would delimit the system, taking into account the possibilities for steering. This may yield a new roughly delimited system which includes even more people. Admittedly, the process involves some redundancy, but may be expected to lead to a system with clearly defined boundaries; this is the delimitation of the scale of the system in space and time (4). Ideally, this will precede the description of the system in terms of its valued characteristics and its resources (5). Thus five iterative steps are recognized in the definition of a *SEnS*; these will be elaborated below (Figure III.2).

III.4.1 Identifying the system

The reasons for designating an object and defining it as a system may be different, but the party that initiates the designating process should be fairly confident that aiming at discussions on development will solve present problems and prevent future problems from arising. Obviously this approach cannot solve large, acute problems, such as famine.

Different kinds of objects can be identified, provided that (i) they are concrete, i.e., part of the biophysical world; (ii) they contribute to economic, social, cultural, and ecological goals; and (iii) they are changeable by deliberate human actions. Identification implies describing the object as a system in space, time, and organizational structure. It may be interesting for purposes of historical analyses to situate a system in the past. However, in the search for sustainable development, the focus will generally be on a concrete steerable system as it appears now, or will appear in the near future.

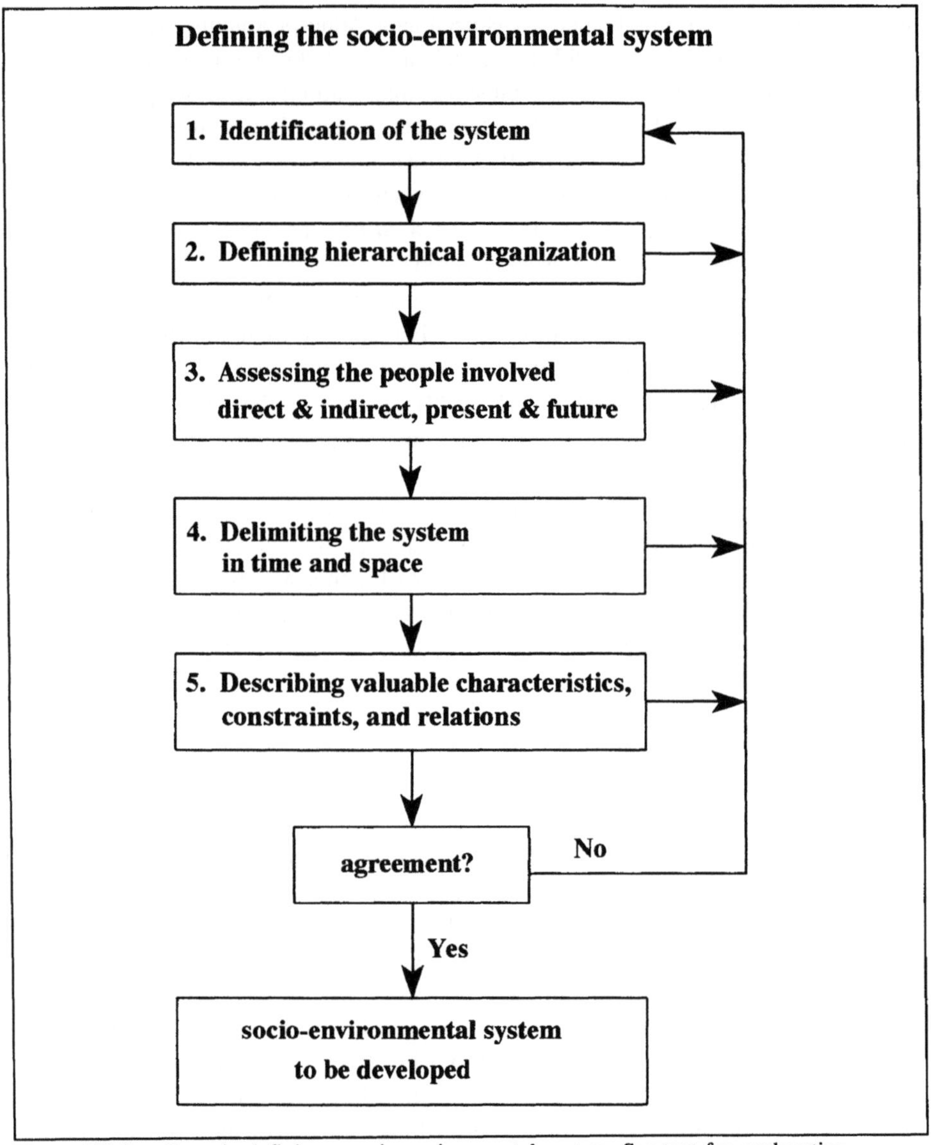

Figure III.2: Steps in defining a socio-environmental system. See text for explanation.

III.4.2 Hierarchical organization of the system

In system theory, it is often said that the boundaries of a system must be chosen in such a way that the relations between elements of the system are stronger or more numerous than those between the elements of the system and outside elements (e.g., Simon, 1962; Platt, 1969). But where the relevant relations are not yet known - as is the case when the system is first identified - this principle cannot be applied. Therefore, the initial identification of the system is tentative, based on what at first sight is presumed to be a steerable entity. Systems that are extremely open, i.e., whose development is assumed to be controlled primarily by processes outside the system, should not be selected and defined as a *SEnS* for purposes of sustainable development. For example, developments in North Sea fishing grounds probably depend so heavily on developments and activities in other parts of the North Sea that it would be useless to strive for their sustainable development on that scale level. Nor should systems that develop largely as a result of uncontrollable processes within the system be selected. For this reason, the possibilities for steering the development of the entire earth are limited, despite the fact that since the Brundtland report (WCED, 1987) the concept of 'Sustainable Development' has often been discussed as if it were a global issue.

In identifying the system, one also identifies the level of scale at which the system is to be studied. Different typologies of levels of scale can be applied. The Dutch National Environmental Policy Plan, for example, recognizes five such levels: global, continental, national, regional, and local (VROM, 1989). Clayton and Radcliffe (1996) employ three levels: the global, regional, and local.

As we have seen, the hierarchical organization of human activities is one possible starting point for placing a *SEnS* within its context. Governments are recognized in this way. Thus when a spatially defined region has been selected as the object to be described as a *SEnS*. For example, the national government can be selected as the higher-scale system. This provides the higher-scale frameworks for the level on which the *SEnS* is defined.

The system itself can be defined hierarchically as a set of subsystems within the system. The system is divided into subsystems in order to identify steerable units that can be manipulated by altering their inputs and outputs, number, or relative spatial or temporal positions. All subsystems are to be described on the same space and time scale. In effect, they are regarded as complete systems, containing structure elements and related sets of processes. This makes it possible to distinguish their economic, social, cultural and ecological aspects and identify their need for human, financial, cultural and biophysical resources.

After it is has been decided which subsystems are to be distinguished, the subsystems can be described in terms of their boundaries, relative positions, and number. They can then be described functionally, according to their inputs and outputs. In a rural setting these subsystems can be land-covering systems like agriculture area, habitation area, or nature conservation area, together with their related processes; or they might be systemized by their processes, such as food production, recreation, and trade, together with the related land, where relevant.

Although subsystems are regarded as entities that can be steered, they do not indicate to what extent the system can be steered and how steering is organized. As noted in Section III.3.3, when relevant characteristics cannot be steered due to the chosen boundaries, it may be necessary to redefine the *SEnS* and its subsystems. For example, when the downstream part of a river basin must be steered in order to solve certain water quality problems, the higher scale can be represented by the province or country it belongs to, and the downstream part can be divided into subsystems with related activities. However, since the quality of surface water and aquifers downstream is dictated largely by activities upstream, the question is whether the downstream part can be adequately steered by steering its subsystems. This question can be dealt with in two ways.

- First, the *SEnS* is defined as the part downstream and the quality of the incoming water is regarded as input that acts as a constraint on the economic, social, cultural, and ecological development in that selected area. In the search for sustainable development, the people of that area will try to steer the area downstream. They may, however, regard the option for steering as too constraining, and decide not to focus on steering their subsystems. Instead they might try to convince the districts upstream that something should be done to reduce the pollution of the river. This is the way in which such problems are usually tackled (e.g., the initial policy of the Dutch government on the pollution of the Rhine: Dieperink, 1995).
- Second, because of the limited steerability the conclusion may be that a new *SEnS* must be defined, one that consists of the total river basin. Consequently, steering must be organized within the enlarged boundaries of the system. The people downstream may take the initiative, but they will have to convince the people in the other part of the basin that the definition of a common *SEnS* is necessary, and equally advantageous for all parties concerned. In such a situation, not only water quality may be at stake, but also economic development and employment.

It must be stressed here that in either case (i) the boundaries of the *SEnS* are defined by the organization of a set of subsystems based on human activities within the river basin, which form the steerable units; (ii) decision making must be carried out by representatives of all the people involved in the *SEnS*; (iii) the higher-scale level is an organization of human activities which includes all the provinces or countries to which the different parts of the basin belong.

III.4.3 *Identifying the people involved*

Changes in *socio-environmental systems* may be autonomous or intentional. The decision about whether the direction of development is acceptable should be made by the people involved (Doyal & Gough, 1991). This means that they must shape the decision making and try to reach consensus on the development desired. However, in actual practice, this ideal situation will rarely be achieved (Van Mierlo, 1988). Thus a distinction must be made between the people involved and their representatives in decision making. But inviting the right representatives to take part in the decision making process depends on knowing who the people involved are, and how they can be involved, directly or indirectly, in the decision making process. Another question is who decides who they are and who invites them to take part (Chapter II). This latter point pertains to the management of decision making and is therefore beyond the scope of this volume. Here the focus is on identifying the people who are by definition involved in the system and who are therefore involved in decision making; at the very least they should be consulted.

Everyone who values, provides or uses - whether positively or negatively - the inputs, outputs, or structures of the defined system is regarded as *directly* involved on the level of the system. Where a region is regarded as a system, this will mean that not only inhabitants, employees, and visitors are involved, but also those who depend on its water supply, use the land to generate income, enjoy its natural beauty, or have other physical or emotional bonds with that area. Not only individual people, but also those organized in institutions may have a direct involvement with the system. A factory in a country that uses the resources of another country has a direct involvement with that other country. The input, output, and structure elements, as well as the relations between the inputs, outputs, and structure elements of a system which can be related to the people or other systems involved are referred to as the '*valuable system characteristics*'.

There are also people and systems that are *indirectly* involved in a specific system. For example, if a lumber-exporting country lowers its prices, then other lumber-exporting countries will probably experience a drop in market share. They are not affected directly by the output of the first country, but by its role on the international market. In the initial analysis of the economic, social, cultural, and ecological relations between one system and another, this '*indirect involvement*' will be hard to detect. But it can be made more apparent by regarding the system as a subsystem within a higher-scale system. For this reason, higher-scale systems should also be represented in the decision making process on system definition. This will at least guarantee that important changes in the context of the system are taken into account, and that the system does not secretly export its problems to its neighbours.

It is highly improbable that all indirect involvement can be assessed in the initial stages (Clayton & Radcliffe, 1996). This underscores the responsibility of representatives of the higher-scale system. They must be alert to changes in the context of the system that may have unexpected effects on the system, as well as to unexpected negative effects flowing from the development of the system itself.

Above we defined the people directly or indirectly involved. But the search for sustainable development is also aimed explicitly at "*the ability of future generations to meet their own needs*" (WCED, 1987). They will have to use the *SEnS* as it is passed on to them, and are therefore involved in today's decisions on the development of the system. Since they are clearly absent, it is the job of today's representatives to speak for them. However, today's representatives can only represent what they assume to be the needs and wants of future generations, on the basis of what they see as the primary life-improving and life-supporting properties of the system.

Thus there are four groups of people or systems involved in a *SEnS*: those involved now directly or indirectly, and those involved in the future directly or indirectly. Recent experience in regional development projects shows that direct involvement is not difficult to assess (Glasbergen, 1995). When important changes in development are at stake, the people directly involved - or their representatives - will make themselves seen and heard in an open society like that of the Netherlands. This does not mean that in all cases all those directly involved also participate in decision making on steering. Sometimes powerful parties succeed in excluding others from the planning process (Van der Meij *et al.*, 1997a). How the representation of involvement should be realised will depend on circumstances. In many cases, representation already exists in the form of political parties, nature conservationists, trades unions, etc. However, Susskind & Cruikshank (1987), advocate *ad hoc* representation, because, they claim, traditional representative groups either focus on a limited number of issues, or are not flexible enough to solve socio-environmental problems. Again, the difficulties in managing participation are beyond the scope of these volume.

Indirect involvement is represented by the higher-scale system to which the system under consideration belongs. Because indirect involvement is harder to assess, higher-scale decision makers may try to escape responsibility, permitting the representatives of the people directly involved to take a chauvinistic attitude, and concentrate only on the values of the system which are of importance for the people directly involved in the *SEnS*, without taking into account the impact on the context (Van der Meij *et al.*, 1997b).

III.4.4 Delimiting the system in space and time

Delimitation in space and time has three aspects: the outer boundaries, the level of resolution of the information, and the boundaries of the context of the system. In ecological terms, the first two aspects are known as the 'extent' and the 'grain' of the variable-set which describes the system (Kolasa & Rollo, 1991). These two aspects are also relevant in the search for sustainable development, because together they delimit the information on the system, and thereby affect the exploration of opportunities for development. The third aspect describes the higher-scale system to which the system belongs. Since the higher-scale system not only puts forward the needs and wants of the people indirectly involved, but also provides the constraints for steering on the lower levels, it is important to establish the nature of this higher scale level.

Extent

The term 'extent' refers to the scope of the definition, the aspect that determines the space or period of time covered. It may be seen as the horizon of steering. As noted above, these horizons are tentatively selected when the object to be steered is identified (Section III.4.1).

As a rule, the determination of the extent of the system in space and time presents no insurmountable problems, since the party initiating the discussions on sustainable development will have a specific object in mind, including the concomitant boundaries with respect to space, time, and organization. However, there are cases where determining the outer boundaries of a *SEnS* is more problematic. For example, the expansion of cities into rural areas can give rise to serious conflicts. Bailey (1990) has pointed out that often boundaries are not open to debate. They are set by higher-scale authorities, who disregard the fact that the people directly involved may want other boundaries.

The spatial boundaries of a system may be subject to change, and in that case the people affected by the changing size of the system will become directly involved. A new system may be created that has not yet been precisely defined, such as in the river basin example (Section III.4.2). Another example is a nature conservation area that has been defined as a *SEnS*. If the national authorities decide that it should be enlarged, then the farmers around that area become directly involved, and their views should be taken into account in the redefinition of the system.

The temporal aspect is usually referred to as the time horizon. When a specific time horizon is selected, it is on the implicit understanding that significant developments can be steered during this period. The time horizon defines which characteristics of a system are constant and which are variable.

Grain
Grain is the resolution of a variable-set; it determines the detail of the knowledge on the system. For example, if only the relation between the yearly nutrient supply and the yearly production of wheat is known, we cannot determine how to increase production by changing the distribution of the fertilizer supply over the seasons.

Dutch regional planning has seldom given explicit attention to grain. Consequently, the goals of development are described at very different levels of detail. In the 'Duin- & Bollenstreek', a region in the coastal area of the Netherlands known for its dunes and bulb farms, "*strengthening the north/south landscape structure*" was one of the supposedly equivalent goals on the system level. But it was on a list that also included "*planting trees in farmyards, that do not attract insects*", which is on the subsystem level (Van der Meij *et al.*, 1997b). The system level can include the subsystem level, but focusing on subsystems does not guarantee that the goals on the system level will be reached. A planning process - in the example above, one designed to change the landscape - can be considerably improved by specifying the level of resolution. Operationalizing the landscape demands on the level of resolution should then lead to a specification of all elements needed on the subsystem level, which includes not only planting tree in farmyards, but also such aspects as restoring hedges or constructing cycle paths. And if such specification is lacking, the items on a list of goals become arbitrary.

Context
While the spatial boundaries of a system are usually easy to assess, the spatial boundaries of the context, i.e., the higher-scale system, may be problematic. What is the next level of scale? What if the next governmental level of scale does not coincide with the next economic, social, cultural, or ecological level? Such higher scales must be defined, as will be clear from the following example (see also Chapter X). The Duin- & Bollenstreek is part of the province of South Holland. But geographically and demographically it lies within a conglomerate of urban and rural areas in the Dutch provinces of South Holland, North Holland and Utrecht known as the Randstad. The other two provinces were not represented during the planning stage of the development of the Duin- & Bollenstreek, because the higher-scale system and the representatives of the people indirectly involved were not explicitly defined. Consequently, without discussing the matter with that larger group of people, plans were drawn up to export part of the bulb-raising activities to other areas within North-Holland (Van der Meij *et al.*, 1997b). What was needed in this case was a governmental entity that represented the conglomerate of urban and rural areas. However, although it has a name (Faludi & Van der Valk, 1994), the 'Randstad' is not a governmental entity that can be made responsible for the needs and wants of the people indirectly involved. Thus while it may be possible to

identify the higher scale, that does not necessarily mean that it can be operationalized.

All the developments that change at a rate slower than it is possible to detect within the time horizon belong to a temporal scale of a higher order. The large-scale regeneration of a forest, for example, cannot be accomplished within a single decade. In steering developments, decision makers should take into account the developments of the higher-order temporal scale, i.e., the slower processes. Thus if one cannot regenerate a forest, the focus must be on certain key factors that can be influenced, such as the agricultural practices in or near the forest. This shows that desirable large-scale changes may argue for effective, small-scale steering.

III.4.5 Describing the system

The process of describing the system can been divided into three parts, dealing with the (i) valuable characteristics, (ii) resources, (iii) the relationship between valuable characteristics and resources with respect to steering.

The valuable system characteristics

In searching for sustainable development, the whole system must be assessed on the basis of the characteristics regarded as valuable to the people involved. These valuable characteristics will include only those inputs, outputs, or structure elements of the system that are related to either people or other systems. Variables that measure or estimate these characteristics are part of the variable-set that describes the system. In other words, the variable-set contains a subset of variables that measure as completely as possible the *valued system characteristics*. Several methods for assessing the economic, social, cultural, or ecological characteristics of a system already exist (e.g., Henderson, 1991; W.T. de Groot, 1992; Cobb *et al.*, 1995). In Chapter IV the problems presented by these valuations are discussed and an alternative method is proposed. The variables that measure the valued characteristics may be seen as defining a special type of *phase space* (Section III.3.3), i.e., the space that contains the set of all *value states* of a system. This space will be referred to as the '*value space*' of the *SEnS*.

The resources

Since sustainable development is regarded as a development within economic, social, cultural, and ecological constraints, the variable-set also contains information on those constraints.

Constraints may be the result of deliberate decision making at other, usually higher, levels of scale. By translating goals for global change from one level to another, it may become possible to constrain the development of regional systems with a view to global sustainability. Since these constraints are determined by goals on higher scales, they must be translated into *valued system characteristics* related to higher-scale systems (Chapter IV).

The development of a system is constrained not only by higher-scale systems, but also by internal characteristics that act on changes in subsystems. The set of internal constraints is called the '*resource space*' (Chapter V), the space that contains all the potential *resource states* of a system. Chapter V examines various ways of assessing the *resource space*. The variables that measure the resource characteristics of the system are regarded as the dimensions of the *resource space* of the *SEnS*.

Consequences for steering

Now the *feasible phase space* can been defined as that part of the *phase space* accessible by steering the system and the *feasible value space* can be defined as the set of *value states* accessible by steering. *Feasible value states* can now be redefined as those states which can be accessible by acceptable ways of using the available *resource space* of the system.

If a desired *value space* of the system lies within the *feasible value space*, then this state can in principle be attained. The challenge is then how to change the system in order to do so, in other words, how to move from the present state of the system to the desired state. However, a desired state of a system may lie outside the *feasible value space*, and such a state can never be attained by steering. This conflict can be solved by adapting the norms, standards, or goals of development, i.e., by selecting another desired state.

It is preferable to assess the *feasible value space* before discussing the *desired value states*, in order to avoid the frustration which may arise among participants in the debate as a result of unrealistic sets of norms, standards, or goals. The debate should be guided by what is actually possible, and steering should be focused on staying within the constraints imposed upon the system. In this way, the search for sustainable development will become a procedure for defining as explicitly as possible economically, socially, culturally, and ecologically acceptable states of the system within its *feasible value space*. Chapter V discusses how the *feasible value space* can best be explored.

Will defining these spaces make the choice of clear, attainable goals for system characteristics superfluous? On the contrary, carefully selected quantitative goals are more likely to motivate the participants in a development than abstract spaces.

The participants in the debate on decision making related to the development of a *SEnS* may have different views on what constitutes constraints and the norms, standards, or goals to be realized. Furthermore, one must be aware that an *acceptable feasible value space* may prove to be non-existent, in other words, the debate on the development of a *SEnS* may lead to a set of constraints and goals that are fundamentally irreconcilable.

To be able to steer effectively, it is necessary to know the relationship between the *resource space* and the *value space*. This relationship must be defined by means of a set of rules describing the variables that measure the *valued system characteristics* as functions of the variables that measure the *resource space*. However, the problem is that this kind of knowledge is not usually available (Steverson, 1995). This may interfere with the construction of overall computer models - as opposed to smaller sub-models - but it should not impede the search for sustainable development. In discussions on preferred states, exact predictions of the state of the system may not be necessary (Munda *et al.*, 1994). Consensus on the feasible states of the system and the direction of development, together with some knowledge of the actual state of the system, will in many cases be sufficient to guide steering.

III.5 Concluding remarks

In the foregoing analysis of concepts, the definition of the object to be steered as a *socio-environmental system (SEnS)* was proposed as the starting point of the debate. It will be clear that the following sequence of methodological steps depends on a clear definition of the system. This definition must comprise the following five parts:

1. Identification of the *SEnS*, i.e., the first tentative designation of the object to be steered.
2. Description of the hierarchical organization of the *SEnS* as a set of subsystems designed to identify steerable units and the place of the *SEnS* within its context.
3. Identification of the people involved. Ideally, all these people should be represented in the decision making; at the very least, their needs and wants should be kept in mind by responsible decision makers.
4. Delimitation of the system in space and time. The spatial extent of the system, the time horizon of the developments, the detail level of the information, and the context of the system should be clearly defined.
5. An empirical description of the system in both structural and functional terms, including the valued characteristics, the resources, and the relationship between them.

Defining the system serves not only to clarify the debate on the development of a *SEnS*, but also to facilitate the collection of adequate information on development opportunities based on verbal or mathematical models. However, it has been demonstrated that even this first step toward a sustainable development of a *SEnS* involves a great many choices, and a specific sequence of questions to be answered. The following major questions have been discussed, while the italized concepts have been discussed and defined:

- May the *system* be regarded as hierarchical?
- What kind of *hierarchy* is assumed?
- Can the system be described both *functionally and structurally*?
- Which *characteristics of the system* are steerable?
- Who are the *people involved*?
- What are the *boundaries of the system*?
- What should the *grain* of the system description be?
- What is the *context* of the system?
- What are the *subsystems* of the system?
- What are the *valuable characteristics* to be considered?
- What are the *resource characteristics* for development?
- What is the *relationship* between *valuable characteristics* and *resources*?
- What is meant by *value states*?
- What is the *feasible value space*?

All these questions serve to underline the importance of definitions of concepts and methodology and of being as explicit as possible on the different aspects of defining the system and on the uncertainties that one is prepared to accept. This approach will facilitate the later steps in the search for sustainable development, and will also ensure that participants in the debate do not use confusion concerning the definition as an excuse to go back on previous agreements. For this reason, the procedures should lead to a formal document that embodies the results of the debate on the system definition.

CHAPTER IV

A Framework for Identifying the Valuable Characteristics of a Socio-Environmental System

CONTENTS

IV.1 Introduction

The political will to achieve 'sustainable development' is not, of course, the ultimate solution to all our environmental problems. This is more obvious now than it was ten years ago, when the concept was first introduced as an issue in science and politics. Even then it was clear that the concept had been constructed primarily as a work-in-progress attempt to take account of equivalent, but different, human goals when making decisions on steering developments. In other words, economic, social, cultural, and ecological goals must be coordinated as closely as possible, to satisfy the needs of both present and future generations (World Commission on Environment and Development (WCED, 1987).

The interaction between economic, social, cultural and ecological development should encourage decision makers not to consider the economy, society, or the biophysical environment in isolation, but rather to see the '*SocioEnvironmental System*' (*SEnS*) as a whole (Chapter II; Wallace and Norton, 1992; Szakál & Illés, 1993). The search for sustainable development may be seen as a debate about new opportunities arising from the integration of the economic, social, cultural, and ecological aspects of such systems.

As explained in Chapter III, defining the *SEnS* is the second major step towards sustainable development, following the decision to aim at sustainable development on a national, regional or local scale. This step involves delimiting the system in time and space, as well as its subsystems and its context, and identifying the people involved. In this way the object to be steered can be identified and accurately described, together with the time-horizon for changes, and the options for steering.

However, the question 'What are the specific goals of sustainable development?' has not yet been answered. This issue must be addressed before examining the question of how those goals can best be achieved (Proops et al., 1996). Therefore, the next step is to assess the values which the system has for the people involved, i.e., to specify the needs and wants of the people involved in relation to the system, and translate them into relevant system characteristics. A method for identifying all the values that are crucial to sustainable development would clearly be a useful instrument (e.g., Norton, 1994: Bingham *et al.*, 1995; Lockwood, 1997).

Could a universal list of values, like that of Max-Neef (1992), meet such an objective? No, because it cannot be specific enough. The people involved or their representatives, as identified in Chapter III, should be asked to indicate what they consider to be the valuable characteristics of their *SEnS*: the *valued system characteristics*. The description of these characteristics will obviously depend upon both the specific system and the people involved, and such specificity can never be universal.

Thus instead of searching for a universal list, some type of instrument, procedure, or framework must be developed that allows the people involved to come up with their own list of *valued system characteristics*. Such a framework can also serve as a monitoring device for assessing the valuations people encounter in local, regional, and national planning and development policies.

As yet, there is no such procedure or framework for selecting the *valued system characteristics* of a *SEnS* which can be used in the search for sustainable development (Bingham *et al.*, 1995). The aim of this chapter is to develop such an instrument. The discussion will focus on the fundamentals for valuing a *SEnS* in relation to sustainable development, and constructing a framework based on these fundamentals (Section IV.2). This framework will then be compared with four applications from the recent literature on existing valuation approaches, in order to identify any gaps in the framework (Section IV.3). These four approaches emphasize economics (Cobb *et al.*, 1995), ecology (R. S. de Groot, 1992), basic human needs (Doyal & Gough, 1991), and human wants respectively (Henderson, 1991). None of these approaches, nor their applications, presume to generate a complete list of all the valuable characteristics of a system related to sustainable development. However, by analyzing the applications, it may be possible to demonstrate the power of using the framework for verification. This is discussed in Section IV.4; the Chapter ends with concluding remarks on the steps involved in a search for the *valued system characteristics* of a *SEnS* (Section IV.5).

IV.2 A typology of valuable system characteristics

IV.2.1 Needs and wants, values, and system characteristics

The 'values' of a *SEnS* are related to both the 'needs and wants of the people involved' and the 'characteristics of the system' (Figure IV.1). These terms have not yet been clearly defined. To begin with the latter, the 'characteristics' may be seen as the properties of the system that are measurable in some way or another. By relating 'values' to 'characteristics', the focus is directed towards those values that are determined by system characteristics such as inputs, outputs, structures, processes, or activities. These values have a traceable relation with the system, and include all economic, social, cultural, and ecological values. The 'needs and wants of the people involved' are the reasons to preserve the existing values of system or add new ones. These reasons may be highly diverse; they should not be delimited by a short-sighted anthropocentrism (Westra, 1997), but should include long-term needs and wants. Thus values - or valuations - are the links between 'needs and wants' and 'system characteristics'. Consequently, the basic requirement which any search-and-check instrument must meet is that it must relate the 'needs and wants of the people involved' directly to the 'characteristics of the system'.

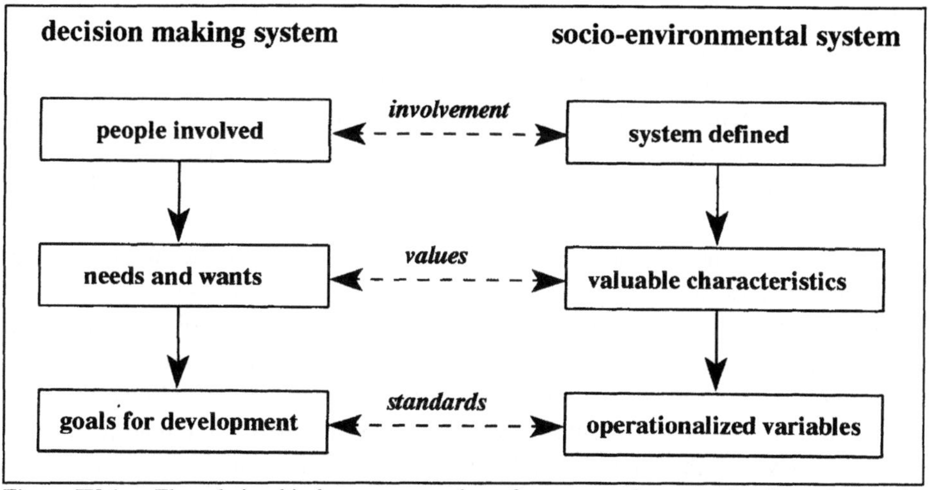

Figure IV.1: The relationship between a number of concepts relevant to the discussion on the sustainable development of a system.

However, a 'value' does not always refer to specified system characteristics. What people call a 'value' can range from the abstract to the concrete, and from the vague to the precise. But in the context of the search for sustainable development, the values of a system must be described as precisely as possible, to guarantee that the discussions on development will lead to a definition of goals before the commencement of steering.

But how does one deal with values such as 'trust' or 'solidarity', which have difficult-to-define relations with the system's characteristics, but are important or even crucial to the striving for sustainable development? These values may be considered properties of the *SEnS* that must be specified, but the question is how. At least, the people involved should indicate what they consider relevant characteristics for 'trust' and 'solidarity'. Since these values may also be based on human perceptions of the quality of government, management, or policy making, they should also be taken into account in organizing the debate on the goals of the search for sustainable development. In other words, they should play an important role in the decision making process (Sagoff, 1988).

Thus not only the *SEnS* has values, but also the system which is instrumental in the decision making process. In Chapter II (Figure II. 3) a distinction was made between the *decision making system* and the *SEnS*. The former is characterized by the processes of information-gathering, goal setting, steering and evaluating. The latter is the object for which the goals are to be set. In this chapter the discussion on the valuation of a system centres on the *SEnS*, rather than the *decision making system*.

The development of a framework for valuing a *SEnS* takes as its starting point the necessity to explicitly relate the 'demand side', i.e., the needs and wants at stake in the development of a *SEnS*, to the 'supply side', i.e., the *valuable system characteristics* (Dodds, 1997; Gregory & Slovic, 1997). But the present aim is not the construction of a typology of values, but rather a typology for system characteristics related to needs and wants. The reasoning is that the aim of the research intended to relate 'demand' and 'supply' is not to determine whether values are, say, 'instrumental', 'optional', or 'intrinsic', but rather to (i) determine the characteristics of the *SEnS* as a whole, and to (ii) make decisions on acceptable changes to those characteristics. In this respect, this chapter differs from much recent literature on environmental valuation that focus on typologies of values (e.g., W.T. de Groot, 1992; Aldred, 1994; Perrings, 1995; Crowards, 1997; Lockwood, 1997). The following subsection first discusses a typology from the point of view of the 'demand side' (Section IV.2.2). This is followed by an analysis of the required typology from the point of view of system changes, i.e., the 'supply side' (Section IV.2.3.). Then the need for more types of characteristics is discussed (Section IV.2.4), and a framework (Section IV.2.5) is constructed that will be used for a comparison with existing valuation approaches (Section IV.3).

IV.2.2 Characteristics related to sustainable development

Future generations and the system

If 'sustainable development' is to retain credibility, the needs and wants of future generations must be considered, as well as those of the present one. The needs and wants of present users can be characterized by assessing today's goods and services. But future generations cannot tell what their needs and wants for goods and services will be. Thus these must be estimated to the best of ones ability and translated into present characteristics. These characteristics cannot be regarded as conditions with regard to the present demand for goods and services, but should be interpreted as conditions that make future use possible.

Thus the first fundamental requirement for a typology of characteristics is that it must distinguish between the present characteristics related to goods and services needed by present users, and the present characteristics related to the goods and services for meeting the needs and wants of future users.

People indirectly involved in the system
Every *SEnS* that is smaller than the Earth is imbedded in a global context. The people involved in this context may not be directly affected by the system under consideration, but they could be involved indirectly (Chapter III). They can produce the same products, compete on the same markets, make use of the same social networks, use the same biophysical resources, or depend on the same environmental conditions as the people directly involved. Seen from this perspective, every living human and non-human being on Earth is part of the context of some system. Norton & Hannon (1997) propose a 'place-based' approach to environmental values. They argue that it is important that people not only value their 'place', but also the 'space' that surrounds that place. If this is done by all people, a system will be valued not only by the people directly involved, but also by a large number of people from the context of the system. This means that the valuation of a system is in danger of becoming so complex that it is almost unmanageable. To avoid such an impasse, one can choose to define any given system as part of only one higher-scale level. On that higher-scale level, the needs and wants of the larger communities must be consolidated into a single set of needs and wants to be met by the system. It must be assumed that representatives of those communities will adequately put forward the collective needs and wants of both the direct and indirect users, and formulate what the system should contribute to higher-scale needs and wants (Chapter III).

This leads to the second fundamental requirement of a typology of valuable system characteristics: is must distinguish between goods and services which are relevant to direct users and those which are relevant to defined higher-scale needs and wants.

Combining these two fundamentals provides a matrix that describes four types of valued characteristics of the present *SEnS* as the 'demand side' of sustainable development: (i) system characteristics related to present use; (ii) the characteristics related to future users; (iii) the characteristics related to higher-scale communities; and (iv) the characteristics related to higher-scale communities in the future (Table IV.1).

IV.2.3 Time and scale characteristics related to a system

The four types of characteristics described above cover all the relevant characteristics on the 'demand' side of sustainable development. However, the search for such development involves a debate in the decision making process on the wanted and acceptable changes in the *SEnS*, i.e., on the changes designed to

Table IV.1: Fundamental types of valuable characteristics of a *socio-environmental system* from the point of view of people involved in a sustainable development (see text for explanation).

Present characteristics related to the people involved		**people involved temporal**	
		present users and communities	future users and communities
people involved spatial	direct users	present characteristics related to present direct users	present characteristics for future direct users
	indirect via higher-scale communities	present characteristics for present higher-scale communities	present characteristics for future higher-scale communities

obtain, maintain or enhance *valued system characteristics*. The various types of characteristics may change in different ways, thus 'supplying' different possibilities for meeting goals. This means that separate information must be collected on these types of 'system supply' characteristics. Have valuable characteristics to be divided into new types of characteristics that are fundamentally different?

Addressing this question means reconsidering the way systems can be described. Chapter III showed that both a structural and a functional description of the system are relevant for sustainable development. It has been argued that a structural descriptions of a system cannot be equated by a functional or vice versa, in other words, a known change in the structure of a system does not immediately tell us what the consequences of that change are for the processes. For example, the number of acres of nature reserve available today and in the future does not tell us the rate at which acres are added or subtracted; it shows only the results of the two processes, i.e., the net increase or decrease. Thus changes in processes are fundamentally different from structural changes.

A similar kind of dichotomy is reflected in the difference between describing a system as a the sum of inputs, outputs, and elements and describing it as the relationship between inputs, outputs, and elements. Known changes in relationships will not immediately reveal how the sum totals change and vice versa. Changes in totals are fundamentally different from changes in relationships.

Obviously, one can design a typology of system characteristics based on these two fundamental dichotomies. However, does such a typology also specify the types of 'demand' characteristics, which are based on the needs and wants of the people involved? In other words, does a further specification of the valuable characteristics based on system description also provide a relevant specification of the needs and wants - or values - at stake? We will now examine this question, making use of the two fundamental requirements highlighted in the previous section: time and scale (the temporal columns and the spatial rows of Table IV.1). The italics below refer to the items which appear in Table IV.2.

Table IV.2: A typology of the characteristics of a *socio-environmental system* related to types of needs and wants of the people involved, now and in the future, both at system level and higher-scale level (see text for explanation).

Present values for:			the present		the future	
	derived from:		processing	use	reserves for future use	capacities for future processing
		type of system charac-teristic:	processes	structures	structures for future use	structures for future processing
direct users	direct users	separate				
	direct users' activities	combined				
higher-scale com-munities	higher-scale activities	combined				
	higher-scale com-munities	separate				

Time-specific system characteristics
In the context of sustainability, the various ways in which needs are satisfied is of crucial importance. There are some goods which lose their specific quality when they are used, such as food and oil. Other goods remain more or less unchanged, such as durables that can be used repeatedly within a certain time horizon, or 'goods' that are conserved for their 'non-use' value, like nature reserves (Booth, 1994; Crowards, 1997). Obviously, in some cases there is a need for *processing*, which results in a transformation of system elements, while in other cases, there is a need for *use* and no transformation occurs. The distinction that Becker & Ostrom (1995) make between 'subtractable' and 'non-subtractable' is based on the same principle. However, not all transformations lead to need satisfaction. On the contrary, some transformations within the system frustrate need satisfaction, and are therefore unwanted, while the lack of transformation - say, recycling - can also frustrate need satisfaction, i.e., the need to conserve fossil energy. Thus system transformation - or the lack of it - can be either wanted or unwanted.

In effect, this difference between processing and use with regard to the way the system changes is reflected in the difference between *process characteristics* and *structure characteristics*. *Processing* refers to inputs and structure elements that can be 'used up or subtracted', and to outputs and structure elements that can be 'produced or added' by the system, and all these processes can be either wanted or unwanted. *Use* refers to the system as a set of structure elements. Although their number and position may change within a certain time horizon, the elements themselves do not change. They are present - i.e., available and accessible - and can be passed on for future use; what is used can be used again, and what is not used may be used in future. As noted in Chapter III, both processes and structures have values.

In principle, only structure elements and their relationships can be passed on to the future. After all, inputs and outputs are gone after transformation, and changing rates are by definition fleeting characteristics. This means that all conditions for the future may be seen as the preservation of present structures and structure elements for later use. Do these present conditions need to be further specified for the future? Yes, they do, since people may have different reasons for passing on those elements, and consequently may want present structure elements for the future that differ from those wanted for the present.

First, they may feel the need to set aside structure elements for the future, in order to provide the opportunity for later use: the *'reserves for future use'*, These elements for direct use are reservoirs consisting of stored material, energy, or information.

Second, people may want to preserve for the future structure elements which can ensure transformation in the future: *'capacities for future processing'*.

Structure elements of this type are characteristics that are conditional for future processing.

In summary, the time columns in Table IV.1 - goods and services for the present and for the future - are divided into four new types of characteristics, which relate the goods and services 'demanded' by the people to the system's present 'supply' characteristics (Table IV.2):

i. The needs or wants concerned with *present processing* are related to the *process characteristics*, i.e., to inputs, outputs, and rates.

ii. The needs or wants concerned with *present use* are related to *structure characteristics* that must be presently availability and accessibility.

iii. The needs or wants concerned with *reserves for future use* are related to present *structure characteristics* that are reservoirs for *future use*.

iv. The needs or wants concerned with *capacities for future processing* are related to the present *structure characteristics* that are conditional for *future processing*.

Scale-specific system characteristics

A system can be described by (i) listing and measuring the relevant separate process characteristics in terms of separate inputs or outputs - such as the food production, energy loss, earned money, etc. - or by (ii) listing and counting the number of separate structure elements - such as number of houses, actors, workers, or available skills. These are the system characteristics at a low level of scale, which determine the 'basic entity level'. This type of system characteristics can be directly valued, when people describe their needed amounts of separate goods and services: the *direct user demands*.

However, when people satisfy their needs and wants, they do not use only one kind of good or one kind of service, but make use of certain combinations of goods and services. When cooking, people make use of food, energy, and water; in their daily lives they make use of houses, shops, roads, schools, etc. To be able to engage in certain activities, people - and organizations - need not only inputs, outputs, or structure elements, they also need those inputs, outputs, or structure elements together, at a certain time, and in a certain place. This results in the need for *combined characteristics*. The possibilities for making adequate use of the separate characteristics depend on a combined supply. What is relevant in these combinations is determined by the actual activities needed and wanted: the *direct users activities*. The needs or wants concerned with these activities results in the need for combined goods and services.

In system theory, this concurrence refers to the definition of systems as an interconnected set of elements (e.g., Clayton & Radcliffe, 1996). In the case of the description of processes, a system cannot be described only as a list of inputs and outputs, or changes in the number of structure elements. A description of the temporal relations between them is also needed. In the case of a structural description, a list of structure elements is not enough, the distributional relationship between the structure elements must be known, i.e., their coherence.

Not only the direct users' groups, but also the indirect users represented on a higher scale may value the combined provision of separate goods and services related to their activities. In the same way their indirect involvement is described, these activities must be described at a higher scale: the *higher-scale activities*. For example, when from the point of view of the higher-scale needs and wants, the focus is on the sufficiency of national employment and habitation. This may be translated into the demand on a system level to realize a certain employment and housing ratio. In other words, there is a need to know the ratio between separate system characteristics: i.e., employment and housing. This ratio is a newly valued characteristic that measures the adequacy of the system to provide combined contributions at a higher-scale level - the *combined characteristics* relevant for higher-scale needs and wants.

In addition to the need for certain combinations of goods and services, the higher-scale communities defined may value the amount or number of goods and services as a contribution to processing and structuring on a higher scale: the *separate characteristics* for a higher scale. One example would be the amount of employment within a region and its contribution to national - higher-scale - employment. But the contribution of the system to indirect interests on the higher level also has to be formulated, for example, its CO_2 emission in the context of the greenhouse effect, or the production of resources for the international market: the *higher-scale demands*. Thus the separate amounts are also relevant characteristics for the higher-scale.

In summary, the scale rows in Table IV.1 - goods and services for direct users and higher-scale communities - is divided into four more specified types of characteristics which relate the people's 'demanded' goods and services to the system's present 'supply' characteristics (Table IV.2):

i. The needs or wants concerned with *direct users' demands* are related to *separate characteristics*.

ii. The needs or wants concerned with *direct users' activities* are related to *combined characteristics*.

iii. The needs or wants concerned with *higher-scale activities* are related to *combined characteristics* which describe the systems' contributions to higher-scale communities.

iv. The needs or wants concerned with *higher-scale demands* are related to *separate characteristics* which describe the systems' contributions to higher-scale communities.

IV.2.4 *More types of characteristics?*

Up until now, the analysis has disregarded the possibility of further subdividing the rows and columns. Is a further subdivision necessary?

If one considers the needs and wants, there is a major argument for extra divisions, since the framework must be tailored to search for characteristics that are valuable to the people involved in their daily lives. From this point of view, it is important that the people involved are able to express their needs and wants in terms of both wanted and unwanted characteristics, and that they can express them as members of a certain group, as members of the community, or as members of a higher-scale community. In order to better accommodate existing distinctions, each type of characteristic could be divided into these types. But does this result in new types of system supply characteristics?

- *Subdividing the cells into wanted and unwanted characteristics.* The difference between wanted and unwanted characteristics is the result of practical alternatives for translating needs and wants into system characteristics. For example, 'poverty' is regarded as a quality of life that the people involved do not need or want. The characteristic related to this demand might be 'number of people with a sufficient income', which is positive. But the alternative characteristic could be 'number of people without income', a characteristic which one will strive to reduce. Both characteristics refer to the same type of characteristic - income - but are expressed in different variables. So here no subdivision of characteristics is necessary.
- *More rows for more levels of scale.* Only one 'higher level of scale' has been assumed, in order to keep the assessment of valuable characteristics manageable. However, it may occasionally be preferable to recognize more than one higher scale. For example, Norton & Hannon (1997) propose three levels: individual, community and global. Westra (1997) has argued that the interests of the national community are not the same as those of the global community. Moreover, recognizing more levels of scale may be necessary when such higher-scale interests are not put forward on the level of the higher-scale community, due to the lack of representatives. But once these issues are identified, they can be operationalized by translating them, first, into needs and wants at the level of the higher-scale community, before being related to system characteristics. In other words, there is no need to relate them directly to system characteristics, resulting in a new row. Only the set of higher-scale needs and wants will be enlarged.
- *More columns for valuable characteristics for the future.* There is a

> comparable problem with the number of columns. There may be differences between needs and wants of the users of tomorrow and the 'future generations', but the type of characteristic remains 'conditions for the future'. Therefore, only one future column will be recognized here.

Neither splitting-up the cells of the proposed framework into wanted and unwanted, nor expanding it with new rows for differently scaled communities, or columns for different time-horizons, would add new types of system characteristics to the framework. Therefore, no such expansion is presented here, although it is possible to expand the framework, and this may be practical, where searching for valued characteristics is the issue and it is necessary to be more specific about the 'demand side'.

IV.2.5 Constructing a framework for system characteristics

A framework is constructed by using one axis to describe the process and structure characteristics for the present and the future and the other axis for the scale-specific characteristics in terms of separate and concurrent combinations of entities on two levels of scale (Table IV.2). Each of the 16 cells of the framework stands for a combination of a time-specific and a scale-specific interest described as a type of system characteristic. Therefore each one defines a type of system characteristic that may be valued by the people involved. Each cell can provide a set of *valued system characteristics* pertaining to sustainable development, and in each of the cells a type of negative or positive effect of a development can manifest itself.

IV.3 A comparison with existing valuation approaches

The framework can be used to determine whether an actual valuation of a *SEnS* includes the types of valuable system characteristics which are important in relation to sustainable development. To search for omissions in the framework proposed, it was compared with existing valuation approaches. This was done by filling it with the actual items involved in different approaches.

Four well-known valuation approaches for policy-making were selected for this purpose. The first is an *economic* approach (Cobb *et al.*, 1995), a monetary index for nations in which the environmental costs of development are taken into account in order to assess the overall health of their economies. The second is an *ecological* approach focusing on the functions of ecosystems and their economic and social values for human beings (R.S. de Groot, 1992). The other two examples are based on defining human needs and wants (Doyal & Gough, 1991, and Henderson, 1991). One takes as its starting point a typology for the basic *needs*, resulting in a list of indicators applied to countries (Doyal & Gough, 1991). The other is a socio-cultural approach concentrating on the *wants* of a local society and the perceptions of the people of that society (Henderson, 1991). It describes a list of social indicators which assess the quality of life in a town.

We are aware that restricting the analysis to these four approaches means that many others have been left out. However, the scope of the present volume does not permit a review of all existing valuation approaches. Therefore, those approaches were selected that are most often referred to in the literature and are applicable at the national, regional or local level of scale. They cover essential types of valuable characteristics that people will actually recognize in existing *socio-environmental systems*. This is confirmed by the fact that economic, sociocultural, and ecological aspects are generally considered of primary importance for sustainable development (e.g., Munasinghe, 1993; Goodland & Daly, 1996).

As noted in Section IV.1, neither the four approaches nor any of their applications presume to generate a complete list of all the *valuable system characteristics*. However, a comparison of the framework with conventional valuation approaches will clarify the relevance of the distinction made in the proposed framework and may bring to light certain omissions.

IV.3.1 Procedure for filling in the framework

The applications of the four approaches entail widely differing systems and levels of scale, and in terms in which the valued items are described may differ considerably. This means that the applications may not all be validated in an equivalent manner. To solve this problem, the following procedure for filling the framework was used:

- All the items are described according to their original designation in the literature, and none of the items have been omitted without explanation.
- The analysis of each application starts by distinguishing items which are clearly related to a lower-scale than the system as a whole from those which are clearly related to a higher-scale community, and fitting the lower-scale ones into the first two rows of our framework.
- The higher-scale community represents 'public' needs and wants, and thus the characteristics valued by the entire community are considered higher-scale needs and wants. These are fitted into the last two rows.
- If an item refers explicitly to a level of scale higher than that of the

system community, this is indicated by an asterisk (*).

- If items are expressed in terms of percentage or rate, they are considered combinations of characteristics and will be entered in the second or third row (combined characteristics).
- Each item is checked to see whether it describes an actual process, i.e., whether the item itself can change within the time horizon. These items are placed in the first column (processes). But the items that do not themselves change - only their number or place - are considered a description of structure characteristics and are therefore placed in one of the other columns (structure elements).
- The items categorized above are checked to see whether they are relevant for the future; if so, they are placed in one of the last two columns. It has been verified that the items in the first two columns are also essential for future use, and thus may be regarded as system elements passed on to the future. These latter items are to be found in the last two columns, between brackets (..).
- Where items could be slotted into more than one cell, they have been placed in the one that seemed most appropriate. The item appears in the other cells between brackets (..).

IV.3.2 Economic: the 'genuine progress indicator'

In environmental economics aggregated indexes are often proposed to assess the quality of developments, and much effort has been put into 'greening' the gross national product (GNP), by defining shadow prices or assessing 'willingness to pay' (e.g., Daly & Cobb, 1989; Hueting & Bosch, 1991; Common *et al.*, 1993; Freeman, 1993; Adger *et al.*, 1995) The main argument in favour of this strategy is its simplicity and power: it dramatically reduces the amount of information presented to decision makers, and compels them - for the sake of their own credibility - to take action when a previously agreed index moves in the wrong direction.

The 'genuine progress indicator' (GPI) is a recent attempt to capitalize the social and environmental costs of developments, so that the actual total consumption of a country can be made more transparent (Cobb *et al.*, 1995). It is a modification of the well-known 'index of sustainable economic welfare' (Daly & Cobb, 1989), and the items taken into account are the same kind of socioeconomic characteristics that must be identified when searching for sustainable development. More than 20 of the aspects of human economics are taken into account in the GPI that are disregarded in the GNP. Let us look at the valuable characteristics which

Table IV.3: The Genuine Progress Indicator aspects as types of valuable characteristics of the United States as a *socio-environmental system*; + = adding to personal consumption, - = subtracting from personal consumption (after Cobb *et al.*, 1995). * = valuable on both community level and higher-scale levels, (..) items also in other cells (see text for explanation).

Characteristics related to direct users

	processes	structures	future use	future processing
separate	+value of household work +value of volunteer work - cost of family breakdown - loss of leisure time - cost of commuting - cost household pollution abatement - cost of auto accidents	(+household work) (+volunteer work) +services of consumer durables +services of highways & streets - cost of under employment - cost of consumer durables	(consumer durables) (highways & streets) (farmland)	- loss of farmland
combined				

Characteristics related to higher scale:

combined		income distribution		
separate	personal consumption - cost of crime - cost of water pollution - cost of air pollution - cost of noise pollution (- cost of auto accidents)		(- water pollution) (- air pollution) - depletion of non-renewable resources	- loss of wetlands * - long-term environmental damages - cost of ozon depletion * - loss of old-growth forests *

are identifiable in the GPI.

The procedure proposed above resulted in Table IV.3. This figure does not include structure items or combinations of characteristics for the higher-scale community, except for 'income distribution', which is a combination of characteristics. There are also very few items for the future. Although the higher-scale items refer mainly to characteristics on the community level in the United States, they may also be related to the context of that community, i.e., the world. This can include 'net investments', 'net lending and borrowing' and such environmental items as 'ozon depletion'.

Some items were difficult to fit into the framework, such as 'auto accidents'. These may be seen not only as an unwanted characteristic for the sub-population of road users - as we do here - but also as an unwanted characteristic for the whole community (it must be noted that the GPI does not take into account the costs of all other accidents which affect the community). In neither case, however, can the item be included in a row reserved for combinations of characteristics. Another example is 'work-related items'; which are also difficult to fit into the framework. For example, if the item 'volunteer work' or 'household work' is described in terms of the number of 'potential jobs', it should be placed in the structure column. But it can also be described as labour which is needed or provided, and in that case it would be placed in the process column; that is what has been done here, since it is being presented in a monetary figure. However, such items cannot be placed in the columns for the future, because they are not presented as a labour market, or as a reserve for the future.

Thus although some items in the GPI may be placed in other cells, moving them does not influence the overall picture presented by Table IV.3. Obviously, the GPI contains no information on combined characteristics and very little on conditions for the future or contributions to the higher scale. This confirms that it is primarily a monitoring instrument, which can be used to assess whether an actual development is desirable or not. It is not designed for exploring opportunities nor, when used on its own, is it helpful in the search for sustainable development.

IV.3.3 Ecological: functions of nature

The biophysical characteristics of a system form the starting point of an ecological approach; their relevance for human beings is described as functions. An environmental function may be defined as 'the capacity of natural processes and components to provide goods and services that satisfy human needs (directly and/or indirectly)' (R.S. de Groot, 1992). There are two ways in which these

Table IV.4: Functions of tropical forests (after R.S. de Groot, 1992) categorized as types of valued system characteristics. * = valuable on both community level and higher-scale levels, "-->" = a copy of the functions described in the 'present' cells in the same row (see text for explanation)

Characteristics related to direct users

	processes	structures	for future use	for future processing
separate	production functions: - food/nutrition - genetic resources - medical resources - raw materials for manufacturing - biochemicals - fuel and energy - fodder & fertilizer - ornamental resources	information functions: - aesthetic information - spiritual/ religious information - historic information (heritage value) - cultural/artistic inspiration - educational & scientific information	-->production functions -->information functions	-->production functions -->information functions
	regulation functions: - watershed protection & catchment - erosion prevention - storage/recycling of human waste	carrier functions (use of the forest) for: - habitation for indigenous people - cultivation - recreation (- nature protection)	-->regulation functions -->carrier functions	-->regulation functions -->carrier functions
combined				

Characteristics related to higher scale

combined				
separate	regulation functions: * - buffering of CO2 * - climate regulation * - fixation of (solar)energy * - storage/recycling of organic matter - biological control *	carrier functions: * - nature protection * regulation functions: * - migration habitat * - maintenance of biological diversity *	-->carrier functions * -->regulation functions *	-->carrier functions * -->regulation functions *

environmental functions can be lost: first, due to changes in environmental properties such as erosion or the dumping of toxic industrial waste, and second, due to more stringent demands that make what was acceptable yesterday unacceptable today or vice versa. For example, water that was 'drinkable' yesterday might be regarded 'undrinkable' today. In the Netherlands, several typologies for environmental functions have been proposed (an overview is given by W.T de Groot, 1992). Most of these are based on the work of Van der Maarel and Dauvellier (1978). The following functions are included: (i) regulation (of ecological processes); (ii) carrying (the environment as a carrier for human and other activities); (iii) production; and (iv) information. Functional approaches of this kind have also been winning support outside the Netherlands (Bingham *et al.*, 1995; Myers, 1996; Christensen *et al.*, 1996; Costanza *et al.*, 1997).

Because functions are 'possibilities for use', they refer implicitly to characteristics that are valuable for the present as well as the future. The function 'production of goods', for example, is of value for the people now involved, but should also be passed on to the future.

In this chapter the focus is on the functions of a tropical forest and its socioeconomic values, as discussed by R.S. de Groot (1992). These functions were entered into the proposed framework, and the results are shown in table IV.4. Only the concept of 'nature conservation' was difficult to fit into a single cell, because of the problems presented by interpretation. Not only does it have special importance for specific local users, at a lower-scale level, but it is also important for the higher-scale community. The latter has been adopted here. It should be noted that the regulation functions are described as *processes*, and placed in the process column. However, some of them, such as migration habitat are expressed in terms of structure elements, which should put them in the structure column. The global level of scale is explicitly recognized.

Table IV.4 contains no information on combined valuable system characteristics. This is probably because the function approach is based on systemizing the potential use, rather than structuring the actual use demanded, and deriving from that the function demands and the way the function must be described. The focus is on showing what a system is worth, rather than what is needed or available for actual use. In that sense it places the focus on the assessment of natural capital (Costanza *et al.*, 1997). As long as the separate system characteristics demanded correspond to the defined functions, the function approach can be used to analyze what is needed to obtain or maintain the selected valuable system characteristics. However, when the focus is on the combined valuable characteristics, such as the combination of food production and habitation, the separate functions listed do not provide the information needed to determine whether this demand can be met. For that, the function demands must first be

Table IV.5: Based on "Substantive need-satisfaction in the Three Worlds" (Doyal & Gouch, 1991) categorized as types of valued system characteristics. Characteristics valuable on both community level and higher-scale levels were not found, (..) items also in other cells (see text for explanation).

Characteristics related to direct users

	processes	**structures**	**future**
separate			(education)
combined	survival/health: - infant mortality before one year of age/1000 live births - mortality of children under 5 years of age/1000 live births - proportion of babies under 2500 gm reproduction: - % of married women of childbearing age who are using, or whose husbands are using, any form of contraception, whether traditional or modern methods - annual number of deaths of women from pregnancy-related causes per 100,000 live births (education)	autonomy: - literacy, % of persons aged 15 and over who can read and write education: - proportion of adults who have ever entered secondary education - idem: post-secondary ed. - number of secondary school pupils of all ages as % of children of secondary school age - number of students enrolled in all post-secondary education, schools and universities, divided by population aged 20-24	

Characteristics related to community

combined	gross National Product/head gross Domestic Product/head at "purchasing power parities" water/nutrition: - calories (daily calorie supply/head as % of requirements) security: - homicides/100,000 population	survival/health: - life expectancy at birth in years water/nutrition: - access to drinking water, as defined by WHO housing: - % of housing units with more than 2 persons per room health services: - population (thousand)/physician - % of population with access to health services as defined by WHO security: - War deaths between 1945 and 1985, as % of population - % of population with incomes below that where a minimum nutritionally adequate diet plus essential non-food requirements is affordable, as estimated by World Bank.	
separate		total population	physicians

defined - suitable for habitation and food production - before the presence or absence of this function can be established. This will lead to the definition of a new function derived from the functions 'habitation' and 'food production'.

IV.3.4 Human needs: satisfaction in the 'three worlds'

This approach focuses exclusively on basic human needs and is based on a typology described in the *Theory of Human Needs* (Doyal & Gough, 1991). The authors present a number of need-satisfaction 'indicators', i.e., measurable characteristics of a system in the 'first, second and third world'. Only a small number of these indicators are actually used in the application they describe, due to lack of information. Those that are used are categorized by means of the framework.

The results of the analysis are given in Table IV.5, which clearly shows that there are few items for the future, and few items that describe amounts; there are no global aspects at all. While not presented as such, three items can nevertheless be seen as important for the future: 'literacy', 'education level', and 'number of physicians'. The 'education' indicators were difficult to place, because the description does not specify whether they refer to yearly changes - a process characteristic - or the overall education level of the total population - a structure characteristic. However, placing these items in both columns does not influence the overall picture.

As with the GPI, the human-needs approach is designed to monitor developments, as witness the term 'indicator'. The economic, social, and biophysical characteristics of the system that determine the value of the social indicators, however, are not identified. As regards the filling in of the framework, this approach emphasizes the listing of combined characteristics, but does not specify the conditions necessary to achieve these characteristics.

IV.3.5 Human wants: quality indicator for progress: Jacksonville, Florida, USA

Because we are looking for characteristics of systems that are valuable for human beings, it is appropriate to focus attention on methods which describe the human wants and the system characteristics necessary to meet them. Such an approach was found in the 'quality indicator for progress' (QIP) for Jacksonville, Florida, USA, which claims focus on 'social cohesion' (Henderson, 1991). The QIP is a set of indicators, based on the issues which the people involved see as important. These indicators were compared with the proposed framework, and the

Table IV.6: The categories of the Quality Indicator for Progress in Jacksonville, Florida, chosen to measure its Quality of Life. Data have been recorded in these categories since 1983 (Jacksonville Chamber of Commerce) (after Henderson, 1991). Categories as the types of valuable system characteristics of a *socio-environmental system*. (..) items also in other cells (see text for explanation).

Characteristics related to direct users

	processes	structures	future
Separate	1. revenue generated by the Bed tax 1. taxed taxable value of real estate 4. total student enrolment 4. academic degrees awarded 5. new sewer permits issued 6. total weekday commercial flights in and out of JIA 8. employment discrimination complaints filed with the Jacksonville Equal Opportunity Commission 9. annual expenditures of major art organizations 9. days of bookings of major city facilities	5. sign permits issued 6. direct flight destinations to and from JIA	
combined	(2. *motor-vehicle accidents/1000 population*) (3. *death-rate indicators*) 4. standard Achievement Test Score 4. student dropout rate 4. education expenditures/student 5. frequency of compliance of St. John's River with water quality standards 5. frequency of compliance of tributaries with water quality standards (desolved oxygen) 6. average weekday ridership on JTA buses 6. average miles of JTA bus services/weekday 6. commuting time from downtown along boulevards 7. % of registered voters who voted 7. % households purchasing local Sunday newspaper 7. % who can name accurately two City Council members 7. rate of Planning Department/City Council concurrence on rezoning petitions 7. % who rate the quality of local government leadership good or excellent 7. days from arrest to disposition of criminal cases 8. child abuse and neglect reports /1000 children under 18 8. % of white persons who believe racism to be a problem 8. % of non-white persons who believe racism to be a problem 8. % who volunteered time during the past year 8. licensed day-care spaces/1000 children under 5 years	1. total unemployment rate 1. black unemployment rate 1. teenage unemployment rate 1. average price of previously owned, single-family home 1. cost of 1000 kw hours of electricity (DEA) 3. % who exercise three times/week 3. % who rate the medical health care system good or excellent 3. % who rate their own health good or excellent 4. average public school teacher salaries 4. % of teachers holding advanced degrees 4. % of faculty holding terminal degrees 4. average faculty salaries at public institutions 7. % of population eighteen and over registered to vote 7. % of City Council members who are blacks 7. % of City Council members who are female	(+) (+) (+) (+)

Table IV.6: continued.

Characteristics related to higher scale

	processes	structures	future
combined	1. effective buying income/capita 1. annual retail sales/capita 2. index crimes/100,000 pop. 2. % who feel safe walking alone at night 2. rescue call response time 2. fire call response time 2. police call response time 2. motor vehicle accidents/1000 population. 2. fire and rescue operating expenditures/capita 2. law enforcement operating expenditures/capita 3. age-adjusted deaths/100,000 population 3. deaths due to heart disease/ 100,000 population. 3. suicide/100,000 population 3. deaths due to cirrhosis of the liver/100,000 population. 3. packs of cigarettes sold/capita 5. per capita tons of solid waste 7. city capital outlay expenditures/ capita 7. total city revenues/capita 8. city government human services expenditures/capita 8. contributions/capita to the United Way and its member agencies 9. public-library book circulation/capita 9. zoo attendance/1000 population	2. % who have been victims of crime 4. % who are high school graduates or above 4. % who are college graduates or above 5. environmental public employees/100,000 population 9. public-park acreage/100,000 population 9. city park and recreation facilities/100,000 population 9. public-library materials/capita	(+) (+) (+)
separate	5. days when the Air Quality Index is in the 'Good' range		

1. The economy:	-	includes individual economic well-being and community health.
2. Public safety:	-	includes the perception of public safety, and the quantity and quality of law enforcement, fire protection and rescue services.
3. Health:	-	refers to the physical and mental health of residents and the local system of health care.
4. Education:	-	includes the system of public education (kindergarten through 12th grade), higher education, including adult education, and the overall literacy and educational attainment of the population.
5. Natural environment:	-	includes the natural elements of the earth's ecosystem, the quality and quantity of water, air, green space, and landscaping and visual aesthetics.
6. Mobility:	-	refers to the opportunities for people to travel freely within Jacksonville and between Jacksonville and other locations.
7 Government/politics:	-	includes an informed and active citizenry, professionalism and performance of local government.
8 Social environment:	-	encompasses collective or group concerns such as equality of opportunity, racial harmony, family life, human services, philanthropy and volunteerism.
9 Culture/recreation:	-	includes the available supply and use of sports and entertainment events, the performing and visual arts, public recreation, and leisure activates.

results are shown in Table IV.6. We see that there are almost no characteristics for the future, or totals for the community, or characteristics for a level of scale higher than the community level.

The 'death-rate indicators' were difficult to fit in. Because the type of cause refers to specific 'user' groups at risk of, say, 'cirrhosis of the liver', it could be located on the direct-user level. However, the rates are expressed per population and not per specific group. They refer primarily to the quality of the system in terms of the total community, and thus are placed on the higher-scale community level.

Since most of the items describe present activities, it was not always easy to decide which of these activities could also be interpreted as system characteristics which are relevant for the future. The exceptions were 'educational activities' and the use of 'public facilities'. The difficulty of assessing the relevance of indicators for the future shows that supplementary information is needed in order to arrive at an accurate appraisal of certain items.

The human-wants approach contains very little information on conditions for the future or the total scale of the system. However, this fourth approach clearly shows that a search for valuable characteristics based on the ideas of the people involved can lead to a list of items covering a wide range of issues, which can be used to fill at least part of the framework.

IV.4 Discussion

The analysis of the applications shows that none of the four conventional valuation approaches seems able to deliver information on all the types of valuable system characteristics of the framework, but also that all items of each of them can be related to one or more types. For purposes of the search for sustainable development, it is important to know whether combining approaches can provide the relevant information.

Combining the four approaches into one framework shows that in many of the cells at least one item of one application occurs. In Table IV.7 we see that the first two approaches - the economic and the ecological - emphasize the valuation of separate goods and services both in the present and the future. The function approach likewise emphasizes the global level of scale. Table IV.7 also shows that the items of the two human needs-and-wants approaches are expressed mainly as combinations of system characteristics. The explanation for this distribution is obvious: the first two focus on 'supply side' aspects of a *SEnS* and value them separately, while the other two start by describing what people want to have and do

Table IV.7: Filling in the framework: cells in which items of the approaches occur as described in the Table IV 3-6. 1=GPI (Table IV.3); 2=Functions (Table IV.4); 3=Need satisfaction (Table IV.5); 4=Human wants (Table IV.6). * = valuable on both community level and higher-scale levels, (..) items also in other cells ((see text for explanation).

Present values for:			present direct users and communities		future direct users and communities	
	derived from:		processing	use	reserves for future use	capacities for future processing
		type of system charac-teristic	processes	structures	structures for future use	structures for future processing
direct users	direct users	separate	1 2 . 4	1 2 . 4	1 2 (3) (4)	1 2 . .
	direct users activities	combined	. . 3 4	. . 3 4	. . . .	. . . .
higher-scale com-munities	higher-scale activities	combined	. . 3 4	1 : 3 4	. . . .	. . . .
	higher-scale com-munities	separate	1* 2* . .	. 2* 3 .	1 . (3) (4)	1 2* . .

- the 'demand side' - and derive characteristics from that. The latter will of course lead to a demand for combinations of system characteristics. In this light, it is remarkable that four cells that describe combinations of characteristics are not filled in at all. These are combinations of characteristics not for present needs satisfaction, but for future use or future processing. They are empty because none of the human needs-and-wants approaches take future conditions into account. They are indeed "short-sighted and anthropocentric" (Westra, 1997). However it is also possible that information on combined characteristics for the future is already present in other cells, so that the cells reserved for the future are redundant.

The conclusion is that this is not the case, because the kinds of combinations of characteristics valued for the future may differ from those valued for the present. When focusing on a sustainable development of a system, there are two major objectives: to identify and solve the problems of the people now involved, and to minimize the risk of creating need-frustrating situations at all levels in the future. That means that the aim is to provide and maintain the conditions for the future use of the *SEnS*. One of the options is to increase the efficiency of resource use. However, such increased efficiency may not be in the interest of the present users, for example, because it would result in higher costs. If the framework lacks the cells that force us to focus on combinations of characteristics in relation to the future, then efficiencies that are not profitable at the moment might be missed.

The above discussion leads to the conclusion that the proposed framework cannot be reduced to any one of the four approaches. Nor does any combination of the four deliver sufficient information on all the interests at stake.

While all the items of all the applications were entered into the proposed framework, some items were difficult to place. This was not because potential cells were difficult to find, but because, depending on how each item was interpreted, they could be entered fit into two or more cells. The item 'auto accidents', for example, was expressed as 'costs', but could also be regarded as (i) costs for the community; (ii) risks for direct users of roads; (iii) loss of capital and durables; (iv) a characteristic reflecting one's perception of safety; and numerous other things. It is suggested here that all these aspects of 'auto accidents' should be expressed separately in measurable variables, because they are valued differently by different groups of users. If the costs are valued, the focus is on reducing the financial impact of auto accidents. If the risk of injuries due to accidents is valued, then improvements in the safety aspects of cars and roads can be considered.

In order to assess the extent to which any specific item covers values, more information is needed about the reason for placing the item on the list in the first place. The fact that the proposed framework exposes the need for a distinction of this kind demonstrates its advantages. It forces users to specify values, in order to find valuable characteristics by which to assess developments.

The above examination also hints at a method for using the framework as a search instrument. Whenever a valued characteristic is slotted into one particular cell, one can also check to see whether it can be placed somewhere else. For example, if 'amount of food production for direct users' is placed in the first cell, the question is whether food reserves are needed, whether food is part of a certain valuable combination of goods, whether food production requires a certain efficiency, whether conditions for food production are to be conserved, what the scale of food production should be, etc. In our view, the ability to reveal this kind of highly relevant questions is the beauty of the proposed framework.

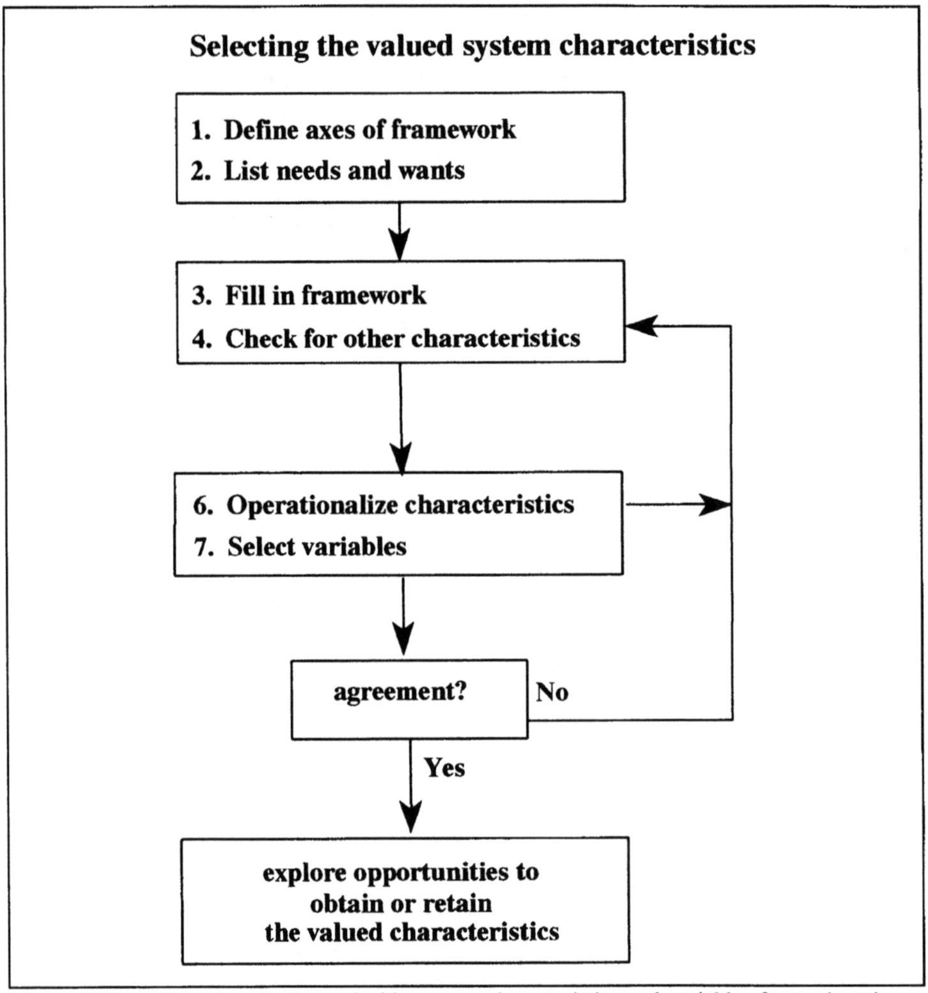

Figure IV.2: Steps in selecting valuable system characteristics and variables for goal setting

IV.5 Concluding remarks

The above discussion leads to the following conclusions with regard to the methodological steps involved in the search for valued characteristics. The search for *valuable system characteristics* is based on decisions made during the defining of the *SEnS* (Chapter III), where the object to be steered is described by its boundaries, its subsystems, the people involved, and its temporal and spatial scale. The selection of the *valued system characteristics* can be based on a framework that categories the characteristics (Table IV.2).

The steps to be taken in the search for valued characteristics are as follows (Figure IV.2):

1. Define the time and scale axes of the framework on the basis of the *SEnS*.
2. List the needs and wants or issues of the people involved
3. Fit the needs and wants described as system characteristics into the cells of the proposed framework
4 Check to see whether characteristics can also be placed somewhere else, to refer to another type of needs and wants

Although not discussed above, this valuation must be followed by the steps:

5. Operationalize the content of the cells through reduction via one or more techniques, such as partial lumping, using indicators, or prioritizing and selecting valuable characteristics
6. Select the variables to be measured in order to assess developments.

Thus far two of the four activities mentioned in Chapter II have been operationalized. The next activity, measuring the possibilities to achieve the *valued system characteristics*, is discussed in the following chapter.

CHAPTER V

Measuring the Resource State of a Socio-Environmental System[1]

CONTENTS

[1] *Based on: Musters, C.J.M., H.J. de Graaf, M.A.W. Noordervliet & W.J. ter Keurs, (1994). Measuring environmental utilisation space: Can it be done? Milieu 9 (5): 213-220.*

V.1 Introduction

In Chapter III the focus was on defining the object to be steered as a *social-environmental system* (*SEnS*). It was stated that the steering of a *SEnS* by the decision makers of that *SEnS* usually means influencing the inputs and outputs of subsystems, their numbers, and their positions. In Chapter IV a framework was designed to select the characteristics of a *SEnS* valued by the people involved, or their representatives. That framework includes characteristics for direct users and higher scale communities, for the present and the future. These are valued characteristics of a *SEnS* as a whole. The application of the framework can ensure that economic, social, cultural, and ecological goals - both present and future - are taken into consideration, for the benefit of those involved either directly or indirectly. However, while the framework can show which aspects of a development are important - and therefore which information should be available to judge developments - it does not make clear what properties of the *SEnS* determine the realization of goals. In steering developments, the latter information is just as essential as the former. In order to make clear the type of the information which is needed on the properties of the system that determine the realisation of goals, we will look at a concrete example.

Costanza *et al.* (1997) set out to determine the total monetary value of the world's ecosystem services and natural capital. According to their calculations, the entire biosphere has a value of somewhere between 16 and 54 trillion dollars per year, on the basis of the potential value of its use for various purposes. This means that the largely uncounted ecological capital and infrastructure is at least equal to - and may be as much as three times higher than - the global gross national product of approximately 18 trillion dollars. This calculation is a valuable exercise for two reasons. First, as the authors emphasize, it will help to focus attention on ecological concerns. Evaluating ecosystem services and natural stocks in a single currency is an attractive option, because it forces people to take a stand when its value changes. Second, the calculations behind this overall figure can provide insight into what are regarded as the individual services that make up ecological capital. But do they also show the opportunities for using that capital?

The services of a tropical forest that are recognized by Costanza *et al.* (1997) are climate regulation, disturbance regulation, water regulation, water supply, erosion control, soil formation, nutrient cycling, waste treatment, biological control, production of food, raw materials, and genetic resources, recreation, and cultural services. In the valuation of these services, it is assumed that they are used in such a way that the forest is not altered. However, what would happen if the forest was changed by such use? A forest that is intensively used for hunting may lose part of its genetic resources, while retaining all its other services. Thus the overall value of the forest may increase, since the loss of value due to loss of genetic resources may be compensated for by the gain in terms of food. But if wood is harvested on a large scale, the forest may change so dramatically that many of its services are lost. In exploring the opportunities of development, this kind of information is crucial. What information is exactly needed to bring opportunities to light?

The problem is, of course, that Costanza *et al.* (1997) are treating a tropical forest as a single resource that generates many different values, while it is in effect a complex of resources which can be used independently, although using one will affect the availability of the others. In exploring opportunities, information is needed on the availability of separate resources, and the relationship between these resources when they are actually used.

As noted above, the steering of a *SEnS* means influencing the inputs and outputs of the subsystems, their numbers, and positions. To be able to apply adequate steering, information is needed on the resources that can be used by the subsystems. Since changing one resource may affect the availability of other resources, the relationship between resources need to be known. Consequently, for purposes of steering, information is needed on the availability of interrelated resources.

In this chapter the focus is on measuring the available interrelated resources which are relevant for the subsystems and the relations between them. Here resources are taken to mean the financial, sociocultural, and biophysical properties of the system which can be used at the subsystem level. The availability of all these resources together will be called the *resource state* of the system. All the options for using each *resource state* constitute the *use space* of the system at a given moment. It is assumed that the opportunities of a *SEnS* arise from modifying the *resource state* as well as the *use space* of the system. But how can the *resource state* be measured in such way that its modifications can be explored? And how can this knowledge be used to explore the opportunities for development? These questions will be examined in this chapter. First, the concepts *resource state* and *use space* will be specified (Section V.2). Then the literature will be examined to find ways of measuring the *resource state* (Section V.3), after which a procedure for measuring the *resource state* will be proposed (Section V.4), and aspects of exploring the opportunities for sustainable development discussed (Section V.5).

V.2 The resource state and use space of a socio-environmental system

Sustainable development is a development with a high potential for continuity, i.e., a development not likely to cause problems for its environment or for society itself, because it meets economic, social, cultural, and ecological goals for the present and the future (Chapter II). However, such a development is subject to various types of constraints. Some of these constraints are a direct consequence of the system boundaries chosen, such as the available area (Chapter III). Others are imposed on the inputs or outputs of the system by higher-scaled decisions, such as the amount of CO_2 production permitted (Chapter III). Still others are a direct consequence of the set of the system characteristics to be attained (Chapter IV), such as the need to reserve a certain non-use area. But developments can also be constrained by the financial, sociocultural, and biophysical properties of a *SEnS* and by the processes or activities that take place within the system. The opportunities for the sustainable development of a *SEnS* are determined by all these various constraints. Section V.2.1 focuses on the properties of the system itself, while Section V.2.2. examines the options for activities and processes.

V.2.1 The resource state

As we have seen, in exploring the opportunities for the sustainable development of a system, all *valued system characteristics* must be taken into account together. Therefore, to steer the system, it is necessary to know how the delivery of this complete set of valued characteristics depends upon the system, i.e., the internal subsystems and their structures and processes. These structures and processes of the subsystems depend upon the *resources* of the system. For example, if 'availability of drinking water' is a valued characteristic, 'drinking water production' is the relevant process, and 'water for the production of drinking water' is one of the *resources*. Thus, '*resource*' is defined here as any property of a system that is needed for the delivery of one or more *valued system characteristics*. Note that by defining *resources* in this way, 'sinks' are regarded as *resources*.

A list of *valued system characteristics* can be used to access the relevant internal structures and processes of the subsystems, and the *resources* needed. The quantity of these *resources* can be measured, and in effect this is what is done in conventional resource management and land-use research, as will be discussed in Section V.3 (e.g., Barbier, 1989; FAO, 1984; WRI, 1996). But for an exploration of the opportunities for the sustainable development of a *SEnS*, a knowledge of the quantity of individual resources is not sufficient. To obtain or maintain *all* the valued characteristics selected, it is necessary to know the relationship between these characteristics and the possibilities for modifying any property of the system, whether resources, structures or processes. And this means that apart from the availability of all resources, the relationship between them must be known.

In the case of the *resources*, this means first of all that the relationship between the quantity of the various *resources* must be known. For example, if more drinking water is to be delivered by the system, the quantity of the resource 'water for the production of drinking water' may decrease. But the same system needs water to produce food. The next question is, whether the quantity of the resource 'water for the production of drinking water' is related to the quantity of the resource 'water for irrigation'? If the two *resources* are measured independently, this question cannot be answered, because it will be unclear whether 'water for the production of drinking water' and 'water for irrigation' are to be found in the same water reservoirs.

But it is not sufficient to know the availability of interrelated *resources* at the level of scale of the system. After all, modifications to the system take place at a level much lower than that of the system as a whole (Chapter III). The availability of interrelated *resources* must be known at a level that is relevant for the processes and structures within the system. This level has been called the resolution, or 'grain of information' on the *SEnS* (Chapter III). Thus the system must be divided into 'units' and the availability of the *resources* and their relationship within these units must be determined.

To overcome the problem of seeing *resources* in isolation, a *resource* must be described in terms that are relevant for different structures or processes. Something can only be used as a *resource* when it has the right qualities, and is available at the right place and time. Thus to be a resource, a part of a system - whether it is a structure element like a reservoir or a flow of material, energy, or information - has to have certain characteristics. Some of these characteristics may have relevance for another resource, for processes or for structures. In the case of drinking water production, for example, the resource has the characteristics 'accessible' and 'nitrogen-free'. This characteristics make it possible to use the same water for irrigation as well, if the resource water for irrigation is characterized as 'accessible'. These are the characteristics of interest here, because they relate resources to one another and to the different structures and processes. To take another example, a nature conservation area that needs the inflow of clean water. In that case 'inflowing water' must be regarded as a resource for the nature conservation area. This resource can be specified as 'unpolluted inflowing water at a certain location within the system'. But when it is known that in the same system agriculture is a major polluting activity as a result of overmanuring, the resource can also be described by the required characteristic 'inflowing water not polluted by fertilizers'. In the latter case, the relationship between agriculture upstream and nature conservation downstream can in theory be established through the description of the availability of a resource at a certain location. Areas that are situated downstream from agricultural areas may be unsuitable as nature conservation areas, because they lack the resource 'inflowing water not polluted by fertilizers'.

Assuming that each resource can be specified by describing it as a set of system characteristics that relate it to other resources, structures, or processes, all the descriptions of resources will together result in a set of '*resource system characteristics*'. By desribing the presence of these characterisitcs in the units of the system, it is possible to describe each unit as a combination of potential resources, and the system as a set of available interrelated resources at the appropriate level of scale. This can be regarded as a measurement of the *resource state* of the system at the relevant level of scale. Thus in theory, it is possible to measure the internal properties of the system in a way that is relevant to our exploration of the opportunities of sustainable development. Parallel to the *value space* (Chapter III), all possible resource states of a system together form what can be referred to as the *resource space of the system*. The dimensions of this space are indicated by the *resource characteristics*. Before discussing how this *resource space* can be used for the exploration of opportunities for development, the concept of '*use space*' must be clarified.

V.2.2 The use space

The *resource state* of a system at a given moment enables the structures and processes of the system to give it its present *value state* (Chapter III). From the human standpoint, human management and activities 'use' the *resource state* to produce the *value state*. It will be clear that the production of the *value state* depends not only on the *resource state* of the system, but also on the inputs and outputs of the system. For the sake of clarity, however, the discussion will focus on the use of the *resource state* to produce the *value state*.

A given *resource state* may be used in various ways. However, in the case of the *present resource state*, the number of ways is not unlimited. It will depend on such factors as the standards for resource use, resource growth rates, techniques available for using resources, social and cultural constraints on management and activities, the willingness to change present use, and potential for financial investment. Each way a *resource state* can be used will result in a different *value state*. All the different ways of using a particular *resource state* must be seen as the '*use space*' belonging to that *resource state*. The *use space* can lead to a set of potentially attainable *value states*, which together form the *feasible value space* related to a specific *resource state* (Chapter III; Figure V.1).

The *resource state* of a system is not constant. It changes autonomously, it changes as a result of present use, it will change under the influence of new ways

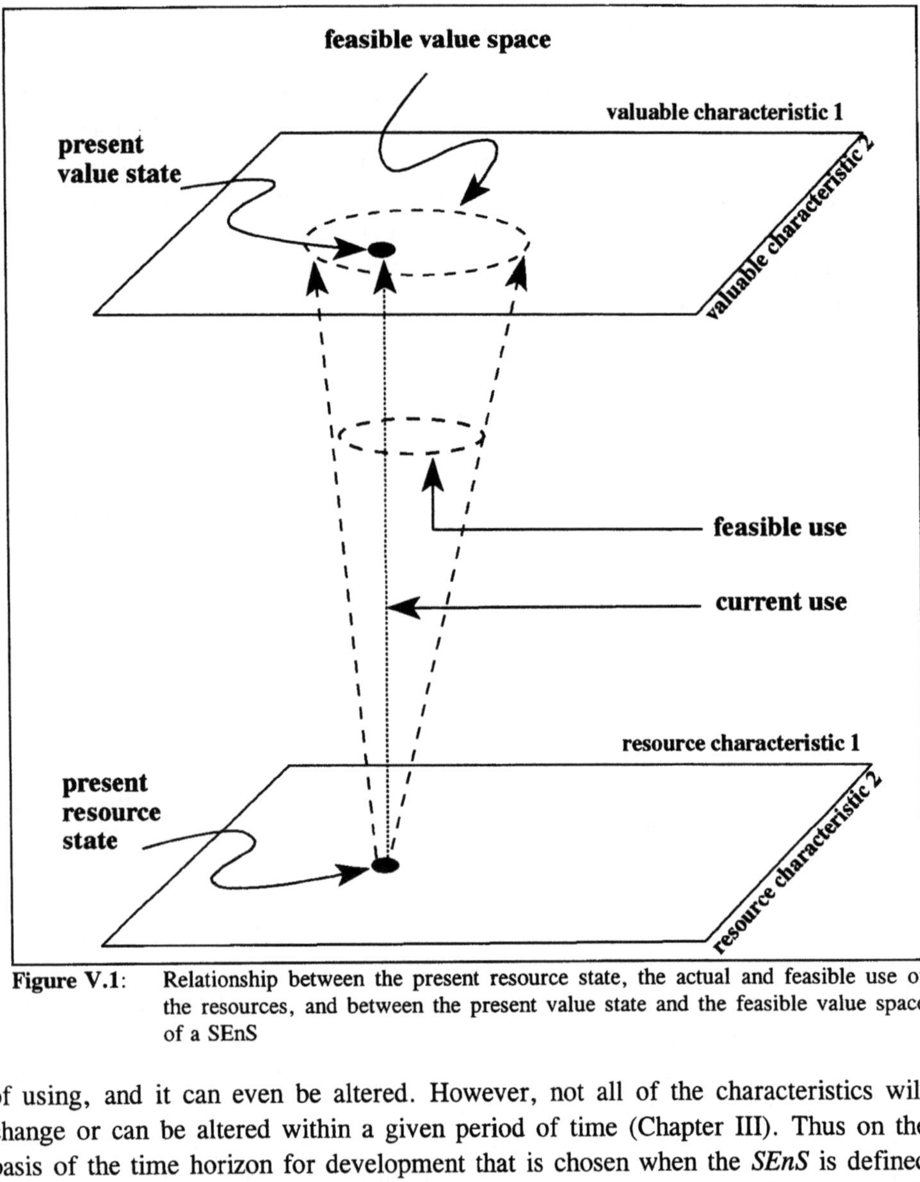

Figure V.1: Relationship between the present resource state, the actual and feasible use of the resources, and between the present value state and the feasible value space of a SEnS

of using, and it can even be altered. However, not all of the characteristics will change or can be altered within a given period of time (Chapter III). Thus on the basis of the time horizon for development that is chosen when the *SEnS* is defined (Chapter III), the system can adopt a limited set of *resource states*, which here will be called the *feasible resource space*. This will increase the use space, but it is still limited due to the limited resource space.

The complete set of *value states* that can be attained on the basis of the available *feasible resource space* and the *feasible use space*, may be termed the *feasible value space* of the system within a certain time horizon (Figure V.2).

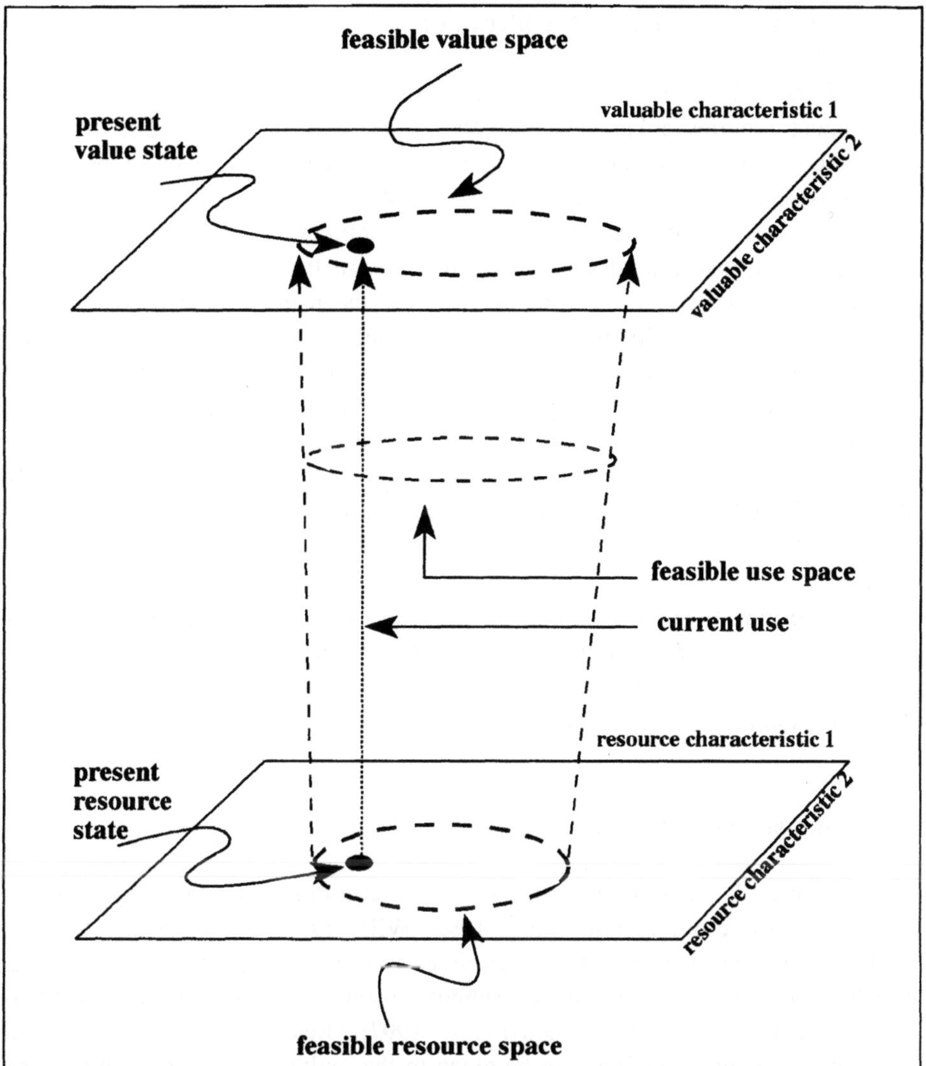

Figure V.2: Relationship between the feasible resource space, the feasible use space, and the feasable value space of the SEnS

Exploring the opportunities for the development of the system can now be defined more precisely as assessing the *feasible value space*. The assessment is based on the exploration of the *feasible use space* and *feasible resource space*. This should result in a set of *feasible value states* of the *SEnS* which can in principle be attained by modifying either the *use state*, the *resource state*, or both. How such explorations can be performed will be discussed in Section V.5.

V.3 The resource state in the literature

It will be clear, then, that an exploration of the opportunities of a development requires a knowledge of the *resource state* of the *SEnS*. As noted above, the *resource state* can in principle be measured at any given point in time. But how is such a measurement to be carried out?

The search for, and management of, resources has a long tradition (e.g., Pointing, 1991), and it is possible that techniques for the measurement of resources have been developed that can also be used for the measurement of the *resource state* as defined above. For this reason, part of the literature on the subject will be analyzed here. The literature on the operationalisation of 'environmental utilisation space', 'human carrying capacity', 'land-use planning' and 'integrated planning' will also be discussed. As a complete review of these areas is beyond the scope of this chapter, a selection from recent overviews will be presented here. Key references include Barbier (1989); FAO (1984), Opschoor & Weterings (1994), Fresco *et al.* (1995), and Rotmans & De Vries (1997). Further references are given below.

The aim of the discussion which follows is to find a way to measure the *resource state* of a *SEnS* in such a way that the quantity of one resource can be related to that of the other resources and to the structures and processes from which the system's valued characteristics flow.

V.3.1 Resource management

There is a considerable body of literature on the depletion and degeneration of resources (Heywood, 1995; WRI, 1996; Brown *et al.*, 1998). In this literature, the measurement of resources is not the issue; the data are simply recorded and discussed, usually to support examples of 'dooms day scenarios' (e.g., Brown *et al.*, 1998; Anonymous, 1997). Resources are always treated separately, although attention may be given to the multiple use of certain resources (Brown *et al.*, 1998).

More may be expected from the research in which resource use is modelled in order to find solutions directed towards sustainable development (e.g., Barbier, 1989; De Vries, 1989; Van den Bergh, 1991; Heijman, 1991). However, in most of these studies the resources are not specified, although different types of resources are distinguished (e.g., depletable vs. non-depletable; renewable vs. non-renewable: Heijman, 1991). The focus is on the changes in 'utility', or resource availability which take place over time, according to different models of resource use (Barbier, 1989; De Vries, 1989; Heijman, 1991). Some studies examine the loss of functions due to resource deterioration (e.g., Barbier, 1989). It is noteworthy that several models assume a closed system. An exception is the work of Van den Bergh (1991), who explicitly examines the behaviour of a 'regional sustainable development' model based on an open system. Van den Bergh also states that attention should be given to aspects such as the 'internal structure of the region'; "*features of the resource base*" which is a "*complex of resources*"; and "*interactions*". However, it remains unclear how the author operationalizes these statements in his models. The conclusion from this sample of studies on modelling resource use is that this type of literature is not very helpful in finding ways to measure the *resource state* of a system.

In recent studies, attention has been directed towards the multifunctionality of certain resources, especially ecosystems (Barbier, 1989; R.S. de Groot, 1992; Costanza *et al.*, 1997). These approaches recognize the relationship between resources and the many different valuable characteristics of a system. The 'resource state' measured by Costanza *et al.* (1997) is the area of all the ecosystems in the world. Each ecosystem is regarded as a resource producing various values. These resources are treated separately, as either present or absent and mutually exclusive. The aim of these studies is to measure not the *resource state*, but rather the *value state*, i.e., the economic value, of ecosystems (R.S. de Groot, 1992; Costanza *et al.*, 1997), and any overlapping of resources would hamper such valuations. Yet, as observed, this takes no account of the fact that ecosystems are actually complexes of resources which can be used independently. Thus describing a system as a set of separate ecosystems is no answer to the question of how to measure the *resource state* of a system.

V.3.2 Land evaluation

Until recently, conventional land evaluation as carried out by the FAO has been based on an assessment of the suitability of the resource 'land' for different forms of land-use (FAO, 1976; 1984; 1993). Although this approach deals exclusively with agriculture, some studies describe ways of measuring a *resource state* that may be used to explore opportunities for sustainable development. Recently the same techniques were used for urban land-use planning (Ofomata & Umeuduji, 1994). Developments within this field include a move towards more integrated planning (Fresco *et al.*, 1995); these will be discussed in Section V.3.3.

At present, land evaluation for a particular region is generally carried out as follows (FAO, 1984):

- select types of land-use
- select land qualities or land characteristics

- identify land-use requirements
- asses land qualities or characteristics per land unit
- draw up a classification for land suitability
- examine land improvements, and environmental, economic and social impacts
- draw up a final classification of land suitability per land unit for each type of land-use.

This procedure will result in a set of maps showing the distribution of land suitability for each separate type of land-use. These maps describe the present state of the resources. By assuming certain 'major improvements' in the future, such as drainage, it is also possible to draw up a potential land suitability classification.

Land evaluation offers a number of features that could make it suitable for measuring the defined *resource state*, since it deals with characteristics that make possible different types of use, thus interrelating resources and various forms of agriculture.

However, in a number of respects land evaluation is not entirely satisfactory for our purposes (for a more extensive discussion see Van Diepen *et al.*, 1991).

- First of all, it focuses on agriculture, disregarding all forms of land-use outside agriculture. Thus the land qualities or characteristics dealt with are relevant only for agriculture, while agriculture itself is reduced to land productivity. Social, cultural, and environmental issues are seen as side-effects.
- Second, the procedure is designed to explore opportunities for land-use shifts, rather than opportunities to change the land or the types of land-use. As a result, the possibilities for modifying land characteristics or land-use requirements are not brought to light. Moreover, both land characteristics and land-use requirements are hidden within a suitability classification. This makes it difficult to access the potential of the land under other circumstances. The procedure is not flexible and transparent enough to explore opportunities that are really new.
- Third, land qualities do not include characteristics that are directly related to actual human activities. As a result, the consequences of current or planned activities, such as pollution, for the suitability classification in the future or for other areas are not explored. This is in accordance with the tradition of investigating options for development by calculating maximum land productivity (e.g., WRR, 1992b; Rabbinge & Van Ittersum, 1995).

In conclusion, the techniques of conventional land evaluation can be used for measuring the *resource state* of a *SEnS*, but they must be adapted and expanded. They must be made suitable for all types of goals and all types of human activities (i.e., more types of resources must be incorporated). Moreover, they must be made flexible and transparent in order to explore opportunities, and they must incorporate the effects of human activities on resources.

V.3.3 Integrated approaches

Since the aim of this section is to find ways of measuring the *resource state* of a system that take into consideration the interrelation between resources and the use of resources, the next logical step would be to examine approaches to development problems that try to reach different goals at the same time. Such strategies are known as 'integrated approaches'. In effect, the search for sustainable development is in itself an integrated approach (Chapter II), which make integrated approaches especially appropriate for measuring the *resource state*. Three different types will be discussed: (i) operationalisation of integrated concepts, (ii) integrated modelling, and (iii) integrated planning.

Operationalisation of integrated concepts

A concept that is closely related to *resource space* as well as *use space*, as defined above, is the 'environmental utilisation space' (EUS) (Opschoor & Van der Ploeg, 1990; Opschoor & Weterings, 1994). This was defined by Opschoor & Weterings (1994) as: "*the locus of all feasible combinations of environmental services that represent steady states in terms of level of relevant environmental quality and stocks of renewable resources*". It is obvious that the concept is intended to capture various uses of the environment, together with their resources, within a single 'locus'. It has been related to such concepts as 'carrying capacity' and 'ecocapacity' (W.T. de Groot, 1992; Opschoor & Weterings, 1994; Tjallingii, 1996), and it has been discussed extensively in Dutch literature (e.g., Opschoor & Van der Ploeg, 1990; Ten Brink *et al.*, 1994; Dietz & Van der Straaten, 1994; Latesteijn *et al.*, 1994; Musters *et al.*, 1994a; Opschoor & Weterings, 1994; Sips *et al.*, 1994; De Vries, 1994; Weterings & Opschoor, 1994; Hueting & Reijnders, 1996; Hofkes, 1996; Stolwijk, 1996; Verbruggen & Dellink, 1996). One major point of disagreement between the various authors is whether or not the EUS can be assessed objectively, rather an odd question considering that the use of the environment is actually part of the definition of EUS, and consequently the EUS will depend on decisions on use (Hueting & Reijnders, 1996; Hofkes, 1996; Stolwijk, 1996; Verbruggen & Dellink, 1996). Up until now no attempts have been made to measure the EUS (except for Musters *et al.*, 1994a). Some authors define it as the selection of a set of standards for resource use (e.g., Weterings & Opschoor, 1994; Moffatt, 1996), but such a selection is obviously not the same thing as measurement.

As noted above, the concept of EUS is closely related to what is known in ecology as 'carrying capacity' (e.g. Kormondy, 1969; May, 1981; Sutherland, 1996; McLeod, 1997). Can the concept of 'Human Carrying Capacity' (e.g., W.T. de Groot, 1992; Cocossis & Parpairis, 1992; Kessler, 1994b; Cohen, 1995; Fearnside, 1997) be used to measure the *resource state*? Ludwig *et al.* (1993) argue that decision making based on estimates of 'Maximum Sustained Yield, one of the supposed measurements of carrying capacity, often fails. Recently Cohen (1995) discussed in greater detail the concept of carrying capacity in relation to environmental issues. The author concludes that "*none of the existing concepts of carrying capacity [...] turns out to be adequate for the human population*" (Cohen, 1995). The reason is, as Haila & Levins (1992) note, that the "*carrying capacity of nature relative to human populations is not a constant but a historical variable. It changes in two ways. The resources that are used change as a result of use or for extraneous reasons, and human needs also change*". The Human Carrying Capacity can be measured 'after the fact', but that would be useless for exploring opportunities. Therefore, it is concluded that carrying capacity is not a concept that leads directly to the measurement of the *resource state* as defined here.

Integrated modelling

In integrated modelling, one tries to design models that integrate different types of goals and the use of different resources in order to forecast the future (e.g., Bockstael *et al.*, 1995). The TARGETS model of the Dutch National Institute of Public Health and the Environment (RIVM) is a recent example of such a model. It is based on 'Integrated Assessment' (Rotmans & De Vries, 1997), and is designed to study global changes by projecting past changes into the future, on the basis of three scenarios.

The TARGETS model consists of six submodels, five of which have a 'state & dynamics' module (for population, energy, water, land, and biochemical cycles respectively). The handling of resource characteristics and resource state can be derived from these modules. It is impossible to discuss all the different aspects of the modules here, but the following issues pertaining to the measurement of a *resource state* must be mentioned:

- Population and capital are regarded as resources.
- Stocks of fresh water are divided into functional quality classes: Water of the highest quality has all functions, water of the lowest quality is unsuitable even for agriculture or industry.
- In general, land-use is dealt with in the same way as in land evaluation (Section V.3.2), except that land quality for agriculture is affected by human use through changes in soil organic matter. Urbanization and the

drainage of wetlands are not included here. Ecological aspects are measured in terms of unused or recovered land area.

- The emphasis is on dynamics: how one stock is transformed into another. Technological development and other changes in human activities are included in the category 'learning factors'.

The conclusion is that human resources must be included in the measurement of a *resource state*, that functional quality classes may be a workable way of relating resource characteristics to different activities, and that the technique of relating human activities to land quality and ecological values has not yet been fully developed.

Integrated planning

Today, many conventional fields of planning strive for an 'integrated' approach, e.g., water management (Van Rooy *et al.*, 1995), urban planning (Tjalingii, 1996), land-use planning (Fresco *et al.*, 1995; Hoanh *et al.*, 1997), and environmental planning (Cairns & Crawford, 1991; Kessler, 1997b). Since much of the literature on integrated planning may be expected to deal with planning procedures rather than measuring the *resource state* (e.g., Cairns & Crawford, 1991; Glasbergen, 1995; Van Rooy *et al.*, 1995; Tjalingii, 1996; Mazzoti & Morgenstern, 1997; Kessler, 1997), this discussion will focus on recent developments in land-use planning (Fresco *et al.*, 1995).

Land-use planning is evolving towards an "*integrating, interactive process*" (Fresco, 1995), as illustrated by many examples discussed in Fresco *et al.* (1995). As a result of developments in computer technology, land-use planning has become flexible and capable of delivering alternative scenarios (e.g., Huizing *et al.*, 1995; Alfaro *et al.*, 1995; Stoorvogel *et al.*, 1995). The emphasis is not only on productivity and economics, but also on risk, employment and emigration, and nature (Veeneklaas *et al.*, 1995). However, where the measurement of the *resource state* is concerned, land-use planning still focuses exclusively on agriculture; land characteristics are not measured in a way that allows modifications of these characteristics to be examined; land-use requirements would appear to be 'cast in stone', and the effects of current or planned human activities have still not been studied (Fresco, 1995).

V.3.4 Conclusions

As stated above, in order to measure a *resource state* with the aim of exploring opportunities for sustainable development, resources must be measured in such a way that both the interrelation between resources and the relationship between resources and human activities can be described. In the light of the preceding discussion, it will be clear that the measurement the *resource state* of a system has not yet been operationalized. While this operationalization can be based on recent developments in land evaluation and integrated modelling, it must be (i) extended to sectors other than agriculture, (ii) incorporate not only resource characteristics that relate a resource to different activities, but also characteristics that relate a human activity to different resources, (iii) make it possible to explore modifications of resources as well as modifications of activities, and (iv) result in a flexible and transparent instrument for exploring the opportunities of sustainable development.

V.4 Measuring the resource state

The basic concepts for the measurement of a *resource state* have been identified in the previous sections, and a procedure will now be proposed (summarized in Figure V.3). This procedure assumes two steps described in previous chapters: the defining of a *SEnS* (Chapter III), and the selection of the *valued system characteristics* (Chapter IV). The description of the procedure concentrates on the subsequent steps to be taken in order to measure the *resource state*. It bears a certain resemblance to the procedure for land evaluation (Section V.3.2). In Chapter VI, the measurement will be embedded in a procedure for sustainable development.

V.4.1 Assessing the structures and processes

Once the *valued system characteristics* have been listed, the necessary structures and processes to ensure that the system will obtain or retain those characteristics can also be identified. Ideally, this list will be complete and not restricted to one sector, as in the case of land evaluation. In some cases, not all natural processes will be known, but the enumeration can be based on the available knowledge. As the list of structures and processes is needed for the assessment of *resources*, no extensive description of these processes and structures is needed. As a rule, the aim is to measure the '*present resource state*', so that it may be assumed that the description of the processes and structures reflects current situations.

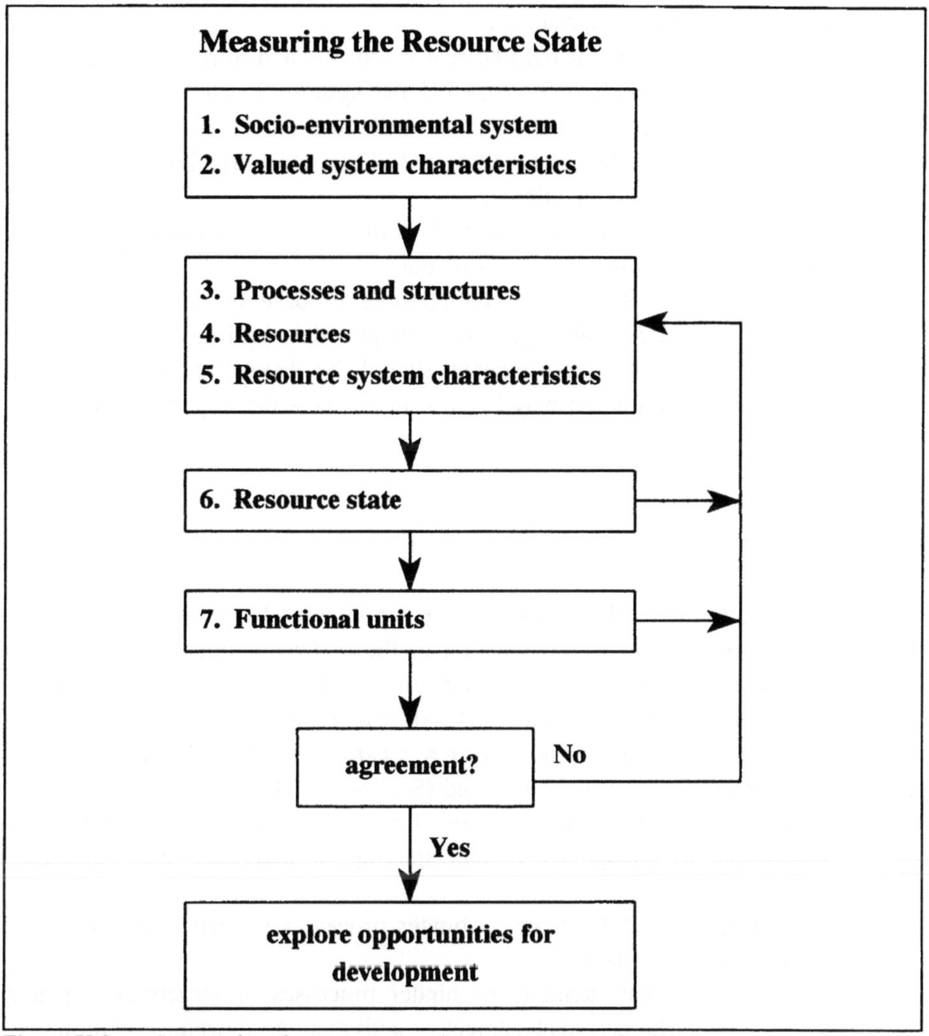

Figure V.3: Steps to be taken to measure the resource state of a SEnS

V.4.2 Assessing the resources

The list of structures and processes is intended to describe the *resources* needed. This can be done by taking the present standards for *resources* as a point of departure.

For example, four types of requirements can be distinguished for the process 'food production on land':

- the biophysical qualities: the soil layer must not be overly thin and should not be polluted; it need not be optimally fertile, since fertilization is

common practice in agriculture
- accessibility: the land should be reasonably flat and in non-built-up areas
- the relationship with other parts of the system: it should be possible to transport fertilizers to, and food from, the location
- human aspects: technical facilities and skills which are important for food production should be available.

The availability of these requirements will define whether a certain part of the system has the *resources* for the process or not.

In many cases, it may not be possible to base the description of the requirements on scientific knowledge. For example, the required sinks may not be known. In that case, the requirements should be based on common sense. Experience with land evaluation may be used to describe the requirements for land-use.

V.4.3 Assessing the resource system characteristics

The requirements that define the *resources* must be translated into system characteristics that indicate whether a part of the system belongs to a *resource* or not. These characteristics have been called the *resource system characteristics* in order to distinguish them from the *valued system characteristics*.

On the basis of the previous section it can be said that the following types of resource characteristics must be selected (Section V.3.4):

- characteristics that facilitate or hinder processes or structures not susceptible to change by human steering or activities, e.g., amount of rainfall
- characteristics that facilitate or hinder processes or structures, but can be steered, e.g., drainage
- characteristics that facilitate or hinder processes or structures which are side-effects of human management or activities, e.g., water pollution.

If possible, each characteristic should be defined in such a way that it is relevant for different structures or processes. 'Functional classes' may be used for this purpose (Rotmans & De Vries, 1997).

Obviously, the characteristics to be taken into consideration will depend on the system boundaries chosen. For example, whether or not a resource characteristic can be changed will depend on the time horizon of the system, and whether groundwater characteristics are taken into consideration will depend on the assumed depth of the system.

V.4.4 Measuring the resource state

The presence of the *resource system characteristics* in each unit of the system can be assessed. What these units are depends on the grain of the information needed, and this is determined during the definition of the system (Chapter III). Thus the available combination of resource characteristics within each unit of the system can be determined. This tells us the present resources of each unit of the system. In other words, the present *resource state* of the system has been measured.

The *resource state* may be seen as a matrix, where the Y axis is formed by the units, and the X axis by the resource characteristics. The cells of the matrix indicate whether or not a resource characteristic is present in the unit. Unless the number of units or resources characteristics is small - which is unlikely when all goals and all relevant processes and structures have been taken into account - the *resource state* of a system will be an observation set that is too large to be suitable for presentation and direct interpretation. For example, the *resource state* of the Dune and Bulb region in the Netherlands, which has a fairly low level of resolution, had a matrix of 641 * 24 cells (Chapter X). Thus the *resource state* must be compressed into a presentable form without losing too much information. The assessment of '*functional units*' is proposed for that purpose.

V.4.5 Assessing functional units

All units of the system that have the same combination of resources belong to the same *functional unit*. Because the required *resources* are derived from the *valued system characteristics*, each *functional unit* can be described in terms of its *resources*, as well as in terms of its potential contribution to the variables which measure the *valued system characteristics*. All units belonging to the same *functional unit* have, by definition, the same potential to obtain or retain the same *valued system characteristics*. It is this reason that the name '*functional unit*' was chosen.

The size of the *functional units* can be determined, but each compartment of a system may require a different type of measurement: land in square metres, groundwater in cubic metres, biomass in kilograms, species in numbers, and so forth.

A table can be drawn up showing all the available *functional units*, the processes and structures that are possible in each of them, their potential contribution to the *valued system characteristics*, and the size of each. Such a compilation represents the *resource state* of the system.

V.5 Exploring the opportunities for development by backcasting

As noted in Section V.2, the measured *resource state* forms the basis for the exploration of the development opportunities of the system, defined as the assessment of the *feasible value space* by means of exploring the *feasible use space* and the *feasible resource space*. This should produce a set of *feasible value states* of the *SEnS* which can in principle be reached by modifying either the *use state* or the *resource state*, or both.

Any exploration of the opportunities for development will depend on the type of development one is aiming for. Here 'type of development' means the field where the initiative for the search for sustainable development was taken. Was the aim of the search regional planning, for example, or legalisation, or technology development? These fields define the type of modifications of both *use states* and *resource states* that should be explored. For exploring the opportunities of development within a field, a backcasting procedure is appropriate (Chapter I).

The first step in such a procedure would be to assume that all constraints on development that might be raised within the field have actually been raised. For example, if technology development is the field of exploration, it may be assumed that all technological constraints on the use of the available resources have been lifted. This would mean that new management and new activities are available that make use of all the resources, and do not hinder each other in any way. Models can be developed that calculate the optimal solution, given these assumptions, and this will determine the potential *value state* of the system. If this *value state* does not meet the preliminary goals that were defined before the exploration started, it must be concluded that these goals cannot be reached, and may need to be changed.

However, if the goals of development can be reached, it is noteworthy to employ a less radical set of assumptions about the possibilities of raising constraints. For example, if it was assumed during an exploration for regional planning that any activity could take place at any location, the *value states* can be examined to see which of them can be reached, assuming that locations are constrained by the availability of infrastructure.

If, in the case of these less radical assumptions, a *value state* is reached that is acceptable, a set of even less radical assumptions can be adopted, and so forth.

By carrying out the exploration in this way, each new step establishes the amount of effort that is needed to reach the desired *value states*. Of course, these efforts cover modifications to the *use state* as well as to the *resource state*; this means that the feasible and desired *value space* is known, as well as the modification needed to reach it. This is in effect an exploration of the *feasible value space*: it indicates which developments are feasible and have the potential to reach a desired state of the system, thus demonstrating the opportunities for development.

V.6 Discussion

In the previous sections a procedure was presented to measure the *resource state* of a system, together with the outlines of a procedure for exploring the opportunities for development. A number of practical problems that may arise must now be discussed.

- First of all, in this procedure the exploration of the opportunities for development is totally dependent on the internal characteristics of the system. This is justified by the assumption that the steering of a system by the decision makers is largely limited to the steering of the internal management and activities (Chapter III). However, it will be clear that the development of a system can actually be governed by external changes. Although knowledge of external changes can be incorporated as changes in the availability of unsteerable resource characteristics, these changes cannot as a rule be anticipated. This is why the focus is on opportunities for development, rather than on the actual development itself. The actual development will always be different from any of the desired developments. However, since the procedure stipulates what the desired and feasible states of the *SEnS* are, it is possible to assess unwanted developments and respond to them.
- Second, it will be clear that the measurement of the *resource state* presently available depends on the assumptions and choices made with regard to structures and processes, resources, and resource characteristics. For this reason, the *resource state* is not an objective, scientific given. It should be discussed with representatives of the people involved, for example, by making use of a table or map of *functional units*, a kind of compilation of the *resource state*. Questions must be answered such as: Have the right structures and processes been selected, which realize the values of the system? Have all the relevant resources been chosen, and the right resources characteristics? The tables and maps of the *function units* can also be used for a quick rundown on the effects of the assumed modifications to the *resource state*.
- Third, in the discussion of land evaluation, it was noted that an instrument intended for exploring the opportunities for development should be flexible and transparent. Does the procedure proposed here (Figure V.3) meet

these criteria? The measurement of a *resource state*, which covers more human activities and values than conventional land evaluation, may require the processing of even more data than is customary in land evaluation (FAO, 1993). But today flexibility depends largely on computer facilities, and is no longer a problem. The transparency is greater than in land evaluation, because information is represented in *functional units* in which information on combined resources characteristics is directly related to the potential to deliver values. Changes in the *functional units* can easily be interpreted as changes in resource characteristics as well as changes in potential values.

- Fourth, the exploration of the opportunities for development depends on the availability of mathematical models that calculate the *value state* of the system on the basis of the available *resource state* and the assumed activities. A great deal of experience has been gained from such methods for land-use planning (Fresco *et al.*, 1990; Fresco *et al.*, 1995). The quality of these models can be tested by starting the exploration with the calibration of the present *value state*, on the basis of the present *resource state* and *use state*. Of course, this calculated *value state* should be close to the actual *value state* of the present system. In addition, the calculated present *value state* can be used as a reference in assessing calculated opportunities for *value states*.

Other practical problems involved in the proposed procedure will be brought to light by means of actual applications of the procedure. This will be discussed in Chapters X-XIII.

V.7 Concluding remarks

It was concluded that the operationalization of the measurement of interrelated resources was still needed. A method for measurement is proposed that consists of the following steps. It takes as its starting point the delimitation of the system in Chapter III and the procedure for selecting valuable system characteristics and variables in Chapter IV. It continues with the selection of the relevant processes and structures on the basis of the selected valuable characteristics. From that the resources are defined. Then the requirements that define the *resources* are established and translated into the system characteristics. This is done in order to determine whether a part of the system belongs to a *resource* or not. These characteristics are called the *resource system characteristics* in order to distinguish them from the *valued system characteristics*. The combination of *resource characteristics* at a certain place and time determines the availability of interrelated resources: the resource state. After that, the units with the same quality can be combined to form *functional units*.

It was concluded that the following types of *resource characteristics* should be chosen:

- characteristics that enable or hinder processes or structures and are not changed by human steering or activities
- characteristics that enable or hinder processes or structures and can be steered
- characteristics that enable or hinder processes or structures that are side-effects of human management or activities.

These selected characteristics must make it possible to explore the modifications of the *resource state* by backcasting in order to find the *development opportunities*.

CHAPTER VI

Procedure for Sustainable Development: Decision Making Phases and Information Requirements

CONTENTS

VI.1 Introduction

In the preceding chapters a number of concepts were presented and defined which are necessary for the construction of a scientific method to facilitate the search for sustainable development. In this chapter the procedure that follows from that method will be outlined as a process that results in information on the opportunities for sustainable development. This procedure follows directly from the strategy proposed in Chapter II for the search for sustainable development, and integrates the concepts developed in Chapters II-V. The essential issues discussed and the steps proposed there will be summarized.

Steering for sustainable development is both complex and uncertain (WRR, 1995). Autonomous developments are by nature uncertain, while there are many values at stake on different levels of scale. Numerous possible measures are conceivable, which may have many different effects, while the results of combinations of measures are unpredictable. All these aspects require decisions the consequences of which are uncertain, but likely to be considerable (Funtowicz & Ravetz, 1991).

If it is acknowledged *a priori* that steering aimed at sustainable development is complex and uncertain, no procedure can presume to provide a 'blueprint for the future'. However, a scientific approach should make it possible to collect information on a given system, and establish the economic, social, cultural, and ecological opportunities in a methodologically correct manner (Faludi, 1986). A procedure for information-gathering for purposes of sustainable development will have many types of applications. In Chapter I, three different ways of applying the proposed procedure for gathering information are distinguished, ranging from applications with low direct impacts on the people involved to application with high impacts. They are described as:

- *Support for scientific research.*
- *Support for parties involved the debate on sustainable development.*
- *Support for sectoral, regional and local planning.*

What will a scientifically sound procedure for sustainable development which is usable for the purposes mentioned above look like? In the following section the rough outlines of such a procedure are described as a set of logical stages, whereby the need to structure each stage is stressed (Section VI.2). Then the information-gathering phases within the initiative and specification stages, and the relationship between them, are examined (Section VI.3). The final section summarizes the major issues involved in embedding a logical procedure in a decision making framework (Section VI.4). In Part Two, the focus is on the analysis of the strengths and weaknesses of the procedure in order to improve it.

The procedure presented here is general and theoretical, so that it covers different kinds of applications. It may be seen as a 'ideal' model of a procedure. To our knowledge, no such 'ideal' procedure for sustainable development exists (Conacher, 1994; Berke, 1995; Bingham *et al.*, 1995). For purposes of real-life application, many practical questions have yet to be answered. Who initiates the procedure? Who is responsible for the outcome? How is decision making managed? How are differences in influence between participants handled? Who represents those people who cannot make known their interests? How does one deal with the needs and wants of people who are not motivated to participate or are not organized? What are the legal aspects of the procedure? What are the governmental and political consequences of the outcomes? These questions will not be addressed in this chapter.

VI.2 Outlines of a procedure for sustainable development

Most people would agree that sustainable development must do more than simply allow people to survive, and that 'quality of life' is one of the central concerns. But it is here that agreement often ends, and debate begins, for people have differing views on what 'quality' or even 'life' actually entails, and what they are willing to pay to achieve it, nor do they agree on how to achieve it and what risks are acceptable (WRR, 1995). Thus the search for sustainable development is as much a political problem as a technical one (Chapter II; WCED, 1987; WRR 1995). It is an ongoing debate about an ever changing society and its environment. It involves a debate on the development of an object defined here as a '*socio-environmental system*' (*SEnS*) (Chapter II & III), as well as on the system's present and future economic, social, cultural, and ecological values, and its human, financial, and biophysical resources (Chapters IV & V). The steering of this system towards sustainable development should be based on agreement concerning the goals, the required changes to the *resource state*, and modifications to the activities that rely on these resources (Chapter II & V; Proops *et al.*, 1996).

Each decision made in the course of such debates has political consequences, whether it pertains to assessing important issues, defining what is to be developed, quantifying needs and wants, determining what should and can be changed in the light of the available resources, exploring the possibilities to reach the goals set, identifying present development opportunities and selecting the measures for steering. This also holds true for such decisions as the organization conducting the debate, the experts to be consulted, the types of changes to be explored, and the types of measures to be taken for steering developments.

Because of these political implications, in a utopian society all the people involved in its development - from individual neighbourhoods to the entire community - would participate in all stages of the debate. While this is not realistic, in discussing a particular *SEnS*, one can strive to approximate full participation in the debate. In any case, if it is to be fruitful, the debate will have to be structured. This will enable conclusions to be reached, decisions made, and actions carried out. At the same time, real-life decision making that incorporates participation may resist structuring (Friend & Hickling, 1987). Thus the challenge is to develop a procedure that can structure the debate on sustainable development, but has the flexibility to cope with the non-linearity in decision making.

VI.2.1 A sustainable development procedure

A procedure for sustainable development can be defined as a formalized working plan aimed at steering the developments in a direction that has the most chance of:

- meeting the needs and wants of the people directly involved in a *SEnS*
- contributing positively to higher-scale interests
- creating the conditions for future generations to meet their assumed needs.

In any application - supporting scientific research, parties, planning processes - the main goal of such a procedure will be to find measures that contribute to sustainable development. Once these measures are found, a process can be initiated to make decisions about executing them: the actual steering of developments. And finally, no procedure can do without a realistic feedback, i.e., the monitoring and evaluation of the impacts of steering measures (Vos *et al.*, 1990).

The primary targets of any procedure for steering developments are decisions about measures to be taken in the present in order to reach the goals which have been set. Before steering can be implemented, at least two types of decisions have to be made: first, decisions during the start-up about how the procedure is to be performed and the overall aim; and second, decisions concerned with the information which is required to design plans for steering (see Chapter I for the main stages of a procedure). Decisions on steering will be based on information-gathering, while decisions about what information is gathered, will be based on the overall aim of the procedure. The procedure shoul consist of clearly-defined consecutive and/or parallel phases which are managed independently, and embedded in a research or planning programme (WRR, 1994; Glasbergen, 1995).

Since the procedure is linear, i.e., decisions have consequences for the subsequent phases, it is impossible to carry out a particular phase before the previous one has been completed. However, as was stated, in real-life decision making linearity may be difficult to attain (Friend & Hickling, 1987). Decisions may be taken within one phase which conflict with previous decisions, for example, in an exploratory phase radical changes in the location of land-use activities may be desired, while in the start-up it was stipulated that locations may not be changed, and that the search is to be restricted to technology within the present distribution of activities. In such cases, the procedure can be restarted. This should be carried out to modify the starting points, constraints and aims, and to establish the possible consequences for subsequent phases. This restarting makes it possible to incorporate cyclicity into the decision making. However, where it is not possible to restart the procedure, because of temporal, financial or political constraints, the best advice may be to suspend the procedure. Alternatively, one may decide to carry on, accepting the inconsistency. This ultimately places the responsibility for continuing of the procedure with those who are responsible for consistency.

VI.2.2 Phases of the information-gathering procedure

Dividing a procedure into phases will lend it structure, which will be helpful for all those involved, especially the people responsible for decision making. This holds true for the research, as well as the strategy development and practical planning, while it also makes the progress of the procedure more controllable (Glasbergen, 1995). Phases should be delimited, so that only a single set of related decisions is considered within each one. Iterations within the process should take place within a single phase, rather than between phases, so that one phase can be concluded and the next one started. Each phase will thus end in a stop/go moment, and decisions made within a phase should not be returned to later on in the procedure (WRR, 1994). In essence, a procedure for sustainable development does not differ from comparable procedures aimed at steering developments (e.g., Werner, 1993; Brinkman, 1995; Rabbinge & Van Ittersum, 1995; Van Rooy *et al.*, 1995; Stinner *et al.*, 1997).

As in all planning procedures, for sustainable development the required information on opportunities and perspectives must be obtained before steering takes place. Moreover, this information should not be based exclusively on the present state of a system established by monitoring, but should also rest on an analysis of the possibilities to modify the system. Five main decision 'moments' can be distinguished, marking five distinct phases (Figure VI.1). Phase 1 is the *Start-up*, equivalent to the 'initiative' stage referred to by Glasbergen (1995). In Phase 2 the *defining of the socio-environmental system* takes place. This will facilitate Phase 3, the *selection of the valued system characteristics*, which is based on the perceived needs and wants of the people involved, expressed in a set of goal

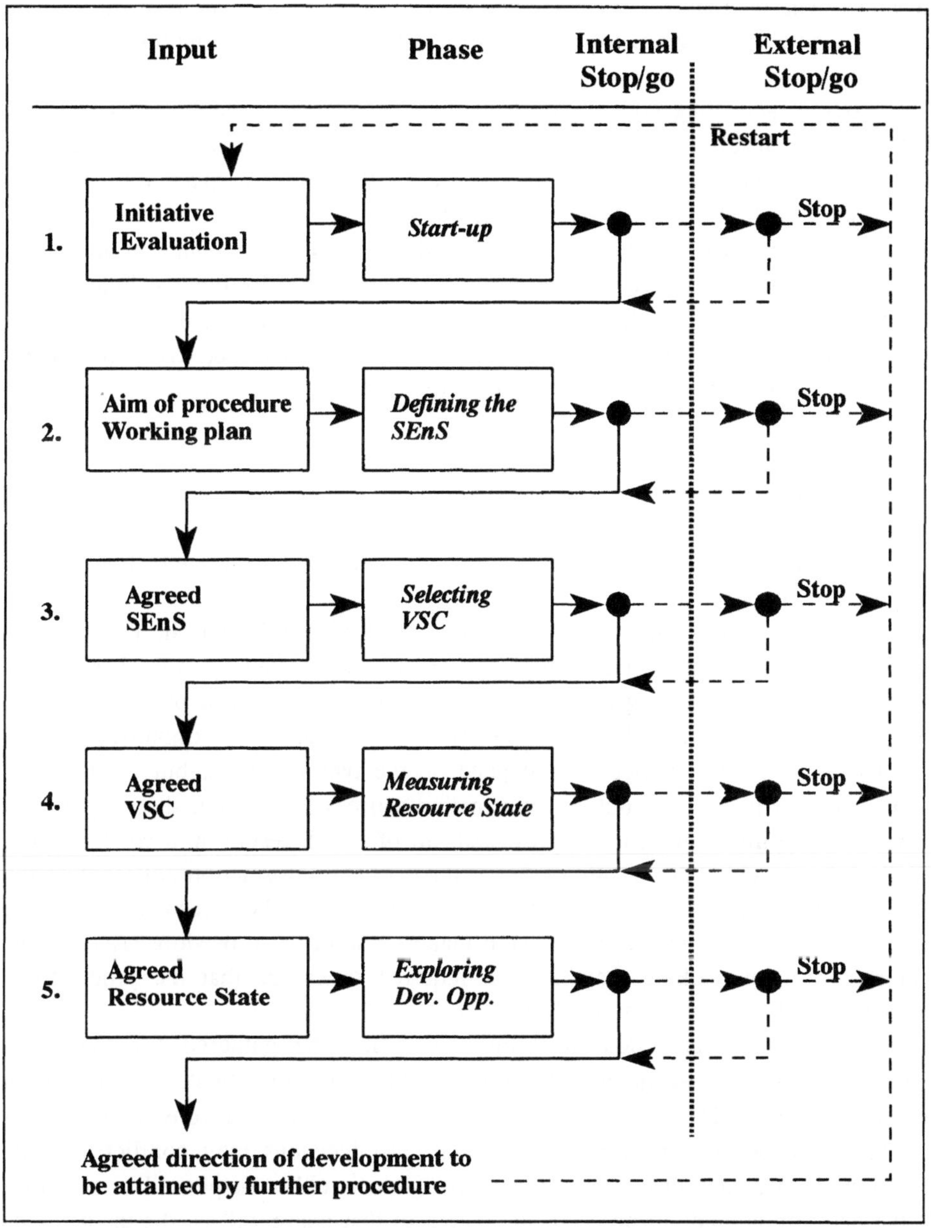

Figure VI.1: Phases in information-gathering in a procedure for sustainable development (see Section VI.2.2 for explanation)

variables. The next step - Phase 4 - will *measure the resource state* of the system, i.e., the available resources for use. Its present use provides information about

difficulties in achieving goals, and about that part of the available *resource space* that is used insufficiently or not at all. In the last phase of Glasbergen's 'specification' stage of the procedure - Phase 5 - the *resource space* and *use space* are explored to find the *development opportunities*.

The information-gathering must facilitate decision making on measures, plans and scenarios for actual steering - the 'operationalization' - followed by 'implementation' and 'evaluation'. These stages are not elaborated here, as they have been discussed in detail by other authors (e.g., Susskind & Cruishank, 1987; Vos *et al.*, 1990; Renn *et al.*, 1995; Glasbergen, 1995).

VI.2.3 How should each phase be carried out?

Before a phase can be carried out, it must be structured. Figure VI.2 proposes such a structure.

Each phase should start with the information collected in the previous phase. This will facilitate consistency throughout the procedure. On the basis of the documented information agreed on in the previous phases, each phase formulates its specific aims. (Action I). The phase must be managed and organized (Action II). The kind of management will depend on the type of phase, but in all phases decisions must be made about the information sources to be used, the people to be consulted, the information to be presented, all of which results in a working plan. Each phase provides the results of the analysis of the information collected in that phase (Action III) and presents a preliminary proposal. These results are verified (Action IV) and legitimated by decision makers within the procedure (Action Va). External decision making may also be required to ensure that the procedure is embedded in legalization (Action Vb).

In each phase adequate information must be gathered for the major decisions to be taken at the end of that phase: the stop/go moments (Action Va and Vb in figure VI.2). As noted in Section VI.2.2, these actions may lead to inconsistency between phases; for example, the decisions may conflict with the decisions in previous phases. If so, the procedure is either suspended or restarted. If the decision is made to proceed, the output of that phase will be the input for the next phase. Those responsible for phase management start with the information from the previous phase, gather relevant new information, analyze it, make the results comprehensible, verify them, and write a proposal for decision making. At the end of the phase, the results of this process are recorded and passed on to the

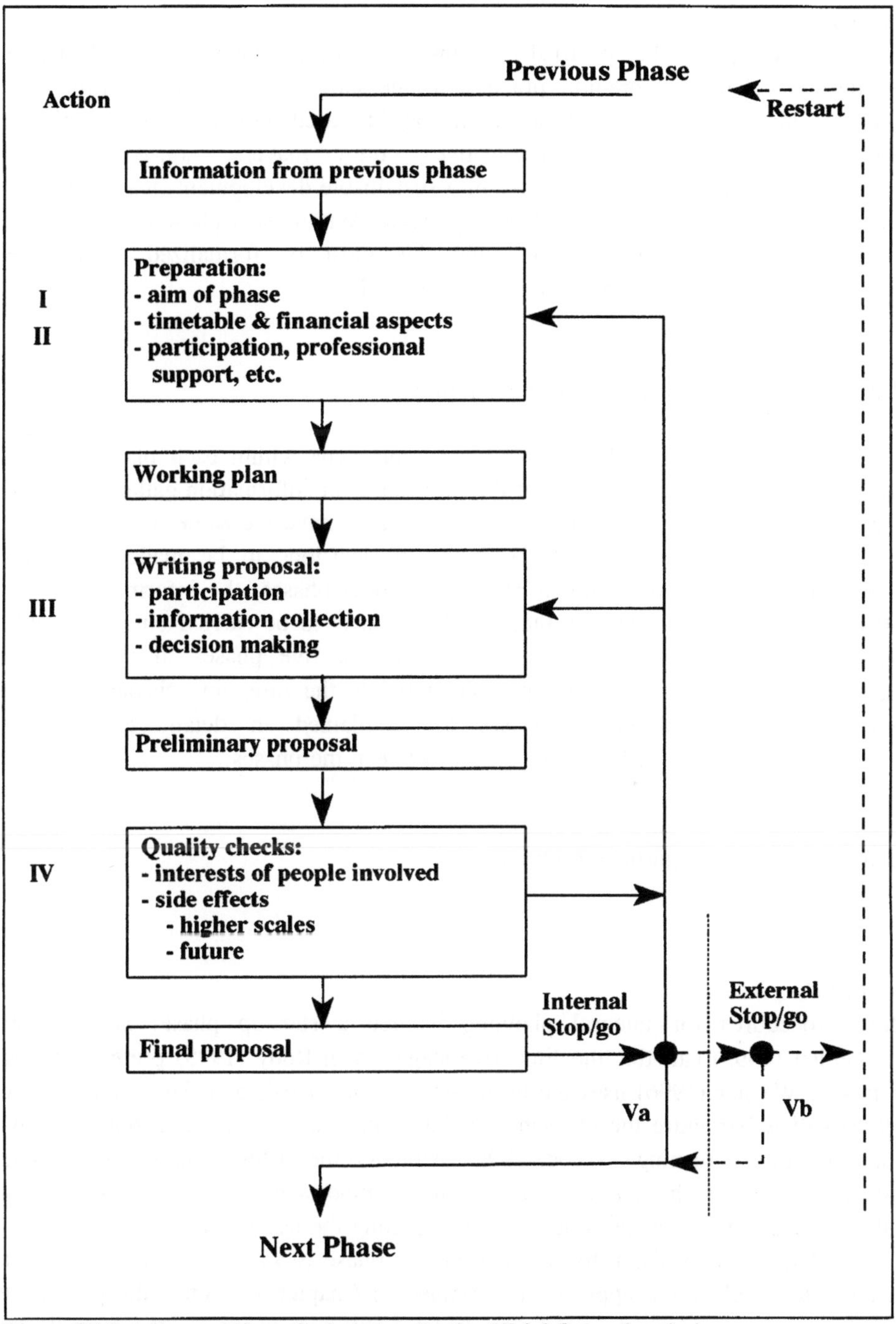

Figure VI.2: Steps within each phase (see Section VI.2.3 for explanation)

next phase.

In each phase the specified opinions of the people involved are taken into consideration. This is common practice in the participative planning processes which are increasingly being employed. It may be based on consultations with the people involved or their representative, policy makers, non-governmental organizations, experts, etc. In the case of a sustainable development, it is important that the set of opinions is as complete as possible. Within each phase there must be ample room for identifying them, and this must be formalized during the organization of the phase (specification in Action II).

VI.2.4 A procedure for information-gathering

The main phases (Figure VI.1) and the steps within each phase (Figure VI.2) are the constituent elements of the procedure for information-gathering aimed at sustainable development. Although all the phases have the same overall design, they differ greatly in terms of the kinds of decisions to be made. This has consequences for the management of the individual phases, the information to be collected, the people to be consulted, quality checks to be carried out, etc. Table VI.1 presents the most important aspects of the five phases in information-gathering, incorporating the concepts elaborated in the foregoing chapters. In the next section the subsequent phases are explained in detail, emphasizing information-processing and the relationship between the phases.

VI.3 Information-gathering phases

VI.3.1 Start-up

Goal of the phase
Most procedures for integral planning include a start-up phase. Driessen & Vermeulen (1995) call this the 'initiative stage'; Van Rooy (1995) prefers 'setting of tasks'; Warner (1996) uses the term 'setting objectives'; and Gregory & Slovic (1997) call it 'structure the problem'; in 'environmental impact assessment' (EIA) it is known as 'screening' (Wood, 1995). Although the different authors emphasize different aspects of this phase, all agree on the importance of clarifying what is at stake, setting the aim of the procedure, and gaining the necessary support.

The major decision to be made in this phase is whether or not to adopt a sustainable development approach (as defined in Chapter II). Since the procedure for sustainable development can be applied for different purposes, the overall aim

Table VI.1 A summary of the important aspects of the five phases of information-gathering (see Section VI.3 for explanation)

Phase: / **Aspects**	**Phase 1.** Start up	**Phase 2.** Defining socio-environmental system	**Phase 3.** Selecting the valued system characteristics	**Phase 4.** Measuring the resource state	**Phase 5.** Exploring development opportunities
Input	[Evaluation] New initiative	Work plan	System boundaries, context and involvement	List of valued characteristics	Available resources and present use and problems
Aims/ Decisions	Start a sustainable development procedure?	Define the system boundaries, time and spatial scale and type of actor, activities and aspects of goals	Assess valuable characteristics of the system	Measure available resource state and its present use	Asses perspective modification of use and resource space
Information	Present problems . correlations . trends Overall . objectives . restriction . constraints	System & context boundaries in time and space: . involvement . the system	Opinions on the values of the system: . people involved . policy documents . scientific documents	Availability resources: . processes . resources . resource characteristics Measurement system characteristics	Quantified goals and assumed changeability of use and resource state Estimates of consequences
Actors in the procedure	Initiator Opinion leaders Policy makers	Policy makers Experts Opinion leaders	Opinion leaders Experts General public	Experts Policy makers	Opinion leaders Experts Policy makers General public
Information for next phase	Work plan: . overall objective . spectrum of choices . restrictions /constraints	Defined system Preliminary list of: . people involved . their interests . the problems	Valued system characteristics Preliminary list of quantified needs and wants	Present resource state: . present use and problems . present (un)used resources	Development challenges: . type of modifications . feasible value space of design space

of the procedure must be made explicit in the start-up. Is it intended to search for scientific solutions to development problems, to modify the use of current resources, or to support parties in the debate?

Is it possible to formulate criteria for choosing between a sustainable development approach and a problem-oriented approach? In the recent literature on integral planning, a number of criteria have been mentioned that also apply to a sustainable development approach. Driessen & Vermeulen (1995), for example, state that one should be capable of indicating where the current situation and the expected developments are not favourable to the parties involved, and that parties cannot independently solving the economic, social, cultural, and ecological problems. It is not yet possible to formulate exact and universal criteria to answer the question Should we opt for a sustainable development approach or not? When developments affect different groups of people - here and elsewhere, now and in the future - and when economic, social, cultural, or ecological issues are at stake, a sustainable development strategy has distinct advantages. However, if the environmental problems are more or less isolated, or call for direct solutions, then a problem-oriented approach is probably more appropriate. If one opts for a sustainable development approach, then it must be decided how the procedure will be organized and executed, and who will be responsible for carrying out the successive steps.

Information needed for the decision to start the procedure

To assess the necessity to start the procedure and the chances of success, overall information should be available on present problems and correlations between problems, as well as trends in development and the problems they may cause. Information on perspectives for solving these problems may show that many groups of people are involved, and that a problem-oriented approach will be insufficient. This information can be collected by consulting experts, policy documents, and opinion leaders.

In order to assess the chances of success, it is important to know which application the initiator has in mind, the overall aim of applying the procedure, and the assumptions and restrictions adopted by the initiator (Faludi, 1986). It must be clear whether the initiator will be using the procedure to identify development options, to support his own strategy, or to design new plans. Other questions are also of importance. How much freedom of choice does the initiator have? To what extent can the initiator change or steer developments? How well will the participation of the people involved be organized? How constraining are higher-scale policies? For example, is present legislation seen as a given set in stone? Or is insight into how existing legislation should be changed one of the motivating factors behind the launch of the procedure? Will radical long-term solutions be chosen, based on the assumed development of new technology? Or is the focus on short-term solutions, based on existing technology?

Once the kind of application is established, more detailed procedural information is needed, including descriptions of the whole procedure as a working plan, the time horizon for completing it, and the cost.

Information for executing the following phases

When a procedure for sustainable development is launched, it must not only be organized, the information for the successive phases must also be recorded. This will consist of a preliminary description of the object to be steered as a *SEnS*, a list of problems and the relation between them, views on the direction and overall aim of development. This information is important because it forms the framework for defining the *SEnS* in the next phase.

In addition, some tentative choices can be made with respect to the definition of the system in the next phase (Phase 2), for example, the tentative boundaries of the system, and a tentative list of the people and institutions involved.

VI.3.2 Defining the socio-environmental system

Goal of the phase

Once the decision has been made to search for sustainable development, the first step is to define the object to be steered as a *SEnS* (Chapter III). The necessary information will consist of the overall aim of the procedure, a list of related problems, the people and policies involved, a broad description of the system, and the assumptions and restrictions adopted by the initiators during the start-up (Phase 1).

The definition of the object includes a description of the boundaries of the system in space, time, and organisation. What are the spatial boundaries of the system? How far does 'the present' extend, and when does 'the future' begin? What are the subsystems and what is its context?

Furthermore, in this phase the system is defined in a social sense, i.e., the categories of people involved in present and future. Who are the people directly involved, and what is the higher-scale community? Who will be classified as present users and future generations? This is important not only because it makes explicit what the procedure is all about, but also because it prevents unnecessary vagueness, later on in the procedure, concerning the present needs and wants and conditions relevant for the future and higher-scales.

In the literature on environmental procedures, there is usually a phase in which the 'objectives' are described. However, no clear distinction is generally made between the definition of the system and the description of the values of that system (e.g., Warner, 1996; Van Rooy, 1995; Mazzotti & Morgenstern, 1997; Kessler, 1997a). In EIA, for example, the 'environment' is defined as well as valued during 'scoping' (Wood, 1995).

Information needed to define the socio-environmental system

Specifications about the boundaries of a system must be based not only on the possibilities to steer the system as a whole, but also on the economic, social, cultural, and ecological problems. However, the boundaries based on the various types of problems may differ. To gain insight into these differences, one must consult experts, policy makers, opinion leaders and the literature. The risk of choosing boundaries that are not suitable for attaining sustainable development can be minimized by ensuring that all points of view on the types of problem together with their spatial and temporal aspects are covered during consultations. Special attention must be given to points of view that are not articulated, for example, those related to future generations, and the people who are involved directly or indirectly, but are not organized. As a result of this iterative process, several options for system boundaries will become available. Thus this information-gathering does not provide the definition of the system, but rather information which facilitates a choice, within the range of choices limited by the framework given in the *Start-up*. The number of subsystems and the categories of people involved must be manageable within the financial and temporal constraints specified during the *Start-up*, and it must be established whether or not the room for steering is acceptable.

Information for executing the following phases

The decisions on system boundaries, space, time and organisation, the categories of people involved, and the type of goals are the input for Phase 3: the *selection of the valued system characteristics* (Table VI.1). Additional information can be derived from the process of defining the system. The list of groups consulted during the 'defining of the system' phase can be passed on to the next phase as a preliminary list for searching valued characteristics. If a comparable list of valued characteristics has already been put forward, this can passed on.

A decision on the definition of the *SEnS* may be in conflict with the preliminary system definition during the start-up phase. If so, this should be discussed with the institution responsible for the procedure to establish whether or not this is sufficient reason to suspend the procedure. It is also possible to adapt the start-up document to the consensus reached on the system boundaries.

VI.3.3 Selecting the valued system characteristics

Goal of the phase

After the boundaries of a system are chosen, the *valued system characteristics* must be selected (Chapter IV). This phase is *not* aimed at selecting quantified needs and wants or goals, but on creating a list of characteristics. Quantification and goal-setting come later on in the procedure (Phase 4 & 5). Such a list is needed in order to guarantee a focus on the prevention of negative side-effects in both Phase 4, when the focus is on the present *resource state* of the system and its *use space*, and Phase 5, when the exploration is aimed at finding options for sustainable development.

Information needed to select the valued system characteristics

The information-gathering in this phase is focused on translating of the needs and wants of the people involved into a set of *valued system characteristics* that is as complete as possible. It must include characteristics desired by the people directly involved, by the higher-scale community and by future generations (Chapter IV). A starting point for this search is provided by the previous phase, as the defining of a *SEnS* includes not only a time and spatial scale, but also a specification of the type of people involved, the subsystems, and the types of goals. This should help to define direct involvement, the higher-scale community, and the present users and future generations. In this way, the axes of the framework for assessing the valued characteristics can be defined (Chapter IV). As a rule, preliminary lists of people involved and characteristics are available, which can be used to fill in the framework. Care must be taken that there are no empty cells in the framework. An empty cell offers possibilities for invisible negative side-effects. For example, if characteristics for the future are missing, today's water collection could then exhaust future water supplies.

At the end of this valuation phase it is important to assess whether the values that people consider important are actually reflected in the proposed list of characteristics. This can be done by presenting a proposed list of characteristics to the people involved or their representatives.

Information for executing following phases

The list of *valued system characteristics* will be the main input for Phase 4, '*Measurement of the resource state*'. However, consulting people and analyzing the system's characteristics provides information not only on the type of characteristics, but also on the quantities needed or wanted. When a nature conservation areas is put forward as a valued characteristic, it will be often accompanied by a quantity demand: the desired extent of the area. Similarly when employment is mentioned, the number of jobs can be included. This quantitative information is important for the next phase, when an initial set of quantified needs and wants are compared with the calculated present contribution to these needs and wants and the extent to which the selected valuable characteristics have been attained or maintained (Phase 4).

When a decision is made on the list of *valued system characteristics*, this list can conflict with the initial lists of the problems and the goals of the start-up phase. In that case, the institution responsible for the procedure can be consulted to determine whether the conflict is serious enough to suspend the procedure, and if not, what changes must be made to the start-up to make it possible to continue with the procedure.

VI.3.4 Measuring the resource state

Goal of the phase

After defining the *SEnS* (Phase 2), the people involved and the *valued system characteristics* (Phase 3), the next step is the *measurement of the present resource state* (Chapter V). The present use of this *resource state* will determine to what extent its use contributes to the quantified needs and wants, and what the side-effects are. This provides information about the problems to be solved by the procedure. This information-gathering should also be directed towards the resources that have not been used yet, and are therefore available for solving the problems. This information can be used in the next phase (Phase 5), in the discussion on goals and modifications to the *resource state* and its use.

In the literature on procedures for steering developments there is usually a phase in which the 'State of the System' is assessed, on the assumption that this will reveal the possibilities of development. Various recommendations for such an assessments are given: "Choose an eco-centric approach" (Van Rooy, 1995); "Take a holistic view" (Magerum & Born, 1995) and "Examine interconnectedness" (Magerun & Born, 1995); "Establish comprehensive base-line data" (Warner, 1996); "Be area-wide" (Warner, 1996); "Be advisory" (Warner, 1996); and "Adopt an anthropocentric approach" (Warner, 1996). However, as far as we know, none of the proposed assessments describes a systematic method for measuring the *resource state* of the system in terms of changeable characteristics that interrelate resources, activities, and values (Chapter V).

Information for measuring the resource state

The information-gathering that addresses the system's possibilities to attain of maintain the *valued system characteristics*, is designed to measure the *resource space* and present it as a set of *functional units*.

Because the availability of resources depends on the quality standards established for resources and processes, and on assumptions about constraints for resource use, the measurement of the *resources state* will be highly normative. This phase includes major normative steps which determine the exploration of opportunities and thus will have to be legitimated at the end of the phase.

While the measured *resource state* must be checked, this cannot be done directly. A model must first be designed that reflects the present use of the resources, and can be used to calculate the expected state of the valued characteristics (Chapter V). This *calculated value state* can then be compared with the *actual value state*. If there are major differences it is important to know whether this is due to the assumptions made for the model, or the assumptions made for the measurement of the *resource state*. Such a comparison will establish whether these assumptions are realistic, and will enable the decision makers to make well-considered assumptions and choices, both on the model and on the measurement of the *resource state*. The comparison will also show how the problems described in the *Start-up* are redefined in terms of model variables.

Information for the execution of the subsequent phases

The decisions on the way the *resource state* is described will determine the possibilities to search for development opportunities in the next phase (Phase 5). Besides the *resource state*, there is important additional information that may be helpful in the execution of the next phase. The redefined present problems, the available but unused resources, and the difference between calculated *value state* and actual *value state* can be used to sketch the goals to be set and the necessary modifications to the *resource state* and its use.

VI.3.5 Exploring development opportunities

Goal of the phase

Once the *resource state* of the system has been measured and the present use implemented in a model, it is possible to discuss the desired changes, using a backcasting approach as described in Chapter V. First, the desired values of the goal variables must be made clear. Then the necessary modifications to the state and use of the available resources must be established, to minimize differences between what is desirable in the future and what is possible at present.

Various types of modifications can be investigated each of which affect the resource and use space differently, by assuming different combinations of number, location and type of activities, and different levels of human, financial, and biophysical resources (Chapter V; cf. the design strategies proposed by W.T. de Groot, 1992). This requires decisions on the kind of modifications to be explored, i.e., the constraints to be removed.

Information about development opportunities

For decision making, it is necessary to explore the possible modification of the system's resources and how it can be used to reach the value set. This cannot be done by focusing on one goal variable at a time, because the assumed modifications will affect other variables as well. Visualizing the benefits of proposed modifications to the system's resource state and its use in establishing the complete set of goal variables is the main aim of this phase. This information will represent the *feasible value state* of the system (Chapter V).

Any proposed modification to the system should be based on a consistent set of assumptions about value priorities, relations between values, autonomic developments in society, technology, and risks (WRR, 1995; Van Asselt & Rotmans, 1997). This can be derived from individual visions on the development needed and the associated goals, or it can be the result of a consensus-building process. In this phase a number of well-designed sets of modifications will be selected, all of which should fall within the 'Spectrum of Choices' determined in the *Start-up* phase. When the emphasis in the *Start-up* phase is on the possibilities of the present technology, the assumptions with regard to new technology will be limited. However, one can deviate from the constraints laid down in the *Start-up* if, for example, there proves to be no *feasible value space*, or if the process of decision making tends towards alternative directions of change acceptable to the participants. A well-chosen set of modifications, each based on a different set of assumptions, can show the boundaries of development and possible conflicts ahead, which will facilitate a discussion of the directions of development. The information-gathering continues with the selection of a set of acceptable value states, which can be seen as a delimitation of the 'design space' for planning purposes. The possibilities for attaining these states can be analyzed by means of such techniques as multiple-goal programming (cf. quantified design , W.T. de Groot, 1992). This will facilitate the selection of the measures to be taken in the next stages of the search for sustainable development.

Information for carrying out the remaining parts of the procedure

The information that is passed on to the next step in the procedure is a set of alternative desired modifications to the system which, according to the calculations, all contribute to a sustainable development.

Because these directions are based on assumed modifications in the *resource state* and resource use, the location and number of activities, and system input and/or output, and all these assumption are known, additional information is now available on the type of measures that can be taken, that is, measures that will realize the assumed changes. Moreover, in the subsequent steps, this information can be used when recommending or selecting measures.

In following stages - which are not elaborated in this volume - decisions must be made on the directions of development to be elaborated. These directions may conflict with the overall goal of the procedure as described in the *Start-up* phase. In that case, consultations should take place with the institution responsible for the procedure to establish whether this conflict is serious enough to warrant the termination of the procedure. The initial aims of the procedure can also be adapted to the consensus reached.

VI.4 Concluding remarks

Coherence between information-gathering, management, and participation is crucial, if inconsistencies in decision making on sustainable development are to be prevented. The application of a procedure such as the one proposed in the previous sections is intended to ensure this coherence. When this procedure is embedded in real-life decision making, its linearity may be lost, but that does not mean that coherence will be lost too. By applying the procedure iteratively, - through the incorporation of restart moments - coherence can be retained. In any case, this should ensure that decisions made during the process that are in conflict with previous discussions, are documented.

Just as the procedure ensures the coherence, the *Start-up* phase should provide clarity about the aims of the procedure. Explicit information on the object to be steered is guaranteed by the *system definition* phase, and the striving for completeness of goals is governed by the steps for selecting *valuable system characteristics*. The measurement of the *resource state* shows the interrelationship between values, processes and structures or activities, and resources, while the exploration of the development opportunities by assuming modification to the state of the interrelated resources and their use make possible the examination of a broad spectrum of opportunities.

It is important to identify the people involved, since they are crucial in collecting information on the relevant needs and wants which must be related to the system. Representatives of the people involved may play a role in the identification of the system, the choice of appropriate boundaries and the identification of the valuable system characteristics. Where the procedure as a whole is concerned, they may even take part in the actual decision making.

Thus, the elaboration of the concepts of sustainable development has resulted in a theoretical procedure for decision making. The next question is whether it can be applied in real-life situations? And if not, what modifications are necessary. This issue will be dealt with in Part Two.

PART TWO

In Search of Improvements to the Proposed Methods and Procedures

In Part One, a methodology and procedure are proposed which are aimed at gathering the appropriate information on the steering of developments. The question to be addressed here is What are the strengths and weaknesses of the procedure in practice, and Where weaknesses are found, what modifications must be made? Two groups of cases were selected and analyzed, in order to find possible improvements to the newly proposed procedure.

The first group of cases selected entails two projects that were carried out before the proposed procedure was developed. The cases focused on the development of a specific sector; Dutch agriculture was chosen as the system to be steered. At the time the projects were carried out, it was felt that the interests of the different groups of people who were directly and indirectly involved should be taken into account. To clarify how the present analyses are related to the cases, the historical background of each case will first be explained, followed by a summary of the research steps and the results. This is followed by a discussion on the choices which would have been made on the basis of the procedure and what the consequences would have been. This brings to light the differences between the case studies and the actual application of the proposed procedure.

The second group of cases selected present simulations of the support for parties involved in the debate on sustainable development. The method chosen to gain insight into possible improvements was to simulate a real-life situation, present the information thus gathered, and compare it with information gathered in existing plans for the chosen situation. If the procedure cannot be carried out within temporal and financial constraints, or no new opportunities are discovered, it may be concluded that the procedure needs further improvement.

CHAPTER VII

The Analysis of the Research Project 'Integrated Agriculture in the Netherlands'

CONTENTS

VII.1 Introduction

In any discussion on steering development, the object - a sector or region - to be steered towards sustainable development must be described adequately. In Chapter II it was concluded that the object should be described as a *socio-environmental system (SEnS)* in order to keep economic, social, cultural, and ecological goals at the forefront of attention. As noted in Chapter III, defining such a system has many aspects, such as setting its temporal and spatial boundaries and the resolution given by its time and spatial scale. This step is important, because it determines which system characteristics can be valued, what the possible goals for development are, and what can be steered.

The core of our procedure is the method proposed for listing *valuable system characteristics* (Chapter IV) and measuring the system's *resource state*, since this provides the basic information for exploring the *feasible value space* (Chapter V). These steps are aimed at connecting the goals of the people involved with the possibilities to meet them. It is from this that the *development opportunities* are derived, together with the measures required to steer developments. The procedure described in Chapter VI emphasizes the importance of a sound elaboration of each step and the relations between them, in the area of research, support for parties in debate on sustainable development, and planning procedures. As stated in Chapter I, the soundness of the steps is theoretically grounded, but there is as yet no answer to the questions: Does it provide relevant information? Does it deliver that information in time while remaining within the available financial resources? Is it appropriate for the actual decision making? Before its role in a real-life application is analyzed, we will focus on the following question: Can the proposed procedure provide adequate information about sustainable development, in other words, does it really have the informative power to support a search for sustainable development? The aim of the present chapter is to analyze this quality. The basic assumption is that the proposed procedure may be seen as effective if it leads to an adequate *definition of the system*, a balanced *selection of valuable system characteristics*, insight into the available related *resources* of the system, and a sound exploration of a broad spectrum of *opportunities for development*. How can the proposed procedure best be analyzed to assess its effectiveness?

VII.2 Method of analysis

To analyze the quality of the proposed procedure, it should be applied experimentally and the results compared with those of other procedures. Since such an analysis was beyond the practical and financial possibilities, it was decided to use the procedure to reanalyze projects aimed at economic, social, cultural, and ecological goals that have already been carried out. In such an analysis, the focus is on the consequences of the choices that could have been made on the basis of the procedure, if it had been available at the time the project was carried out and a comparison of those results with the choices made during the actual project. This forms the basis for a discussion of the strengths and weaknesses of the various steps of the procedure, and the modifications required to improve the procedure.

The case selected was the research project 'Dutch Integrated Agriculture' (Van der Weijden *et al.*, 1984a & b; 1985a, b & c; Van der Wal *et al.*, 1985). It was regarded as suitable for the following reasons.

- First, the project was aimed at the integration of economic, social, cultural, and ecological goals, and in that sense was 'value oriented'.
- Second, it was aimed at providing information for steering, since it centred on new measures for unifying goals without causing negative side-effects, i.e., the integrated steering of development.
- Third, since we had participated in the project, we were able to make use of background information about the *start-up* and the course of the investigation, including the motivation behind the choices made during the process.

In the following section, the project 'Integrated Agriculture' will be summarized, including background, aims, methods, results, conclusions, recommendations, and follow up (Section VII.3). The steps taken during the project and the results will then be analyzed on the basis of the information-gathering procedure proposed in Chapter VI (Section VII.4). The focus is on the consequences of the choices made in the *start-up*, the problems involved in *delimiting the system*, verifying the completeness of the *valuable characteristics* assessed, and discussing the way the *resource state* and *use space* within the agricultural land-use was described and modified to meet the goals set. In each subsection, it will be made clear which steps would have been taken according to the procedure; then the choices made in the project 'Integrated Agriculture' are presented and the differences analyzed. A number of conclusions will be drawn with respect to possible improvements to the proposed procedure.

VII.3 The case: 'Integrated Agriculture in the Netherlands'

In the seventies and eighties there was a great deal of discussion about agriculture in the Netherlands and in the European Union as a whole. In addition to a growing recognition of the economic function of agriculture, as a result of the expanding agricultural production of the sixties and seventies, there was also increasing criticism (e.g. De Jongh *et al.*, 1978). Due to the enormous overproduction, immense sums were required to market the surpluses outside the European Union (EU, 1982). There was also criticism from a social point of view, because of a drop in the number of people employed in agriculture and the relative low income of large groups of farmers (Nieuwenhuize, 1981; Titulaer, 1983; Bauwens, 1983). Moreover, from an environmental point of view, there was criticism of the effects of modern agriculture on the rural environment and animal welfare (SNM, 1979 & 1982; CE, 1982). Alternative strategies focusing on employment or the environment were put forward (Nieuwenhuize, 1981; Titulaer, 1983; SNM, 1979 & 1982; CE, 1982). However, these were open to immediate criticism on the grounds of cost and/or side-effects.

In the seventies farmers, environmental researchers, and students joined in the development of a new concept of agriculture, which did not feature in mainstream discussions (RUU, 1978; Edel *et al.*, 1980, De Graaf, *et al.* 1980). It was aimed at stimulating the development of an agriculture that would contribute not only to economic goals, but also to social, cultural and ecological goals. It was described as 'Integrated Agriculture' or 'agriculture that is in accordance with the current goals of society and the environment' (after Edel *et al.*, 1980). On a enterprise level studies were initiated on the development of integrated arable farming systems, to find new techniques aimed at reducing the environmental risks and negative ecological impacts of arable farming (Vereijken & De Jonge, 1980-1983; Vereijken *et al.*, 1985; Vereijken, 1990).

The Netherlands Scientific Council for Government Policy (WRR) invited the Environmental Biology Group at Leiden University to elaborate this concept of agriculture. The Dutch Agricultural Economics Research Institute was invited to assist in modelling the sector (Bakker 1986).The request of the Council was made within the framework of the Council's recommendation to the government 'Towards a broader perspective. A policy-oriented survey of the future', (WRR, 1983). In that recommendation, two perspectives were elaborated: the segregation and the integration of the functions in rural areas (WRR, 1983). The basic concepts of the latter perspective were described in the study 'Towards an Integrated Agriculture', which was part of the project Integrated Agriculture (Van der Weijden *et al.*, 1984a & b; Van der Wal *et al.*, 1985; Bakker, 1986). In the reports, the possibilities to unify goals were explored.

VII.3.1 Analysis of the problems in the agricultural sector

The project started with a analysis of the major developments in the agricultural sector (Van der Weijden *et al.*, 1984a & b). Since 1950 the sector had undergone a number of dramatic changes, the most striking of which was the increase in the volume of production, particularly in animal husbandry, despite a reduction in the labour force and the land under cultivation. There was a drastic intensification, mechanization and 'chemicalization', stimulated by an increased labour costs and low energy prices. These were regarded as radical changes, leading to an increased use of energy, artificial fertilizers and concentrates, larger farms, fields and plots, and a move towards more animal husbandry. All this resulted in an increase in exports, translating into a greater share of the market for dairy products, meat and potatoes.

The study 'Towards an Integrated Agriculture' made it clear that these developments benefited Dutch and European society in a number of ways. The European Union became self-sufficient in an increasing number of agricultural sectors. Consumers gained trough lower prices for agricultural products; farmers saw their income, social status and working conditions improve, and the whole community benefited from the sizable contribution made by the agricultural sector to the balance of payments.

The environment was the other focal point, and here there were major negative side-effects. The increase in the use of artificial fertilizers was seen as a major disadvantage, since it affected the maintenance of soil quality. In some regions soil was lacking in moisture and organic matter, and the incidence of soil diseases increased. The surroundings of agricultural areas suffered from increased pollution caused by artificial fertilizers, manure and pesticides; this contributed to the impoverishment of the flora and fauna. The study concluded that since production was concentrated in a few major products and farms, the fields and plots became larger, and the countryside more and more uniform.

From an economic point of view, major disadvantages were recognized. There was the enormous overproduction and the accompanying financial burden for the EU; at the time this represented some 40 billion guilders per year, creating a new form of dependence of agriculture on world market prices, United States agricultural policy, and EU policies. Moreover, Dutch agriculture had become vulnerable in another respect, due to its increased dependency on the world market for energy, raw materials, and concentrates. In addition, and EU policy had an increasingly distorting effect on world markets, which affected the possibilities of developing countries to export their agricultural products.

From a social standpoint, the major disadvantages were described as a decline in employment in the agricultural sector; the fact that while farmers' incomes increased, they were unable to keep up with socioeconomic trends; and an increasingly skewed distribution of income within agriculture. The say that farmers had in matters concerned with the management of cooperatives and the environment also declined. The polarization between the farmers and the general public increased, because of intensive animal husbandry, the use of pesticides, and the conditions in which animals were kept.

It was recognized that these disadvantages affect not only the community at large, but also the agricultural sector itself. Everyone was 'passing the buck': consumers to farmers, farmers to animals and the environment, the EU to the Developing Countries, the Netherlands to its EU partners, and the present generation to future generations.

VII.3.2 Proposed alternative: Towards an integrated agriculture

It was argued that a new strategy for agricultural development was obviously needed. In this new strategy, defined as the striving for an 'Integrated Agriculture', widely divergent objectives are integrated in such a way that the burden is not simply passed on from one sector to another, and from one level of scale to another. It was assumed that major changes in agriculture were needed to reach the objectives described. Box VII.1 summarizes the objectives and assumed impacts (The figures between brackets refer to the second column in Table VII.1).

As regards new measures, the emphasis was on creating opportunities rather than imposing limitations. A set of conditions was drawn up, which included the continuation of EU market mechanisms; no substantial increases in government subsidies to agriculture; no drastic changes in patterns of consumption; and no stricter regulations than those in force at present. In land reconstruction projects, conflicts of interests were not assumed *a priori*: cooperation among the various parties and a greater say in the management and design of the rural areas was regarded as a realistic option.

The request of the Council was not only to analyze the agricultural problems and elaborate the concept of 'Integrated Agriculture' in terms of directions of development and measures, but also to quantify the possibilities for unifying goals. However, the aim was not to design a ready-to-use system or a blueprint for Integrated Agriculture, but rather to quantify the widely divergent objectives of farmers, consumers and others. (Van der Wal *et al.*, 1985). The objectives were expressed in terms of the direction of change as well as norms. The values of certain variables would have to increase, such as the production of fodders and energy, while others would have to be lowered, such as energy use per unit or product. The changes desired for other variables were formulated as 'its

	Objectives	Aimed or expected impacts
a	**Maintenance of net levels of employment in agriculture (3)** - no freezing of production structure - no freezing of mechanization - high skilled labour aimed at: . reduction of energy and resources . increasing net production level, . more care for animals (e.g. stable and fodder quality) - more high technology aimed at .waste reduction .recycling	- lower unemployment costs - more social justice for farmers - possible continuation of farming - age distribution among farmers (with a view to continuity of farming) - mechanization: more light machines - improvement: . product quality (1) . environment, nature and landscape . animal welfare (fodder and stable) (9)
b	**Equilibrium demand and supply and self-sufficiency (the EU market)(4)** - Reduction of costs of overproduction: . dairy products . sugar . meat - Increasing the production of . fodder . energy . industrial crops	- reduced EU market risks - improved crop rotation plan - reduced pesticide use - less distortion of competition with regard to developing countries (5)
c	**A more economic use of energy and raw materials (6)** - reducing use of energy per unit of product - reducing of use raw materials per unit of product - no increase in rough manual labour - reducing overproduction - higher prices for: . energy . resources - lower labour taxes - energy production to replace external energy by solar, wind, biogas and crops - more efficient cows - less imported fodders - more fodder crops - efficient use of organic fertilizers - transport of organic fertilizers - storage or organic fertilizers - more processing of wastes for fodder production	- higher income/hour - reduction in use of gas - reduction in use of: . imported fodders . artificial fertilizers . electricity - future generations less vulnerable - fewer resource imported - lower cost of processing human wastes - less waste material - less overfertilization - new employment: . recycling sector . transport . processing etc.

Box VII.1: The concept of Integrated agriculture described as desired developments and expected impacts (after Van der Weijden *et al.*, 1984). (n) = refers to the list of main objectives in Section VII.3.4. '*' = example of issues not operationalized in the model variables in Table VII.1

	Objectives	Aimed or expected impacts
d	**Selective use of auxiliary means** - reducing the use of risky means - replacement by less risky means especially pesticide in . potatoes and . beets - reducing use of copper	- better working conditions (2) - higher product quality (residues) (1) - less soil deterioration - less pesticides - less heavy metals - less risk of resistance - fewer marketing risks (residue) - more labour in agricultural sector
e	**Improvement external production circumstances (7)** - reducing over-fertilization - reducing contamination with heavy metals - improvement soil: . less potatoes and beets . more organic fertilizers - improvement water availability	- lower emissions of: . ammonia . nitrate . pesticides - less erosion - less contamination
f	**Integration agriculture and nature(8)** - integration of nature in the whole agricultural area - improvements for both nature and agriculture through reduction in: . overproduction . imported fodders . over-fertilization water supply * maintenance and use of hedges - special machines - new production methods * volunteers nature protection * functional landscape for both agriculture, nature and recreation - landscape changes allowed: no freezing	- higher quality of vegetation, - more birds especially meadow birds - more mammals - better management of wooden wind breaks and hedges - less acid - less over-fertilization . nitrogen . phosphate * spatial distribution of intensive livestock breeding - more fodder crops in landscape
g	**Decision making: participation (10)** - more say for people involved and less top-down steering - government provides the framework for integration (shift of burdens)	- increase the chance to develop an integrated agriculture.

Box VII.1: Continued

Table VII.1: Relation between groups/sectors, objectives, goal variables of integrated agriculture and variables in the Dutch Integrated Agriculture Model (Bakker 1986)

Groups/ sectors	Objectives	Goal variables	Model variables
consumer	1. reasonable price and quality	- fl products/Dutch consumers	- price/product, internal consumption
farmers	2. reasonable income and working conditions	- land rent - fl/farmer - pesticide use	- rent/ha - ha/farmer - see: soil fertility
farmers and agricultural industry	3. maintenance of net levels of employment	- n-farmers - n-jobs in industry	- labour-time farmer - volume inputs & products
European Union	4. more equilibrium and self-sufficiency on the EU market	- expenditures milk sector - intensity energy use	- fl/milk production, J/fl-product - fl/energy
developing countries	5. more competitive position for developing countries	- milk production	- kg milk production
global users	6. more economical use of energy, raw materials, land	- energy use - use of phosphate - use of fodders	- J/fl-product - import kg P - import kg fodders
future users of the rural area	7. improved quality of the: - production environment	- soil deterioration - soil deterioration in arable farming - soil fertility rotation risk - livestock diseases	- ha potatoes & beets - ha potatoes & beets/100 ha arable land - ha potatoes & beets 1:2 year - gr copper pollution/ha, kg potassium/ ha
	- external environment	- ammonia emission - phosphate accumulation	- kg/ha - kg/ha
nature conservation and tourism	8. promotion of varied wildlife and a varied and attractive countryside	Internal wildlife risks: - high-risk use of pesticides use - manure pastures (+maize)ha - pastures cattle density - low manuring level	 - ha potatoes & beets 1:2 year - kg N/ha - cow/ha - ha 0-kg N/ha
		External wildlife risks: - phosphate emission - ammonia emission	 - kg P accumulation/ha - kg NH3/ha
animal protection	9. improvements in animal welfare	- see: livestock diseases	
national benefits	- no increase in subsidies or costs	- levies on input - unemployment costs	- kg fodders & ng energy - n farmers and jobs
EU benefits		- EU expenditures	- kg milk
scale of production	- maintenance of scale of production	- final output - agricultural area - livestock	- price/product - ha grass, arable land - n cows, pigs, etc

value must be at least maintained' (see 'objectives' in Box VII.1).

VII.3.3 Dutch Integrated Agriculture Model

To explore these possibilities, an activity-based, regional, comparatively static, linear programming model of Dutch agriculture was used. It was developed by Bakker (1979, 1984a &b, 1985a) of the Dutch Agricultural Economics Research Institute, and modified for the purpose at hand (Bakker, 1985b & 1986). The modified model, known as the Dutch Integrated Agriculture model (DIA model), describes the area available for agriculture activities. Regions were distinguished on the basis of location and soil quality, and the potential agricultural crop-growing activities of each soil type were modelled. A certain number of hectares were available for each crop-growing activity, and for each hectare there were various options for agricultural use, within several rotation restrictions. Domestic and imported fodders were used in livestock breeding, and manure was employed in crop growing within biophysical and economic constraints, such as the type and quality of fodder, and manuring levels. The starting point was that current arable land must be kept in production. The DIA model projected activities in various regions, in order to maximize the economic profitability of land-use - the land rent - within a set of constraints imposed upon the model. Areas producing vegetables and fruit were excluded, as were areas consisting of protected horticulture.

In implementing the project, the variables of the DIA model had to be linked as closely as possible to the objectives of Integrated Agriculture and the relevant variables described in Box VII.1. The selected 'goal variables' and their relation to the 'model variables' are listed in Table VII.1. A comparison of Table VII.1 with Box VII.1 shows that not all relevant *goal variables* could be linked to the DIA model (in Box VII.1, examples are marked with a '*'). The spatial distribution of intensive livestock breeding, for example, is not included, nor such aspects as the maintenance and use of hedges, the role of volunteers, and functional landscape for both agriculture, nature and recreation.

Changes in the variables mentioned in Table VII.1 were considered an indication of the extent to which the objectives had been met. The changes desired were imposed upon the DIA model. The analysis of the results focused on the need to formulate new measures that were in accordance with the concept of Integrated Agriculture. In this way, both opportunities to meet goals and sources of conflict were brought to light, resulting in a new run of the DIA model. This continued until the outcome was reasonably acceptable in terms of attainable goals. A satisfactory outcome was translated into recommendations for governmental policy.

To provide a reference for comparison, the present situation was modelled. Although there were a number of major discrepancies between the reference year and the calculated activities - e.g., too much arable land at the expense of grassland - the investigators considered the calculations to be sufficiently reliable to enable the use of the DIA model for further calculations.

VII.3.4 Summary of the results

Six out of all runs that were carried out are presented here. The results are summarized in Table VII.2 (the italics below refer to the definition of the runs in the tables).

The search for Integrated Agriculture started with two runs aimed at preventing the accumulations of *copper* and *phosphate* (Run C and P in Table VII.2). The results were unsatisfactory from the point of view of Integrated Agriculture, because the socioeconomic goals were not met.

The next run of the DIA model was aimed at improving environmental conditions indirectly by increasing the cost of energy through a *levy on all energy used directly and concealed in production means* (Run E in Table VII.2). The run simulated the introduction of an energy levy on the means of production in agriculture which was equivalent to a doubling of the price of fossil energy in the Netherlands. It was found that such a measure would have spectacular consequences, e.g., energy consumption was reduced, while a positive side effect was the reduction in such aspects as over-fertilization with phosphates and copper. However, a number of serious disadvantages were recognized: employment in general would drop, income would decline, and the risk of a deterioration in soil quality remained.

The investigators felt that such results on a run in which the economic screws were tightened was unsatisfactory. Therefore, in a following run of the DIA model compensation was simulated by *reducing labour taxes*, using the revenues from the *energy levy*; in other words, a shift of the burden from labour to energy was simulated, in combination with *phosphate and copper* restrictions (Run LECP in Table VII.2). This run did indeed produce better results, but there was still a loss of income and employment in comparison with the reference run.

To minimize the negative socioeconomic outcome, a new run was carried out. Again the starting point was *reducing labour taxes*, but instead of a levy on energy, an *imported-fodders levy* was chosen, combined with the introduction of *quotas for milk production*. Due to the levy, the prices rose to the same level as domestically produced cereals. The milk production was limited and EU benefits resulting from the decrease in overproduction were used to reduce labour taxes. The main purpose was to make more efficient use of energy and raw materials and to stimulate the use of labour. The modifications in agricultural production were

Table VII.2: Results of the project Integrated Agriculture: changes in % from model reference (+/- =less than 5 %; ++/- - = 5-10%; +++/- - - = more than 10%; p+/n- = not quantified positive or negative changes.

Changes in runs	C	P	E	LECP	LFQ	TLFQ
1 **Consumer expenditures** (fl)	-	-	-	-	-	-
2 **Farmers**:						
rent (fl/ha)	-	- - -	- - -	- - -	+++	+++
income (fl/y)	0	-	- - -	-	+++	++
3 **Agricultural employment**						
farmers (y)	-	- -	- -	- -	- -	+++
agricultural industry (y)	-	- -	- - -	- - -	- -	- -
total (y)	-	- -	- - -	- - -	- -	+
4 **Balance and self-sufficiency** EU						
expenditures milk (fl)	0	0	+++	+++	0	0
intensity energy use (j/fl product)	0	+++	+++	+++	+++	+++
5 **Competition developing countries**						
less milk production (t)	0	+	+++	+++	+++	+++
6 **Use of energy and raw materials**						
energy use (total) (j)	++	+++	+++	++++	+++	+++
use of phosphate (kg)	0	0	n-	n-	n-	n-
use of imported fodders (kg)	++	+++	+++	++++	+++	+++
7 **Internal production environment**						
soil deterioration (ha)	0	- - -	- - -	- - -	- - -	- - -
idem (ha/100ha)	+	+++	+	+	++	+++
ha high-risk rotations (ha)	-	- - -	- - -	- - -	- - -	- - -
9 accumulation copper (gr/ha)	+++	+++	+++	+++	+++	+++
9 potassium above maximum (kg/ha)	0	+++	+++	+++	0	+++
7 **External production environment**						
ammonia (kg/ha)	+	+++	+++	+++	+++	+++
phosphate (kg/ha)	+++	+++	+++	+++	+++	+++
8 **Varied wildlife internal**						
high-risk use pesticides (ha)	-	- - -	- - -	- - -	- - -	- - -
pastures (+maize) (kg N/ha)	0	++	+++	+++	+	0
pastures cow density (c/100 ha)	0	- - -	- - -	- - -	0	- -
low manure (ha with 0 kg N/ha)	0	n-	p+	p+	n-	n-
8 **wildlife external & landscape**						
phosphate emission (kg/ha)	+++	+++	+++	+++	+++	+++
ammonia emission (kg/ha)	+	+++	+++	+++	+++	+++
N **National benefits** (fl)	n-	n-	n-	p+	n-	p+
S **Scale of production** (fl)	-	- -	- - -	- - -	- -	-

C = Copper: reduction of accumulation to zero
P = Phosphate: reduction of accumulation to zero
E = Energy levy on all energy used directly and concealed in production means
LECP = reduced labour taxes, energy levy and reduction of copper and phosphate emissions
LFQ = reduced labour taxes, imported fodders levy and assigned quotas for milk production
TLFQ = idem LFQ + new production technology

similar to those in the previous run (Run LFQ in Table VII.2), the main problems remained unsolved. There was still a negative impact on employment, and the environment for agricultural production continued to be problematic.

Although in the previous runs many positive impacts were recorded, according to the investigators, they all failed in two respects: employment and agricultural-production environment. The last run focused on ways to solve these problems. The question was What would be the impact of a strategy which introduced labour-intensive *new production technology* aimed at increasing the efficiency of input use, combined with a *reduction in labour taxes*, an *imported-fodders levy* and the introduction of a *quota system for milk production*? This run produced the same outcome as the previous one (Run TLFQ in Table VII.2): reduction in the use of energy and raw materials, and in ammonia and phosphate emissions. There was also a reduction in livestock, but the new technology had an important positive impact on employment. Not only the positive results were accentuated, but also the negative impact on the environment for agricultural production. Focusing on the extent to which goals were reached, it became clear that there was still no solution to the risk of pesticide use in crop growing and no improvement in the quality of nature in agriculture (Table VII.2). Although consumer benefits were also negative and the maintenance of productions appeared to be unattainable, the changes were relatively small.

VII.3.5 Policy recommendations

The results of these exercises provided sufficient grounds to recommend a strategy aimed at new technology designed to increase net production by reducing losses, as opposed to the more common strategy of increasing production. The recommendations included continuing EU policy on the quota system for dairy farming and extending the quota system to other sectors, including the meat-producing sector. In addition, there was a recommendation to shift the burden from labour to other means of production, in order to save animal feed, commercial fertilizers, and auxiliary substances. Priority should be given to reducing levels of heavy metals from commercial fertilizers, concentrates, the copper in pig feed, and manure. However, alongside restrictive measures, some relaxation was recommended. The objective of research should no longer be labour-saving measures, but rather the deployment of highly qualified labour which can pay for itself. This presents a completely new challenge to those involved in agricultural research. Agricultural research strategy should be focused on alternative production technology that takes into account environmental and employment aspects, reduces energy and raw material use, and improves efficiency by reducing losses.

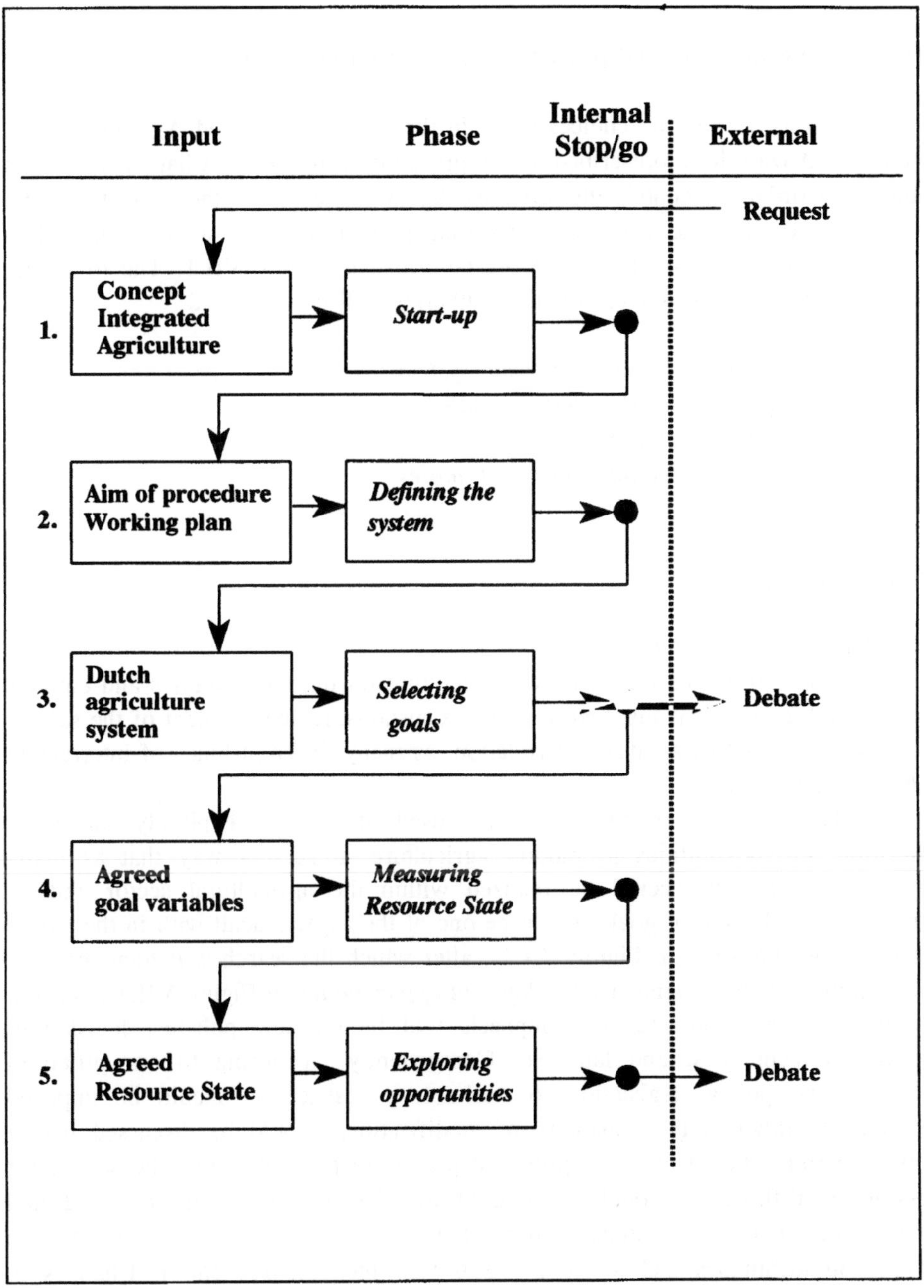

Figure VII.1: The proposed procedure applied to the project Integrated Agriculture

VII.4 Using the proposed procedure to analyze the project

In this section the choices made in the project Integrated Agriculture will be compared with those prescribed by the procedure proposed in Chapter VI of this volume, in order to identify the strengths and weaknesses of the procedure, and discuss the consequences for the proposed procedure. The steps to be taken according to the proposed procedure are described in Figure VII.1. The following phases of the procedure for information-gathering will be compared.

1 *Start-up*
2 *Defining the socio-environmental system*
3 *Selecting the valued system characteristics*
4 *Measuring the resource state*
5 *Exploring the development opportunities*

VII.4.1 Start-up

According to the procedure

In view of the problems facing Dutch agriculture during the seventies and eighties (Section VII.2.2), it was justified to aim at a sustainable development of the sector. Problems could not be analyzed in isolation, as many different kinds of interrelated interests were at stake.

The research procedure as proposed should be explicitly aimed at describing the possibilities to modify agriculture in such a way that as many objectives as possible could be realized within the agricultural sector and its context. The selection of goals should be one of the aspects dealt with in the public debate (*Selecting goals* in Figure VII.1), after which the search will focus on new opportunities for development (*Development opportunities* in Figure VII.1). Since a vision must be elaborated, the emphasis will be on new policies, legislation, production techniques, and land-use. Consequently, according to our proposed procedure no policy, legislation, or technology restrictions should be imposed beforehand, although the feasibility of modifications should be discussed before actual steering takes place. The proposed procedure prescribes that the search for possible modifications should be focused on changes in agricultural land and natural, man-made and human capital (the *resource state*), as well as on the production techniques and location (the use of the *resource state*). Changes in external socioeconomic constraints such as price settings, should also be incorporated, to facilitate the use of new resources.

The project Integrated Agriculture
As noted in the results of Section VII.3.4, the search for possible modifications to the way agriculture uses its *resource space* was based not only on the need to reach the selected goals. It also took as its starting point a set of policy measures that were under discussion at that time. Thus, the project was 'value-oriented', i.e., since it focused on measures that could contribute to the goals set, but at the same time, it was 'problem-oriented', in that it examined the causes of problems, in an effort to establish current economic constraints and the required modifications to the use of arable land. Measures were then derived to solve the problems which had been identified, followed by a calculation of the potential impact of such measures on the goals which had been set.

Discussion of the differences
In the *start-up* of the project Integrated Agriculture there was clearly a dual objective, while our proposed procedure deals with only one aim. This difference must be emphasized, because such a dual objective will have - and should have - consequences for the way development opportunities are explored in the subsequent steps. It can complicate this exploration, especially when it is not made clear in the *start-up* whether only the measures formulated in the *start-up* are to be calculated or alternative measures, where relevant, will also be taken into account later in the proposed procedure. There is a danger that only measures proposed in the public debate will be explored, which undoubtedly reduces the range of possibilities for modifications. The project Integrated Agriculture took this into account. Focusing on the motivation behind the runs, it is clear that the quantification started with a set of measures, but evolved in the end into a strategy of calculating goal-oriented modifications to the *use space* within the agricultural sector. New uses and new technology were assumed, and their potential contribution to the goals set justified the formulation of new measures and research (Policy Recommendations in Section VII.3.5). The question of whether this approach has been fully elaborated will be discussed in Section VII.4.5. In any case, the project showed that such a dual aim can be practical and that the proposed procedure should also be tailored to such use.

VII.4.2 Defining the socio-environmental system

According to the procedure
The choices to be made in this phase are related to the hierarchical organization of the object to be steered, and its temporal and spatial boundaries.

It was clear from the analysis that several *scale levels* were involved, and various interests were at stake, including Dutch consumers, the European Union, developing countries, and farmers. This means that the objectives are also related to very different levels of scale, from 'landscape' to 'EU finances' and 'a more competitive position for developing countries'.

According to the proposed procedure, one scale level must be chosen, and all the higher-scale needs and wants must be described on - or translated to - a level that is one scale higher than the *SEnS*. But which level of scale is the appropriate one for the *SEnS*? Let us assume that we are focusing on Dutch agriculture and the interests of developing countries. The market interests of developing countries are frustrated not only by Dutch agricultural production, but also by the trade regulations of the EU and access to resources such as energy. This complex of relevant factors could be seen as an argument in favour of defining agriculture on a higher level of scale than 'Dutch' or even 'European'. However, that would make it difficult to take into account regional, or even local objectives, such as wildlife and recreation governed by local processes and structures. Thus in delimiting the system, the necessity of adopting a different scale and detail level for the needs and wants of the various people involved causes a problem. Unfortunately, this problem cannot be solved by means of the procedure itself (Chapter VI.3.2.). It says only that during the *start-up* guidelines should be given and that any choice will be highly normative.

The choice of a time scale is less open. Since, one is looking for changes in all aspects of the *present resource* and *use space* of agriculture - location and technology, new measures, and price-setting - the time-scale is obviously one of decades rather than years, although short-term changes are also strived for.

After the scales are chosen, the *boundaries* of the system must be defined. For the sake of argument, let us suppose that regional agriculture in the Netherlands has been selected as the relevant scale for the search for sustainable development. The question is whether the agricultural sector and its land-use should be isolated from other rural activities (cf. Chapter III.4.2, and the remarks about selecting parts of the North Sea), or the rural area should be chosen as the system to be described as a *SEnS*? In Dutch rural areas nature conservation, agriculture, drinking water collection, and other rural activities are spatially strongly interwoven (see for example Chapter XII and XIII). The description of the system as a more or less spatially isolated set of agricultural land-use activities is therefore not sufficient. It does not meet the requirement that the steering within the selected boundaries must not be hindered by the steering of external activities (Chapter III). The system selected for analysis should be an entity that is not overly 'open'. This argues in favour of taking into account the complete set of rural activities (an example is given in Chapters XII and XIII). Thus according to the procedure proposed, the Dutch rural area would probably have been chosen as a

SEnS, recognizing the Netherlands or EU as the higher level of scale. The time-scale selected would have been decades.

The project Integrated Agriculture
It will be clear that the Dutch agriculture was chosen as the *spatial level of scale*, and that consequently the relevant needs and wants had to be translated to this level; the quantification was also aimed at this level. The boundaries of the system were defined as the sum of the agricultural sub-sectors in the different regions: the context of agriculture was also emphasized. The list of objectives (Section VII. 3.2 and Box VII.1) shows that the project took into account direct needs and wants (e.g., farmers, consumers, and animal protectionists) and higher-scale needs and wants (e.g., EU budget and market, and developing countries). Future needs and wants were acknowledged by emphasizing the need for new production methods and new technology, and by the environmental objectives. Neither a *time-scale nor a time horizon* for change was defined. Where the modelling and quantification of goal attainment is concerned, the *boundaries* were the land-use areas of the sectors within selected rural regions.

Discussion
Since, as we have seen, the choice of a certain level is highly normative, the choice made in the project Integrated Agriculture must be regarded as a given. This means that the analyses will be less adequate on all other levels of scale. The discussion will centre on the need for a time scale and the way Dutch agriculture is defined.

No *time-scale or time horizon* was defined in the studies, although the proposed procedure prescribes that one must be selected. Is it really necessary to set a *time-scale*? If research is aimed at predicting the most likely developments and their alternatives, then all processes must be modelled, including the actions of farmers and policy makers. In that case, a time path and a time horizon will be needed. But if the initial focus is on the question of 'what' can be unified under 'which' socioeconomic and biophysical circumstances, and it is not until later that attention turns to 'when', then a time horizon would seem to be less relevant. Moreover, to gain insight into attractive modifications of farming and policies, the actions and reactions of the farmers and policy makers should be excluded. This is a crucial condition for exploring new ways of agricultural production that may provide the necessary motivation to change current agricultural practice and policies. In this light, the decision to set no time horizon is consistent. However, the assessment of what can be unified is not independent of the *time-scale* for change. Possibilities for change on a decade scale are undoubtedly more numerous than those on a year scale.

Does the type of *goal* require that a time horizon be set? For example, does a goal, such as reducing the risk of 'eutrophication with phosphate' or 'disease due to copper in the soil' prescribe a time horizon? The answer must be 'no', if the focus is only on current process characteristics, such as the rate of accumulation. These alter quickly when modifying the breeding of livestock. But this is not an adequate answer to present problems caused by accumulation in the past. In the set of goal variables mentioned in the studies on Integrated Agriculture, goals which entail 'solving existing problems rooted in the past' are not operationalized. Therefore, it was acceptable to refrain from setting a time horizon. However, this also implies that in the studies such goals are considered as not relevant, which actually conflicts with the overall objective to improve the conditions for nature. Assuming that the goals aimed at restoration should be taken into account, then a restoration rate must be chosen, which can only be derived from a time horizon for complete restoration. Thus the requirement that a time horizon be set must not be dropped, and is even essential to the operationalization of restoration goals.

An implicit *time horizon* was set, since new production technology was assumed (Run TLFQ). This would put the time horizon at a period of time needed to develop new technology (see also Part Three). Since the aim of the studies was to identify development opportunities and decide on the necessary measures - rather than to provide a blueprint - this implicit time horizon is not a problem, although there is a chance that the results will be interpreted as developments that may be realized in the near future.

This leaves us with the question: How justified was the selection of the Dutch agricultural sector and its present land-use area as the *spatial boundaries* of the system to be studied? Should it not be the rural area, since many objectives, such as wildlife, landscape, and the external environment, are affected not only by agriculture, but also by the overall use of rural areas, while in the same way, agriculture depend on the development of the activities in surrounding areas? Choosing the agricultural land-use area in the Netherlands as the system *boundaries* implies that the system delimited encompasses only a part of the rural land-use. This may be sufficient and acceptable where there is a clear spatial division between agricultural areas and areas intended for other rural land-use but, as noted above, in the Netherlands agricultural and other types of rural land-use are closely interwoven. Thus an approach in which system boundaries are chosen in such a way that the system covers a complete rural area is probably more appropriate. However, such a change in system boundaries would transform the object from 'Integrated Agriculture' into 'Integrated Rural Area'. Moreover, it would undoubtedly require enormous extra effort to quantify the meeting of goals within these new boundaries. If the proposed procedure had been followed in the project, there would have been a real risk of time and money constraining the implementation of the next steps. As these are good reasons to focus on the agriculture sector rather than the rural area, a conflict arises with the proposed procedure. The demand to select a system that is 'not overly open' is in obvious conflict with the initiative for a study on a spatially defined sectoral development.

VII.4.3 Selecting the valued system characteristics

According to the procedure

The procedure for gathering information for sustainable development prescribes the quantification of the system's potential to contribute to the objectives simultaneously. To that end, the *valued system characteristics* must be expressed in *goal variables*, after which the possibilities of the system to contribute to these goal variables are measured. Thus the selection of goal variables is crucial for the following phases. In the procedure for sustainable development, it is prescribed that the people involved should be identified and consulted when selecting the valued characteristics of the system.

The project Integrated Agriculture

Instead of drawing up an inventory of the people involved, an indirect method was followed. The main production developments in the sector since World War II were analyzed. By relating these developments to the people that may be affected, the people involved could be identified. This resulted in a list of the benefits of the developments for the farmer, the consumer and the European Union. When it came to listing the disadvantages, a different starting point was chosen. The side-effects from an economic, social and ecological point of view were discussed. This resulted in the recognition of the interests of the local nature and environmental-protection agencies, supply and manufacturing industries, and developing countries. On the basis of these interests, the study described the relevant objectives for agricultural production and developments (Box VII.1).

Discussion

The question is whether the proposed procedure could have improved the selection of *valued system characteristics* and *goal variables* within the time and money constraints of the project Integrated Agriculture.

In order to analyze the consistency and completeness of the goals of integrated agriculture, the issues listed in Box VII.1 have been entered into the framework for *selecting valuable system characteristics*. First, the types of 'direct' and 'indirect' users are listed to define the rows of the framework; then the issues are systematized by fitting them into the cells of the framework (Table VII.3).

Table VII.3 shows that Integrated Agriculture is focused on characteristics relevant for the interests of several direct users in the present and the future. Characteristics related to present and future higher-scale interests are also emphasized. It this sense, the project was not hampered by the lack of a systematic framework for identifying different types of valuable characteristics. Nevertheless, for a systematic search for missing items, it might have been better if the investigators had had such an instrument at their disposal. It is noteworthy that for one group of direct users the row is empty. No quantitative food needs at the level of national consumers were identified. The quantities of agricultural products such as milk and meat were only taken into account with regard to the contribution to the EU and the needs and wants of developing countries. Although the investigators apparently regarded self-sufficiency primarily as an EU interest, this standpoint should not have overruled the possibility that national populations have wants with regard to the national self-sufficiency. Clearly, an adaptation could easily have been made within the time and money constraints of the project.

Many of the interests described as objectives in Section VII.3.2 were translated into characteristics that may be seen either as conditions for the future, or as contributions to the higher-scale needs and wants. This is due to the fact that the agricultural sector was chosen as the system to be studied: all needs and wants not directly related to agriculture can only be regarded as externalities, and the same applies to all environmental issues that do not affect agriculture itself. This means that some of the needs and wants or values distinguished have to be translated before they can be related to relevant system variables, while others can be related directly. Thus there is a risk that not all needs and wants are translated in the same way into goal variables on the level of the agricultural system. The extent of this danger cannot be established by studying the framework of valued characteristics, but only by studying the actual choice of variables.

To analyze the *selection of goal variables* in the project Integrated Agriculture, the selected variables in the project are also entered into the proposed framework (Table VII.4.). As the table shows, all the variables for present direct users are socioeconomic ones. It is noteworthy that values such as nature and landscape are not clearly linked to certain user groups, while socioeconomic values are linked to groups such as farmers, consumers, industry, and developing countries. As a result, these values are not elaborated to the same extent as those in the above groups. Although process characteristics such as emissions of fertilizers were taken into account, the natural and environmental structure conditions have been underexposed. From the point of view of improvements to the landscape, variables may be expected which determine the structure of the landscape.

One could argue that the type of model used resulted in the absence of

Table VII.3: Valuable characteristics of the agriculture sector as a socio-environmental system within the context of national, European and global policies derived from Box VII.1 (a-g = the letters in Box VII.1)

Characteristics related to direct users

	Processes	Structures	for future use	for future processing
Separate				
-consumer	-	-	-	-
-labours	c.rough manual labour	a.skilled labour c.employment in: transport, processing, etc. d.labour in agricultural sector	-	-
-production in agriculture	b.fodder production c fodder crops c.energy production	a.production structure a.mechanization technology c.more efficient cows c.storage organic fertilisers	c.efficient cows	a.production structure a.mechanization/ technology d.risky means, risk of resistance, and marketing risk
-environment agriculture	e.soil: organic fertilizers	c.storage organic fertilisers	e.erosion	d.pesticide use in potatoes & beets e.soil: potatoes & beets d.heavy metals (e.g., copper)
.nature	f.use of hedges	f.hedges f.volunteers nature protection	-	f. special machines
Combined				
-consuming	-	a.product quality d.product quality (residues)	-	-
-farming	a.net production c.income/hour	a.social justice farmers d.working conditions c.price resource and energy c.labour taxes	- c.storage organic fertilizers	a.age distribution farmers a.continuation farming b.reduced market risks c.use organic fertilizers c.distribution organic fertilizers
-environment use and nature	-	e.water level f.integrated nature/ agriculture f.new production methods	a.environment, nature landscape quality d.soil deterioration f.integration nature and agriculture f.new production methods	b.crop rotation c.over-fertilization (e) d.soil deterioration e.contamination heavy metals f over-fertilization water f over-fertilization (N and P) f.acidification
-animals	a.fodder quality	a. stables	-	-

Table VII.3: continued

Characteristics related to higher scale

	Processes	Structures	For future use	For future processing
Combined				
-production	c.energy use/product c.raw materials/ product	-	-	c.energy use/product c.raw materials/ product
-landscape management	f.changes in landscape	f.functional landscape f.distribution livestock	-	e.ammonia/ha e.nitrate/ha e.pesticide/ha
Separate				
-socio-economic	a.unemployment costs b.overproduction costs (dairy, sugar, meat) b.developing countries	c.new employment	-	-
-biophysical	a.waste reduction/ recycling b.production fodder, energy industrial crops b.pesticide use c.processing wastes c.resource import c.cost processing wastes c.waste dumps c.waste recycling c.overproduction (f)	a.energy and resources	a.energy and resources	c.vulnerability future generations
-nature/ landscape	-	f.quality vegetation f.birds and mammals f.windbreaks and hedges f.fodder crops in landscape	c.fossil energy use (gas & electricity) c.imported fodders (e) c.use artificial fertilizers	-

Unfitted goals:
- a. more social justice for farmers
- g. more say for people involved and less top-down steering
- g. government provides the framework for integrating

Table VII.4: Model variables (Table VII.1) systematized as valuable system characteristics of the agricultural sector. ( ) = refers to Table VII.2.

Characteristics related to direct users

	Processes	**Structures**	**For future use**	**For future processing**
Separate -consumer	consumer benefits: fl (1)	-	-	-
-farmer	sector production: - agriculture: fl. (S)	employment: number of (3) - farmers - jobs in agricultural industry Arable land: ha (S)	Arable land: ha (S)	environment agriculture: (7) - ha potatoes and beets - ha potatoes and beets /100 ha arable land - ha potatoes and beets in 1:2 rotation wildlife in agriculture: - ha high-risk use of pesticide (8) - ha zero kg nitrate/ha
Combined	farmer income: (2) - fl/ha - ha/farmer	-	-	environment agriculture: - gr copper/ha (sheep) 9 - gr potassium/ha (cow) (9) - ha potatoes and beets/ 100 ha arable land (7) wildlife in agriculture: (8) - kg nitrate/ha pastures - cows/ha-pasture

Characteristics related to higher scale

	Processes	**Structures**	**For future use**	**For future processing**
Combined	energy: - J/fl product (4)	-	-	wildlife in rural area: (7,8) - kg phosphate/ha - kg ammonia/ha
Separate -national: socio-economic	national benefits: (N) - levies: fl - unemployment costs: fl	total employment: jobs (3)	-	-
-EU/world	- milk surplus: fl (4,5) - milk surplus: kg (4,5)	-	-	-
-energy & resources	- phosphate: kg (6) - total energy use: J (6) - imported fodders: kg (6)	-	- phosphate: kg (6) - total energy: J (6) - imported fod-ders: kg (6)	-

such variables, since it only describes agriculture production. But that is no more than a partial explanation, for landscape is also determined by the types and variety of crops, and variables of this kind were included in the DIA model. For instance, many landscape conservationists in the Netherlands consider the growing of green maize as a major disturbance factor in the Dutch landscape, because in the summer it changes the open landscape into a closed landscape. The DIA model made it possible to demand a certain combination of crops, in order to avoid such a disturbance of the landscape, but it was not explicitly valued and quantified by the investigators. Other structure variables, such as farm size, are also missing, but that was caused by the type of model selected. Farm size and farm distribution could not be modelled without radical modifications which would exceed the time for executing runs, and the maximum number of variables allowed in the DIA model.

It is concluded that the selection of goal variables was not complete, due to the model selected and an inadequate use of the possibilities of the model. This lack of landscape-related variables could have been partly prevented by using the proposed procedure within the project constraints.

VII.4.4 Measuring the resource state of the system

According to the procedure

In the procedure, the selected goal variables determine the relevant processes and structures within the system, and these processes and structures in turn the resources needed. The available combination of resources within - in this case - a spatial unit describes the *present resource state* of the system (Chapter V). In this step, the main question is which processes should be included, and which standards should be adopted to describe those processes. If one wants to use groundwater for drinking water, for example, the relevant specific environmental qualities in sandy soil areas are the presence of sandy layers and the absence of high concentrations of nitrate. When the continuity of agricultural production is valued, than unpolluted soils are required. In view of this relationship between the *valued characteristics* and the potential of the system to provide the *resources*, the *resource state* should be based not only on present agricultural land-use qualities, but also on the qualities relevant for future processing, such as soil deterioration, and the characteristics that determine this quality, such as heavy metals. Furthermore, the availability of *resources* for other than agricultural products should be measured, such as the available water and the quality of that water.

The project Integrated agriculture
In the project, the processes and resource description were based on an early version of the DIA model (see Bakker, 1979 & 1985a). This version was modified, but was still based mainly on the availability of resources such as agricultural land and production means. In this sense, the basic version determined to a large extent the type of land-use distinguished and the type of activities capable of using the available land and its products. It describes the land as useful for various crops as in conventional land evaluation (Chapter V), since it made a distinction between soil types and their suitability for several crops (Section VII.3.3).

Discussion
The way possibilities for agricultural production are modelled describes the resources in terms of suitable for agricultural production. The project obviously did not follow the consecutive steps described by the procedure proposed, because it was not constructed on the basis of the *selected valued system characteristics*. Comparing the choices in the proposed procedure with the choice made in the project, it is clear that the *resource state* in the model disregarded the soil qualities that are relevant for future activities, and other actual and potential products of agriculture or rural land-use. Consequently, the exploration of the opportunities was constrained, as will be discussed in the next Section VII.4.5.

Concentrating on the 'resources state' as described in the model, the question is whether it was adequate. Although the DIA model includes major agricultural resource characteristics, it obviously does not cover all the characteristics that determine the relevant environmental problems for the sector itself. Soil type and average production levels were included, but again, not their qualities in terms of the extent of present soil deterioration and the loss of soil fertility which is relevant for future processing. These are important constraining factors that can be changed by organic manuring and by diversification of the cropping, which were steerable in the model. Thus the description of the resources is inadequate with regard to the available steering possibilities of the model. Incorporating such basic information on deterioration and soil fertility in the model would have required a substantial investment in the data collection and model modification, which certainly could not have been done and an extra investment would probably have exceeded the time and money constraints of the project. Thus it was necessary to keep to the available model and its description of the agricultural resource state.

VII.4.5 Exploring development opportunities

According to the procedure

On the basis of the availability of a model that describes only the available agricultural resource state, the question arises of the strategy which should be followed to *explore the development opportunities*. As noted in the *start-up*, the focus should be on the modification of the *resource state and use* by changing location, technology, and price settings.

The project Integrated Agriculture

At the time, it was proposed by members the Netherlands Scientific Council for Government Policy that the technique of multiple-goal programming should be used to study 'Integrated Agriculture'. It was regarded as an adequate method for studies aimed at combining various kinds of goals. However, the selected model was not designed for such purposes. It could have been modified, but that would have required extra investment, leading to new runs, which at the time were expensive. Because of the expense of modification and calculation, this was not a feasible solution. Therefore, it was decided to keep to the model, and choose another strategy which would make it possible to play within the budget. Subsequent modifications of the runs were used to simulate multiple goal programming. It was decided to focus first on the environment, starting with environmental constraints, and then go on to price-setting and compensation strategies. Depending on the results of subsequent runs, the choices for next runs were calculated to attain even more goals. By maximizing land rent, large negative impacts on income were prevented, and by constraining other variables, such as the use of copper and phosphate, the DIA model was prevented from selecting solutions with unacceptable values for other goals. The idea was that in this way the DIA model would come as close as possible to a multiple-goal analysis.

Discussion

It would have been more effective to choose another type of model, but such a choice would certainly have exceeded the time and money constraints of the project. Respecting these constraints, there are two main subjects which have yet to be discussed here. First, was the manner in which the DIA model was used suitable for exploring the development opportunities, and second, did the set of selected runs cover all relevant and feasible development opportunities for 'Integrated Agriculture' within the temporal and financial constraints of the project? And what can be learned from this about possible improvements to the proposed procedure?

First the *exploration of the development opportunities* will be discussed. In the project a linear programming model was selected that aims at maximizing the farmers' income. It was modified to allow the incorporation of several constraints, and to visualize the contribution to *selected goal variables*. However, since the constraints are formulated at the level of emissions, only the costs are visualized, without calculating the external benefits of a sound environment for other production activities.

The impossibility of following a real multiple-goal programming approach explains the failure to reach all the environmental goals within the constraints of maintenance of income and employment. But is this really the only cause? *Were the appropriate runs selected*, and if not, could the newly developed procedure have been helpful in finding appropriate runs?

The set of measures covers many different kinds of modifications - ranging from adding and removing internal environmental constraints via price settings, to changing activities by assuming new technology - but it does not tell us whether all directions of searching have been adequately explored. This may become clear by analyzing the extent to which the goals have not been met, and discussing which runs would have solved the remaining problems.

The results of the runs show that not all the problems identified were solved (Table VII.2). The decrease in total employment was solved by simulating alternative production activities that differed in net production and labour required, but the problems related to the environment for agricultural production could not be solved by any of the modifications (7 in Table VII.2). This is not surprising. When the focus is on, say, the modification of price settings, then it is easy to neglect the possibilities to modify activities as well. This could explain the lack of crop rotation modifications in the final run. The inconsistency here is that a part of the valued characteristics, i.e., crop rotation, is not translated into constraints on the *resource and use space* level. Runs aimed at a diversification of crop-growing should be carried out.

Could this have been done within the temporal and financial constraints? Assuming that the final run was the last one that could be executed - and it was - the question is actually whether the diversification of crop-growing could easily be incorporated into the last run without exceeding the temporal and financial budget limits. The answer must be 'yes', because the basic principles in preventing deterioration problems had already been defined as solvable by modifying rotation. The incorporation of a modified rotation in the model would probably take hours rather than days, and the same 'final' run could have been carried out. This would probably lead to less positive economic results. The question is whether it might even lead to a negative impact, a question that cannot be answered without executing the run and is beyond the scope of this analysis.

VII.5 Concluding remarks: improvements to the procedure proposed

The analysis of the project 'Towards an Integrated Agriculture' using the proposed procedure for sustainable development revealed a number of strengths and weaknesses in both the project and the procedure proposed.

First, the description of the problems and the values at stake, as described in the studies, obviously demanded a sustainable development procedure designed to provide insight into possibilities for unifying goals. This was stated in the introduction, and the analysis leads to the same conclusion. The comparison showed that in the *Start-up* a double aim was formulated, while the proposed procedure implicitly assumed that only one aim would be chosen in the *start-up* phase. By focusing on measures, the impact can be analyzed; by focusing on opportunities, the necessary measures can be found. However, the double aim, while it does not make the execution of the proposed procedure impossible, it does increase the risk that feasible solutions will be missed. When a double aim is formulated, it becomes even more important to be extremely explicit about the type of assumed changes to be calculated. Nevertheless, this is not an argument for excluding such an approach. Consequently, the procedure should admit the incorporation of various starting points, but it should be modified so that at the same time, the steps to prevent the risks can be addressed.

Second, the *system definition*. The agricultural sector was the only system in the project which was possible in the light of the budgetary limits, but this choice had far-reaching consequences. If a different system definition had been chosen, in accordance with the demands of the procedure, the system might have been less open, but it would also have ceased to be the appropriate system from the point of view of the *start-up* of the project. The procedure requirement that the system should not be 'too open' must be applied less strictly than was assumed in Part One. But if it is 'more open', more attention must be paid to the definition, in order to ensure that the following steps will be manageable.

Third, the list of *valuable characteristics* shows that practical research applications can come up with a set of characteristics that includes items corresponding to all the types distinguished in the proposed selection framework. In this sense, one could conclude that the framework is superfluous. On the other hand, the framework proves to be useful in making a quick check to identify missing items. In the re-analysis, such a check was quite easy, and it result in a more explicit decision to refrain from quantifying the need for national self-sufficiency. A comparable check should be made on the level of goal variables, to make sure that all valuable characteristics are taken into account when describing the resource state and its use, and when using a model for calculations. Consulting the people involved might have made it possible to include certain valuable characteristics which were now missing, but applying this part of the procedure would undoubtedly increase the time and money required.

Fourth, the selection of goal variables, related processes and structures, and resources to define and modify the resource state and use were in some respects inadequate. The proposed procedure pointed out the missed opportunities to solve internal environmental problems related to agriculture. This was due in part to the use of a model selected *a priori*, instead of constructing a model based on the goal variables selected. This resulted in an incomplete description of the *present resource state*. Moreover, a number of *valued characteristics* were not operationalized, even though it could have been done within time and financial constraints. Using the proposed procedure would have made it possible to include goal variables which were missed, and would have improved the results.

CHAPTER VIII

The Analysis of the Research Project 'Reduction in Production in the Dutch Dairy Sector'

CONTENTS

VIII.1 Introduction

In Chapter VI a procedure for information-gathering was presented in which the logical relations were established between *Defining the socio-environmental system* (*SEnS*; Chapter III), *Selecting the valued system characteristics* (Chapter IV), *Measuring the resource state* (Chapter V), and *Exploring the development opportunities* (Chapter V). The analysis in Chapter VII made it clear that the selection of valuable characteristics and goal variables must coincide with the listing of the relevant *resource system characteristics*, since they are the basic element for *measuring the resource state* and for *exploring development opportunities*. It has been shown that a description of the *SEnS* and the *selection of valued system characteristics* does not guarantee an adequate *exploration of the development opportunities*. The quality of the exploration, depends on the quality of the intermediate steps taken. Chapter VII also states that a clear description of the research aim formulated in the *start-up* is important, because of its far-reaching consequences for the actual exploration in the subsequent phases. However, accurate guidelines for the way each phase must be carried out cannot ensure consistency between the desired results and the actually attained results. Ultimately, this consistency will depend on the quality of the decisions made at the end of each phase, where the proposal is approved and the guidelines are formulated for the execution of the next phase.

This chapter focuses on problems of this kind. The question is: Can the proposed procedure prevent inconsistencies in the information-gathering, and if not, how can the procedure be improved?

VIII.2 Method of analysis.

The project 'Reduction in Production in the Dairy Sector in the Netherlands' (De Graaf, 1990; De Graaf & Tamminga, 1990) is the second project selected for a comparison with the application of the proposed procedure. The preparation of this research project on production control was started in the early eighties, and the project itself was carried out in the period 1984-1990.

Although the project focused on ways to reduce milk production, the measures to be studied were not defined *a priori*. In order to identify the measures, or variants of measures, a research strategy was chosen in which different parties - politicians, farmers, and environmental organizations - were invited to explain more clearly their vision on production control and the development of the rural areas. Goals were listed, possible measures were worked out, and the socioeconomic and environmental consequences were calculated. The project was aimed at generating a static picture, in order to visualize the possible consequences of each vision. Those who adhere to a particular vision may not be satisfied with this picture, and this may lead to a decision to adjust the vision or design new measures, the consequences of which are then calculated.

Although the aim was to force drastic changes in the dairy sector, where the socioeconomic and environmental side-effects are concerned, the focus was not primarily on the dairy sector. Since it was assumed that any reduction in the production would have major impacts on other agricultural sectors, the socio-environmental impacts were visualized at the level of agriculture as a whole. Consequently, the iterative process of adjusting the visions and measures focused on agriculture as a whole.

The process adopted for this project closely resembled the exploration of the feasible value space of the agricultural sector as part of a proposed procedure for Sustainable Development, as described in Chapter VI: it is comparable to a procedure to design perspectives of development by generating coherent images of the potential changes in environmental resources and their use which fits the values selected by the people involved.

First the project is summarized, to provide the back-ground information for the analysis (Section VIII.3). Then the steps to be taken and the choices to be made according to the proposed procedure are presented and compared with those actually taken during the project (Section VIII.4). Not all the steps are given in detail. Parts of the research methods used in the project 'Reduction in Production in the Dairy Sector' are quite similar to those of the project on 'Integrated Agriculture' discussed in the previous chapter, especially the selection of goal variables. Therefore, the completeness of this selection will not be discussed here. Instead, the question addressed in the present chapter is: How consistent is the project and how effective is the procedure proposed in Chapter VI in ensuring the consistency of the successive steps taken during a project such as 'Reduction in Production in the Dairy Sector'? With regard to the proposed procedure in Chapter VI, the question is: Can inconsistencies arise when using the proposed procedure, and is this due to imperfections in the procedure for information-gathering, or rather a consequence of the embedding in a decision making process.

VIII.3 The project 'Reduction in Production the Dairy Sector'

To protect dairy farmers, the European Union (EU) has for many years guaranteed that all milk produced will be bought. As a result, in the late seventies the EU was faced with increasing costs involved in marketing the surplus milk production (Meester, 1985; Meester & Strijker, 1985). To compensate for these expenditures, a small levy was placed on each kilo of milk produced. In spite of this measure, by the early eighties agriculture in Western Europe was recording a level of milk production far in excess of internal consumption. The costs of marketing the surplus came to weigh heavily on the EU and its member states (Meester, 1985; Meester & Strijker, 1985). Measures were needed to call a halt to the increase in expenditure on agricultural production.

At first glance, a problem-orientated approach appeared to be an adequate strategy for solving this kind of problem: one could focus on the agricultural sectors that were causing the overproduction, and take measures which within a few years would reduce production in these sectors. Such measures were in fact taken. In the Netherlands a quota system was introduced that permitted a surplus production, but production higher than the quotum was frustrated by an extremely high levy (EU, 1983; LV, 1984; Krijger, 1986). In the mid-eighties, a debate was initiated on the desirability of continuing the measures to control production (Oskam *et al.*, 1987). There were discussions on various ways to reduce the cost to the EU of the overproduction of milk (EU, 1985). One option was regulation by means of a product-price policy, while another solution consisted in incentives to wind down production activities. In the discussions on the continuation of current measures to reduce the costs of overproduction, the European Commission favoured towards a market-oriented policy based on production levies (De Graaf & Tamminga, 1990).

Although it could have been predicted in the early eighties that all EU member states would be confronted with a compulsory reduction in production, it was also plausible that within the framework of EU policies national policies would be able to develop their own variants of production control, tailored to internal agricultural and rural demands. These variants were expected to have different consequences for the development of the entire rural areas.

The expectation that major changes in agriculture would have a considerable effect on the rural area, was of great importance for the national Spatial Planning Agency of the Ministry of Housing, Spatial Planning and Environment (Veer, 1983; RPD, 1983). THerefore, the department took the initiative in setting up a study of the consequences of different visions in the debate on production controle, although overproduction in agriculture is actually the responsibility of the Ministry of Agriculture. In day-to-day policy making in the Netherlands, it was unthinkable for a study started by a governmental department concerned with agricultural production to be carried out without the cooperation of the Ministry of Agriculture. Thus a 'steering committee' was formed that took the major decisions during the project. It was made up of representatives of scientific organizations and governmental agricultural and spatial planning agencies.

Research institutes were asked to prepare a detailed research proposal for studying reduction in production in the dairy sector (LEI & MIBI, 1984). The Agricultural Economics Research Institute was asked to model the sector and to calculate variants of production reduction. The department of Environmental Biology of Leiden University was invited to develop an instrument for calculating the impact of those variants on nature and the environment (De Graaf, 1990). Together the two institutes analyzed current visions on agricultural and rural development, described the measures to be calculated, and reported the results (De Graaf & Tamminga, 1990).

VIII.3.1 Aim of the project

Agricultural activities not only take advantage of the spatial qualities offered - infrastructure, water supply, natural resources, etc. - they also have a considerable influence on those qualities. Agricultural activities change their own production conditions, as well as those of other activities. Landscape and nature suffer from the intensification of agricultural production, but also from the reverse: when agricultural activities disappear completely, many values related to agricultural and natural history disappear with them. Therefore, as far as production control in dairy farming was concerned, the focus was on its effect on the variety, coherence and sustainability of the physical environment, and on the production function and social structure of the rural area in the Netherlands (Veer, 1983). In other words, the project was placed within a broader context, because production control in the dairy sector affects not only dairy farmers and their land-use, but also other farmers and other users of the rural area, such as drinking-water producers and nature conservation organizations. So, the consequences of various visions for the agricultural sector as a whole were studied, rather than only the effects on dairy farming and its land-use.

VIII.3.2 Research strategy

One important argument to focus on production control was that it had been recognized by several societal and political organizations involved in the development of national rural areas as a feasible tool for meeting their objectives:

- The EU embraced it as an instrument that could guarantee a reduction in the expenditures of the common market.
- Nature and environmental conservation organizations felt that it could reconcile agricultural production with the conservation of nature and landscape, thus alleviating the existing environmental problems.

- Several dairy farming organizations saw it as an instrument that could improve the income of farmers, provided that prices were stabilized, rather than lowered.

As noted in Section VIII.2, the project opted for an elaboration in which the various organizations concerned with rural development and dairy farming were given an opportunity to explore the impact of their vision and proposed measures on agriculture and rural development (RPD, 1984a). These groups would be confronted with the potential side-effects of putting their vision into practice (RPD, 1987). If the results did not tally with their expectations, there would be a possibility to adapt the measures. The side-effects of the measures were to be calculated, using the same set of values for each vision, in order to facilitate the discussion on measures to reduce milk production. Ideally, this would also lead to a common strategy for agricultural development.

The first step in the project was to list the organizations which had put forward ideas on the development of dairy farming in relation to rural development, or vice versa. Interviews were held with representatives of an agricultural youth organization, a society for nature and environment protection, agencies of the Ministry of Agriculture, Nature Conservation and Fisheries, and the European Parliament (Bakker, 1984c). On the basis of these interviews, several variants of production control were distinguished. To reduce the costs of investigation, similar variants were consolidated; variants that were described on the level of farms were omitted, because they could not be modelled without considerable expenditure.

VIII.3.3 The Dutch Regionalized Agricultural Model

A new version of the Dutch Integrated Agriculture Model described by Bakker (1985 & 1986) and discussed in Chapter VII was used: the Dutch Regionalized Agricultural Model (DRA Model) (Tamminga & Verhoog, 1988-1989). The instrument for calculating the impact of agriculture on nature and the environment was also adjusted where the agricultural module was concerned (De Graaf, 1990). Relationships between agricultural and environmental aspects were further elaborated, in order to incorporate them into agricultural modelling (Van der Meij *et al.*, 1992a & b; Hanegraaf *et al.*, 1992a & b).

As in the project Integrated Agriculture discussed in the previous chapter, the objectives of the different organizations involved were translated into model variables, i.e., variables the value of which could be calculated by means of the model (Table VIII.1). For example, the risk of negative impact on the diversity of

Table VIII.1: The project Reduction in Production in Dairy Sector: relation between groups/sectors, objectives, goal variables and model variables of the Dutch Regionalized Agricultural Model (Tamminga & Verhoog, 1988)

Grou)s/sectors	Objectives	Goal variables	Model variables
Socioeconomic:			
1 - European Union	- reduction in milk production	- net milk production	- kg milk
2 - farmers	- incomes - maintenance of employment	- land rent - labour needed	- rent/ha - labour-time/activity, volume of activities
Environmental:	quality production		
3 - agriculture	environment - grassland - crop growing	- livestock diseases - soil deterioration in arable farming - soil fertility, rotation risks	- kg potassium/ha grass - ha potatoes and beets/ha arable land - ha potatoes and beets 1:2 year/ha arable land
	- continuity of production	- soil deterioration in agriculture	- ha potatoes and beets/ha agriculture
		- soil fertility, rotation risk	- ha potatoes and beets 1:2 year/agriculture
		- copper pollution	- gr accumulation copper/ha
4 - forestry	- production environment	- deterioration by concentrated ammonia emission	- ha agriculture with concentrated ammonia emissions
5 - water collection	- production environment	- nitrate in groundwater	- kg NO_3/ha
6 - nature on farmland	- grassland	- low manuring level - low manuring level - meadowbirds	- ha grass land $N<100$kg - ha grass land $N<200$kg - lapwings/100ha
	- arable land	- soil fertility, rotation risk	- ha potatoes and beets 1:2 year/ha agriculture
	- agriculture	- diversity crop growing	- diversity index
7 - nature conservation areas	diversity of plant species	- ammonia emission	- kg NH_3/ha

plant species was related to the volume of phosphate and nitrogen emissions, which were in turn derived from the overall output of the land-use projected in the model. The values chosen and the related effect variables can be divided into two groups: agricultural socioeconomic and environmental, each further subdivided.

A time horizon was not made explicit. However, the measures were intended to produce the desired reduction in the short term, i.e., years rather than decades. Therefore, certain aspects of development were disregarded. For instance, changes in EU market and price policy other than those concerned with the dairy policy were not taken into consideration, nor were changes in the environmental policy, technical developments, or the daily and seasonal changes in the sector. The yearly averages were used to describe the inputs and outputs of the activities.

Five calculations were carried out: a reference calculation and four calculations which constrained milk production on the basis of four different variants of measures to control production. According to a decision made in the Council of Ministers in 1989, milk production had to be reduced by 15.5% (De Graaf & Tamminga, 1990). For the other sectors, such as cereal and beef production, the relevant autonomous development had to be taken into account as a reference for side-effects. The four variants were described as:

- non-tradeable quotas, a regional reduction of 15.5% (non-tradeable quota: NQ),
- tradeable quotas within the constraints of a national reduction of 15.5% (tradeable quota: TQ),
- levies for regional surplus production to discourage production above 84.5% of the reference year; more production is possible only if a lower price is accepted (levies: LE),
- decreasing milk prices; prices were lowered until the reduction in EU costs reached a level equivalent to a reduction of 15.5% by means of the quota systems mentioned above; an iteration resulted in a 12% reduction in milk production (market price: MP).

In each calculation agriculture was optimized: the difference between benefits and costs was maximized by combining the different types of agriculture in each region, within the constraints of the chosen variant. One could say that current agriculture was optimized only in a financial sense, while it was described not only in socioeconomic terms, but also in ecological and biophysical terms.

The reference calculation reproduced the agricultural situation in 1983, just before the measures were taken by European Union. In each calculation, all values of the variables shown in Table VIII.1 were compared with the reference calculation. These four comparisons, referred to as 'runs', were regarded as pictures of the consequences of the four constraints.

Table VIII.2: Survey of the most important model outcomes of a reduction in the milk production in the dairy sector. Non-tradeable quotas (NQ), tradeable quotas (TQ), levies (LE), and market price (MP) (model reference = 100: + or - = changes less than 5%, ++ or -- = changes 5-10 %, +++ or --- = changes > 10 %).

	Run changes NQ	TQ	LE	MP
Milk production (kg)	-15.5	-15.5	-15.5	-12.0
Land use (ha)				
Arable farming	++	+++	+++	+++
Grassland	++	++	0	-
Green maize	---	---	---	---
Employment (year)	--	--	--	-
Dairy sector	---	---	---	---
Extensive cattle-raising	+++	+++	+++	+++
Regional differences				
Milk production (kg)				
- peat districts	---	0	--	0
- sandy districts	---	---	---	---
- sea-clay districts	---	0	--	0
Land cover				
Grassland (ha)				
- peat areas	0	0	0	0
- sandy areas	+++	++	0	-
- clay areas	-	++	+	++
Green maize (ha)				
- sandy areas	---	---	---	---
- clay areas	0	--	+++	--
Income (change in guilders per ha)	---	---	---	---

VIII.3.4 Summary of the results

Impacts on agriculture

The differences between the four measures in terms of their socioeconomic consequences for agriculture were not substantial (Table VIII.2). Obviously, the milk production decreased in all the runs. A lower density of dairy cattle per

Table VIII.3: Environmental consequences of the runs in percentages of the reference situation: non-tradeable quotas (NQ), tradeable quotas (TQ), levies (LE), and market price (MP). In regions: + or - = positive or negative development in one region; +/- = a positive and a negative developments in two region. Runs: model reference = 100: + or - = changes less than 5%, ++ or -- = changes 5-10 %, +++ or --- = changes > 10 %).

	In regions	NQ	TQ	LE	MP
Agriculture production					
1. grassland (cow-disease risks: potassium: kg-k/ha)	+/-	+	+	+	+
2. arable land (ha rotation risk)					
- sandy areas (%)	++oo	+	++	+	+
- clay areas (%)	+++o	++	++	++	++
3. Continuation of production (ha rotation risks)					
- four sandy areas (%)	++++	+++	+++	+++	+++
- one clay area (river clay) (%)	+	++	++	++	++
Woods					
4. (deterioration by concentrated ammonia)					
- one peat areas (west) (ha)	-	- -	- -	- -	- -
- one sandy areas (north)(ha)	-	- - -	- - -	- - -	- - -
- four clay areas (ha)	++-	+	++	+	+
Water collection					
5. sandy areas (nitrate risks)					
- north (kg-NO_3/ha)	+/-	- - -	+++	++	+++
- east (kg-NO_3/ha)	-	++	+++	+++	++
- central (kg-NO_3/ha)	+/-	- -	+++	+++	+
- south (kg-NO_3/ha)	+/-	++	+	+	- -
Nature (ha low manure level)					
6. grassland vegetation					
- two peat areas (ha)	++	+++	0	+	0
- four clay areas (ha)	+- -	-	+	0	+
7. meadowbird populations					
- two peat areas (pop.)	++	++	0	++	0
- four clay areas (pop.)	+++o	+	+++	++	+++
8. farmland vegetation (ha pesticide risks)					
- four sandy areas (ha)	++++	+++	+++	+++	+++
- one clay area (river clay) (ha)	+	++	++	+	++
9. nature conservation areas (ha eutrophication risks by ammonia)					
- four sandy areas (kg-NH_3/ha)	++++	++	+++	++	++
- two clay areas (kg-NH_3/ha)	(+++-)	+	0	+	0

hectare made it possible to shift from fodder with a high nutritional value (green maize) to grass with a low nutritional value. Moreover, arable farming and extensive cattle breeding had expanded. These changes had a number of consequences for income and employment in agriculture. A smaller dairy sector could be only partially compensated for by increases in extensive cattle-breeding sectors. These larger sectors resulted in lower prices, which had a negative impact on income in all runs. More marked distinct differences between the runs were detectable on a regional scale. In the case of the 'non-tradeable quota', milk production had to be restricted in all regions with dairy cattle, so alternatives for the use of soil were necessary across the board. In the other three runs, there were more possibilities for regional specialization.

Impacts on nature and the environment

The runs had a number of results in common with regard to the consequences for nature and the environment (Table VIII.3). For example, the risk of deteriorating soil fertility and structure declined, due to more crops in the rotation scheme (2 and 3 in Table VIII.3). All runs contributed to higher local problems for forestry in the sandy area in the north, due to the direct impact of concentrated ammonia emissions (4). An overall alleviation of ammonia emissions resulted in better perspectives for nature conservation areas in sandy areas (9). It was not expected that the decrease in emissions as a result of the agricultural shifts of the runs would be sufficient to do away with the negative effects of agriculture on drinking water and nature (5 and 6). Positive effects did occur - tradeable quotas would have a positive effect on water- but where nature was concerned, a great deal depended on the developments in arable farming (8). Only where more diverse arable farming was involved in the policy to control milk production were positive developments to be expected.

VIII.3.5 Conclusions of the project

In the project report it was concluded that in the case of an optimal shift in production, non-tradeable quotas would have either a positive or a negative effect on water quality, depending on the region, and a positive effect on grassland vegetation and meadowbird populations. Tradeable quotas would have a positive effect on water. The effects of levies were about the same to those of non-tradeable quotas. According to the project, a shift from an area-dependent assignment of quotas to tradeability of quotas would lead to a shift of quotas to those areas where the land is less suitable for arable farming.

VIII.4 Using the proposed procedure to analyze the project

The project described above will be analyzed, using the phases of the proposed procedure as a model, i.e.,

1 *Start-up*
2 *Delimiting the socio-environmental system*
3 *Selecting the valued system characteristics*
4 *Measuring the resource state*
5 *Exploring the development opportunities*

The focus is on the consistency of the steps taken. Since the modelling was comparable to that in the project previously discussed (Chapter VII), the completeness of the list of *valued system characteristics* and the *measurement of the resource state* will not be dealt with here.

VIII.4.1 Start-up

According to the procedure.
The proposed procedure states that it is important to be clear about the aim and the working plan. Based on the intention to focus on the visions of agricultural and non-agricultural organizations, the main aim should be formulated as follows: to explore within each vision the possibilities to meet that vision's goals for the rural area in general and the dairy sector in particular, by altering agricultural activities, especially dairy activities. The consequences of modification to all other selected valuable system characteristics will be visualized, in order to provide an opportunity to modify the goals set, and to repeat these steps until the picture is satisfactory. To ensure that all valuable characteristics are taken into account, the steering of the research should be scientific and based on a committee made up of representatives of the visions.

The project Reduction in Production in the Dairy Sector.
As noted above, the initiative for the project was taken by the Spatial Planning Agency of the Ministry of Housing, Spatial Planning and Environment. This body was interested in development of the rural area and aimed the project on the visions to be calculated, including non-agricultural organization and its consequences for rural development goals. The Ministry of Agriculture, Nature Conservation and Fisheries, however, concentrated on calculating the impacts of measures in the area of dairy farming (RPD, 1984b).

Discussion.
This means that there were clearly two points of view from which the analyses could be carried out: a spatial one (land-use) and a sectoral one (agricultural production). As a result, it was important to define beforehand a research strategy that would do justice to both. The first touchstone for the consistency of the project was the way the '*system to be studied*' is defined.

VIII.4.2 Defining the socio-environmental system

According to the procedure.
As regards the dairy farmers, the system to be analyzed could have been restricted to dairy farming. However, if changes in land-use area in this sector are permitted, then from the point of view of sustainable development, the new activities using the same area would also be part of the system. The same holds true for areas which are no longer devoted to dairy farming, but rather to providing fodders. Thus there are good arguments for focusing on the agricultural system as a whole. However, new activities that take over dairy farming areas may also be non-agricultural, so from the point of view of national spatial planning, the system ought to include agriculture as part of the complete rural area in which both agricultural and non-agricultural activities take place.

The project Reduction in Production in the Dairy Sector.
Instead of searching for an acceptable common system encompassing those systems most appropriate for both approaches - the agricultural and the spatial planning approach - the common element of those appropriate systems was chosen. As a result, only agricultural land was taken as the system to be studied, leaving out other land-use activities in the rural area.

Discussion.
As we made clear in Chapter VII, this choice means that agricultural interests are regarded as internal and all others as external, while shifts from agricultural land-use to non-agricultural activities are excluded. This decision has far-reaching consequences for the way needs and wants are taken into account and related to *valuable system characteristics*. These consequences will be discussed below.

VIII.4.3 Selecting the valued system characteristics

According to the procedure.
Selected values must include all needs and wants of the people involved here and elsewhere, both in the present and the future. This results in a set of valuable characteristics that shows which side effects should be prevented when searching for development opportunities. Although a common set of goal variables must be provided, this does not mean that these variables are valued equally by the parties involved in the debate. After all, different visions actually mean different priorities. Nevertheless, the proposed procedure prescribes not only agreement on a common *SEnS*, but also on a common set of *valuable system characteristics* that is used to visualize all impacts, in order to find sound development opportunities.

The project Reduction in Production in the Dairy Sector.
Because the system was restricted to the agricultural sector and its land-use, the valuable system characteristics chosen all refer to the characteristics of agriculture (see also Chapter VII).

Discussion
Defining the system as agricultural land means that some of the values of rural areas automatically became external values, such as the quality of nature areas and aquifers. From the spatial planning point of view, however, these values should be handled in the same way as agricultural values. This can only be done by translating them into *valued characteristics* of the agricultural system. This project made use of the same models, and the same model variables, as those in the project dealt with in the previous chapter. Therefore, the analysis of the selection of the valuable system characteristics performed there (Chapter VII.4.3) also holds for this project: the selection of goal variables was not complete, due to the model selected and to an inadequate use of the possibilities of the model. There are also differences: the needs and wants related to conserving nature and the environment were explicitly explored by calculating the possible effects of agriculture on certain aspects of nature and the environment.

VIII.4.4 Measuring the resource state

In modelling the agricultural sector, the same approach was chosen as in the project discussed in the previous chapter. The basic *resource state* was described as the available agricultural land with its present environmental qualities. This leads to the same conclusions as in Chapter VII 4.4: only the resources for agricultural production were described, thus disregarding the resources for all other activities and the possibilities for altering them. Agriculture is constrained not only by biophysical, but also by sociocultural factors. These were not taken into account, except for an employment constraint in intensive livestock breeding, and the constraint to keep all land in production.

VIII.4.5 Exploring development opportunities

According to the procedure.
Using the backcasting approach, the procedure prescribes that each vision should be taken seriously, by assuming that all goals within the vision can be realized. This means that the goal variables selected in previous phases would have been divided into variables for goal-setting from the point of view of the organization involved (the goal variables related to the vision) and remaining goal variables describing the side-effects that are prevented (the 'non-vision' goal variables). Then the *resource state* would have been modified to match the vision goals, and the impact of non-vision variables would have been calculated. If no satisfactory picture emerged, because of unexpected and unwanted side-effects, a modification of the set of goals might have been considered by the representatives of that vision. The new set of goals would have to be taken as a starting point for a new calculation, and this would continue until a satisfying picture emerged.

The project Reduction in Production in the Dairy Sector.
Although an agricultural model was available for the iterative process proposed, and the groups were asked to explain their vision more clearly, there was no feedback after the runs were executed. During the investigation the original pursuit of an iterative process was discussed and abandoned (De Graaf *et al.* 1987). The aim of the project shifted from 'confronting' the groups with the consequences of their views, to 'analysing four different kinds of measures' proposed for production control. The context of the measures - the visions of the organizations - was dropped.

Discussion
At this point it is important for future investigations and the future use of the proposed procedure to examine the question of why this project did not meet its original aim.

As we know, the steering committee that decided on the major issues of the project was made up of representatives of scientific organizations, the agricultural department, and the spatial planning department. This was contrary to the recommendation of the research committee of the Spatial Planning Agency, which was against the participation of representatives of the Ministry of Agriculture (RPD, 1983). It is noteworthy that later it was explicitly stated that no environmental or farmers' organizations were to be included in the steering committee, because the aim was to form a committee of experts, rather than a political body (RPD, 1984a). As a result, the non-governmental organizations involved had no steering power in decision making. This enabled the steering committee to combine variants of the different measures put forward by the environmental and farmers groups into a single measure (RPD, 1985 & 1987). It was argued within the steering committee that all the proposals for dairy farming were similar and thus could be combined. Where the measures to reduce milk production were concerned, this was indeed the case. But the interviews clearly showed that the organizations had different views on the opportunities for increasing other production sectors to compensate for the loss of production in dairy farming (Bakker, 1984c).

The environmental organizations, in particular, emphasized the impact of side-effects, because compensating for lost milk production by increasing meat production, such as intensive pig breeding, would increase the risk of environmental damage. They proposed a variant designed to prevent such developments (Bakker, 1984c). As the results show, negative side-effect did occur: two districts were confronted with an increase in locally concentrated ammonia emission caused by the expansion of intensive pig breeding (Table VIII.3). The proposed variant was neglected, since the steering committee regarded measures to restrict meat production as unrealistic, in view of the political situation at the time. This position was noteworthy in view of the fact that within a year measures to constrain pig breeding were incorporated into nation agricultural policies (Anonymous, 1985). When searching for *development opportunities* the decision of the steering committee was to reduce the system boundaries *de facto* from 'agriculture' to 'dairy farming'; no conditions were formulated for the other agricultural production sectors. The researchers pointed out the consequences of this departure from the aim of the project, but this did not alter the decisions taken (De Graaf *et al.*, 1987). This had three main consequences:

- The results could no longer be presented as the consequences of the visions of the organizations involved: it would be unacceptable to confront the environmental organizations with more environmental deterioration as a result of increasing meat production, while those same organizations had emphasized from the start that they wanted specific measures to prevent this.
- Because the variants containing a complete vision on rural areas and agriculture had been omitted, it was no longer possible to motivate the

search for coherent packages of measures. The strategy by which pictures of impacts are used to provide insight into attractive opportunities for development was discarded

Because the project now focused on measures for dairy farming and forecast the possible impact of those individual measures, the negative side-effects of the measures could be made visible; however, there was no basis for a societal discussion. The project provided information, but it was keyed to the government policy makers and not the parties involved and their vision.

In short, what remained of the original plan was an estimation of the impact of measures to reduce milk production on a wide range of values.

The procedure that was actually followed in this project is shown in Figure VIII.1. From the start, there should have been consensus on the aim of the project, and that aim should have been reflected in the decision making group, i.e., the steering committee. If the visions are the issue than all of them must be represented or none of them. However, the actual steering committee was not guided by the aim of the project, but rather by government responsibility for agriculture. As a result, one participant emphasized spatial interests and wanted to investigate the possibilities offered by variants of the measures proposed by the groups, while the other focused on economic functioning and was interested only in the effects of measures aimed at reducing production. It is possible to combine these varying starting points, but that would mean defining a common system that included the scope of both participants, such as 'rural area'. Since the common element for steering, i.e., 'dairy farming', was chosen as the system, most of the relevant opportunities for development from the point of view of one of the participants would have been excluded. This problem could have been prevented if the representatives of non-governmental organizations had been involved in the decision making. Since this was not the case, the project cannot actually be regarded as a search for development opportunities based on different visions, because the runs accept certain negative side-effects without any attempts to do away with them.

VIII.5 Concluding remarks: improvements to the proposed procedure

The project shows that although each phase of the proposed procedure for gathering-information for sustainable development was executed correctly, the procedure as a whole may not supply the relevant information where the visions of the participants are concerned, because the consistency of the procedure can be violated by decision making, in this case by decisions of the steering committee.

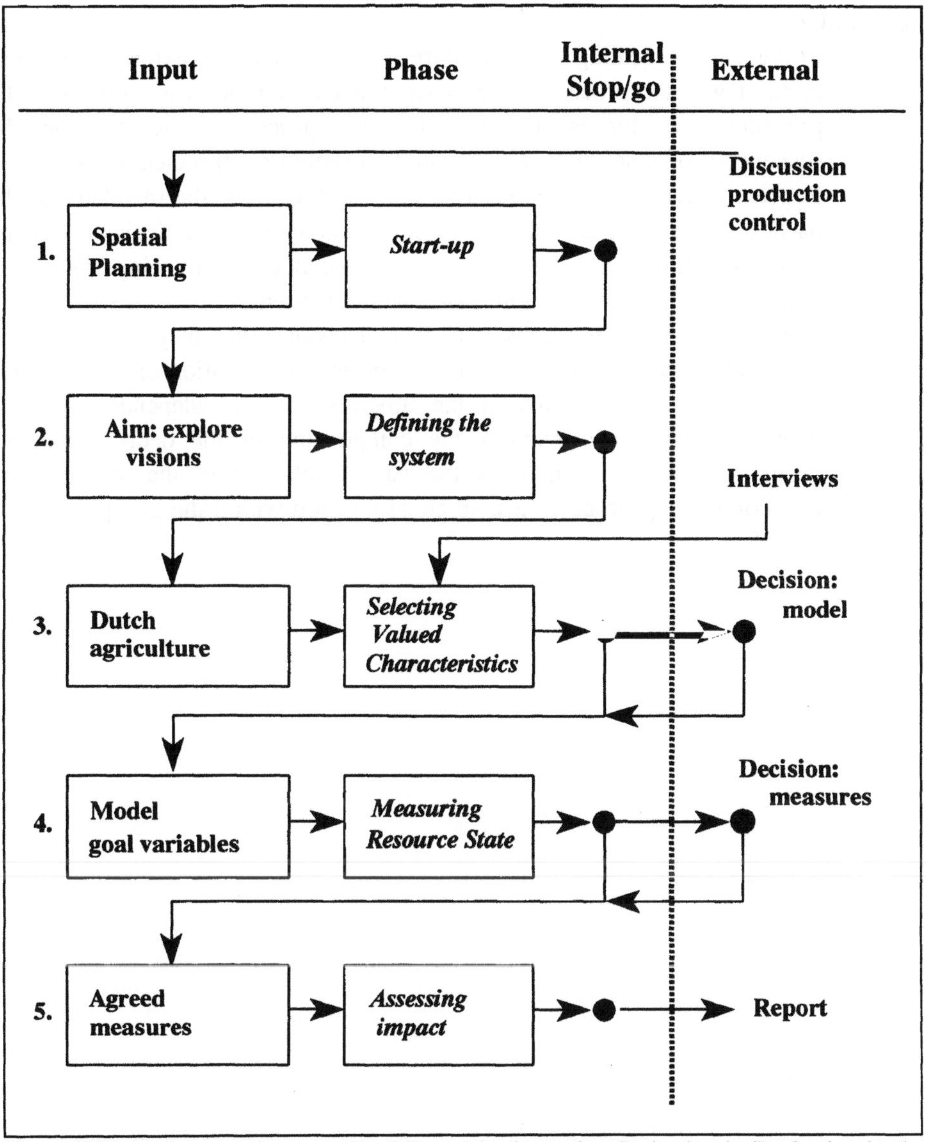

Figure VIII.1: The procedure actually followed in the project Reduction in Production in the Dairy Sector in terms of the procedure proposed in Chapter VI

Therefore, the evaluation of the project does not provide much information on the strength or weakness of the separate steps of procedure. The discussion of the project shows that the proposed procedure as a whole can be used as a model for evaluating existing research. Moreover, evaluation points the importance of the setting in which the procedure is executed. In this respect the following remarks are in order.

- Whoever is executing the procedure should verify that external decisions pertaining to the procedure are in accordance with the original aims of the procedure. If this is not the case, the procedure should either be suspended, or restarted, i.e., the *Start-up* should be executed again.
- Representatives of the people involved - in this case, the members of the parties involved in de debate - should participate in the decision making in applied research. This may prevent decisions from being taken that unilaterally change the original aims of the procedure.
- Governmental organizations may take decisions that they assume to be 'politically realistic', but that hamper an open exploration of development opportunities. This points in the direction of an independent process manager who would monitor the composition of the decision making group and see to it that all the participants within the steering and decision making process have an equal position within the group.

CHAPTER IX

A Simulation of The Selection of Valued Characteristics of the Rijnmond Region of the Netherlands

CONTENTS

IX.1 Introduction

In the previous chapters the procedure for sustainable development presented in Chapter VI was used to discuss the strengths and weaknesses of the procedure as a whole. This chapter deals with an application of the first three phases of the procedure leading to the selection of the valuable characteristics of a *SocioEnvironmental System* (*SEnS*). The focus is on the definition of the system and the application of the newly developed framework for selecting *valued system characteristics* presented in Chapter IV.

In the chapters on projects pertaining to Dutch agriculture (Chapters VII & VIII), the newly developed framework was used only to verify the completeness of the set of goals presented in those projects. Thus the framework has not yet been applied to identify the valuable characteristics of a concrete system. The present chapter examines a simulation of such an application, in which the framework is used to support parties in developing a strategy in the debate on regional planning (Van der Meij *et al.,* 1997).

The starting point of the simulation of the proposed procedure was a government project for the development of the region known as 'Rijnmond' (Dutch Parliament, 1990). The region, which is dominated by the city of Rotterdam and its immense harbour at the mouth of the river Rhine, also includes rural areas. Several islands within the Rhine estuary are home to valuable nature conservation areas. The planning area comprises approximately 1000 square kilometres, covering the city, the harbour and the surrounding rural areas, inhabited by some 1.4 million people (Figure IX.1 & 2).

The Zuid-Hollandse Milieu Federatie (ZHMF), a non-governmental, environmental organization, was one of the parties participating in the development of a plan for the region, in collaboration with industry, business and government groups. To participate adequately in the debate, the ZHMF needed to develop a strategy of its own, independent of the negotiation process aimed at developing a plan for the region. They had to make sure that important values were not missed or ignored, potential conflicts were identified and potential allies were found. The question was whether the proposed procedure would have been able to support this search if it had been available at the time.

Here we simulate the application of the proposed procedure in order to answer the question: Could the procedure have been more useful in identifying the valuable characteristics of the 'Rijnmond' region than the procedure actually used?

The available information and the method of analysis is presented below. In Section IX.3 the proposed procedure is applied and in Section IX.4 it is compared with a current plan for developing of the region: the 'Rijnmond Development Plan' (Dil *et al.*, 1993). The chapter ends with a discussion of the

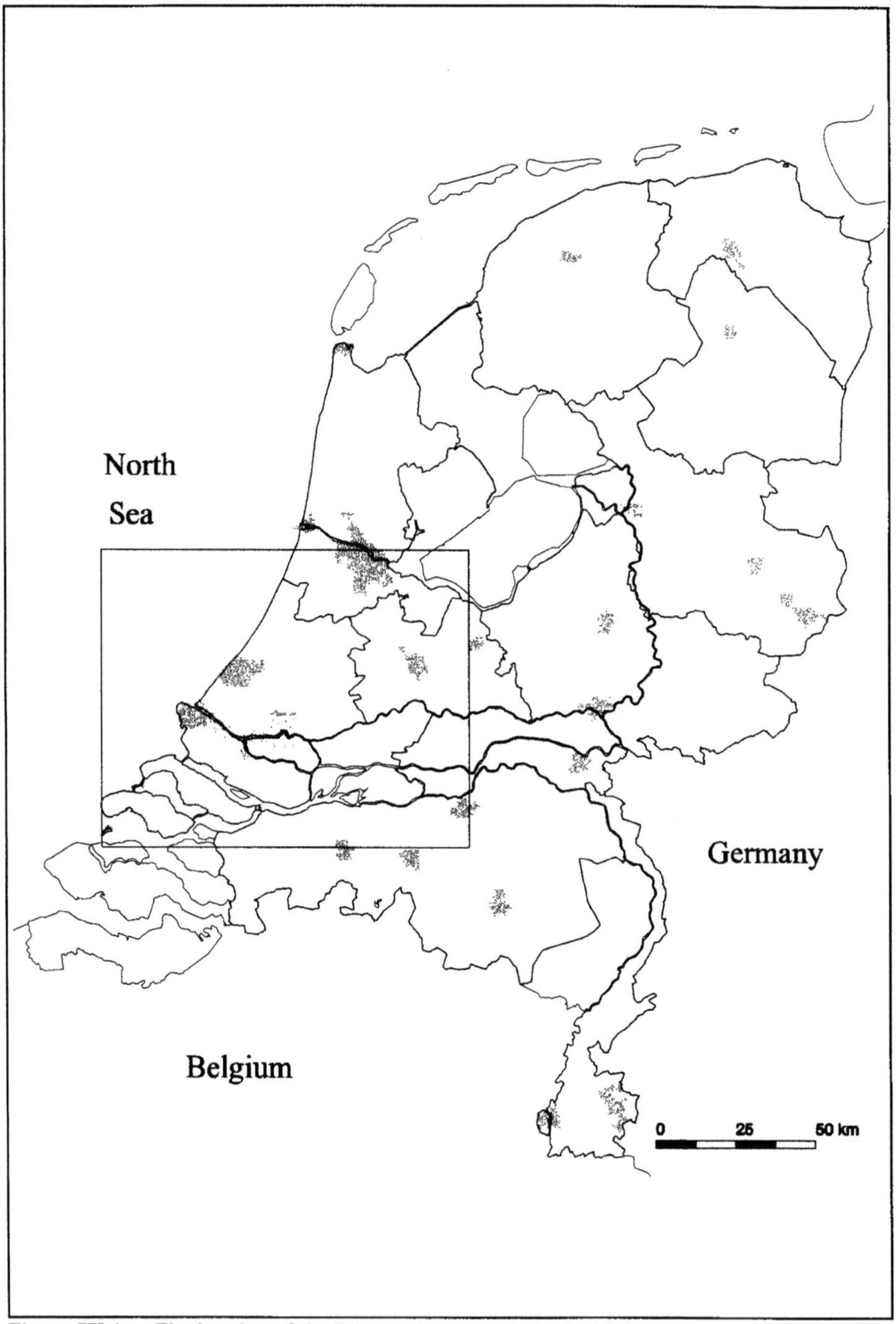

Figure IX.1: The location of the Rijnmond region within the Netherlands

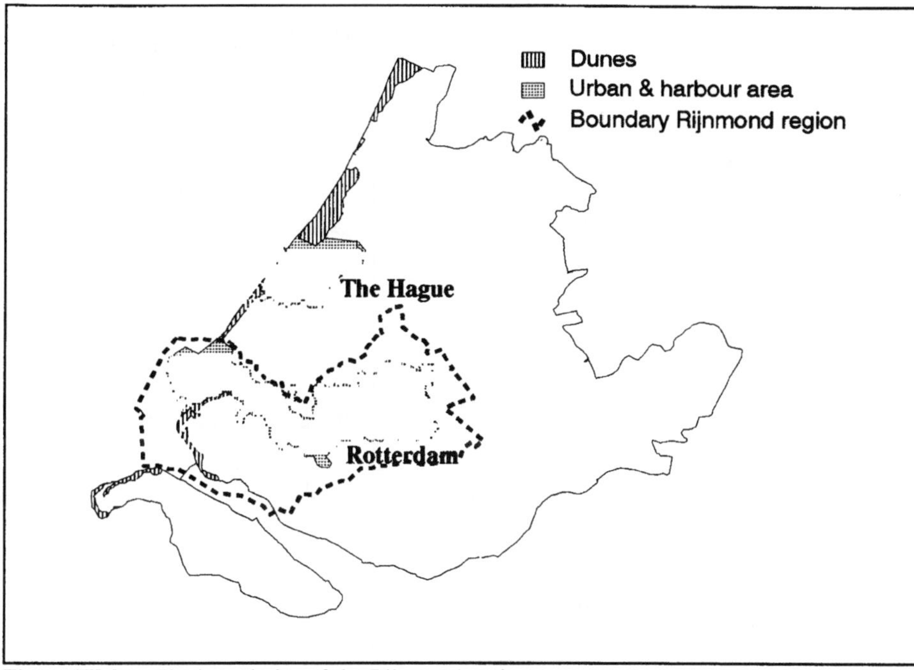

Figure IX.2: The boundaries of the Rijnmond region

application and concluding remarks on improvements to the proposed procedure (Section IX.5 and 6).

IX.2 Method of application

The proposed procedure for information-gathering is only partly applied in this chapter. The chapter is aimed only at the first three steps: to identify *valued characteristics*, without either measuring the *resource state* or exploring *development opportunities* (Figure IX.3).

In order to analyze the efficiency of an application of the proposed procedure, it was decided to compare the simulated application with the results of another method for defining and valuing a *SEnS*. This method of analysing requires two sets of information. First, information must be available on the application of the framework by or on behalf of an organization; it must be a real-life situation constrained by time and cost. And second, information must be available on the valuable characteristics actually selected in that same, real-life situation by decision makers who did not make use of the framework. The working hypothesis is that the application of the proposed procedure will result in a more complete and more

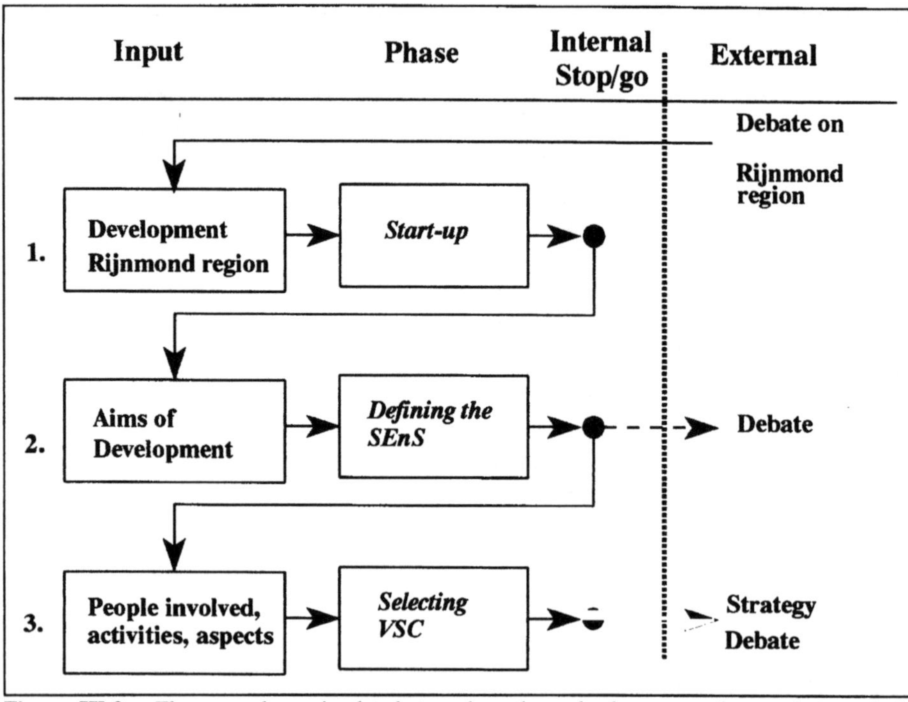

Figuur IX.3: The procedure simulated to select the valued system characteristics of the Rijnmond region, to support strategy development

consistent set of valuable characteristics than the procedure selected by the decision makers.

As is stated above, the 'real-life situation' that was chosen, was the discussion on the development of the Rijnmond region. In cooperation with the ZHMF, the application of the newly developed framework as a search instrument by that organization was simulated (Figure IX.3); this is described in Section IX.3. The aims of the application were the same as those laid down at the beginning of the development process (ROM-Rijnmond, 1992a & b), thus ensuring that discrepancies between the results of the application and the actual valuation were not caused by differences in starting points. By applying the new framework, it was possible to obtain first-hand information on the difficulties of application. This ensured that it was not overly biased by difficulties arising from an inaccurate description of the new framework, or by difficulties in the communication between the designers of the framework and the people responsible for applying it. Such problems must also be studied, but they are beyond the scope of the present chapter.

Information on the actual valuation of the Rijnmond region was provided by the Rijnmond Development Plan (Dil *et al.*, 1993). Both reports list the goal variables to be monitored, i.e., the valued system characteristics and the variables that measure them. The comparison between these lists and the results of the application of the framework is discussed in Section IX.4.

It was assumed that in an actual application it will be difficult to organize an extensive consultation with the people involved and to make use of other information than the available documents. A situation was assumed in which such information-gathering had to be carried out over a period of a few months. Thus the simulation would not exceed the time available for an organization involved in the decision making process.

IX.3 Simulating the proposed procedure

IX.3.1 Start-up

The main governmental objective in developing a plan for the region was twofold: to strengthen the position of Rotterdam as the mainport for Western Europe, and to improve the residential environment (ROM-Rijnmond, 1992a). This meant that not only industrial, but also social and environmental organizations were invited to take part in the discussion (ROM-Rijnmond, 1992b; Van de Klundert & Glasbergen, 1995). This strategy was intended to produce a list of goals which had the support of all the major stakeholders. For those participating in this debate, it would be useful to gain insight into the values at stake. The first step is to identify the problem area.

Problem area

The major reason for devising a new plan for the Rijnmond region was the fact that the region is currently facing a number of serious problems. As a consequence of economic growth, the harbour must be expanded, but suitable locations are limited. The housing preferences of the population have led to a decrease in the number of people per single-family house, and thus an increased demand for housing. Traffic is congested, and this problem will increase as a result of the expansion of the harbour and the new residential areas. Nature conservation areas are under threat from industry and from residential development, and should in fact be enlarged to ensure the survival of endangered species and the landscape. Industrial activities are threatening the environmental quality of the surrounding areas.

None of the above problems can be solved without adding to other problems. Therefore, it was decided by a group of governmental and non-governmental organizations (national, provincial, and municipal governments, business and industry) to formulate a plan designated to produce integral solutions for these problems (ROM-Rijnmond 1992a).

As noted above, the objective was to strengthen the position of Rotterdam as the 'mainport' for Western Europe, as well as to improve the environmental quality. This 'double aim' was operationalized by setting a number of objectives that had to be dealt with. The need for extra space for harbour facilities and improvements to the accessibility of the harbour and the industrial sites, was placed high on the list, alongside the management of mobility. To safeguard the competitive position of the harbour, the transfer volume was also given priority. In addition, the discussion should centre on generating employment and income for the present and future population, as well as creating conditions for the improvement of housing, accessibility and recreation. With respect to the rural area, high-quality nature and recreation facilities were items mentioned. It was considered important to focus on the environmental impact on living conditions and work and, in rural areas, on recreation and nature. Finally, the higher-scale environmental objectives should be stressed, especially the emission of greenhouse gases. Thus the discussion focused not only on economic considerations, but also on social, biophysical and environmental aspects.

The intentions formulated at the start were indeed ambitious. However, from the point of view of the independent participants in the discussions - the social and environmental organizations - the translation of the various items into socioeconomically and environmentally measurable goals is crucial. The major problem is how to increase the likelihood that all these objectives will actually result in a broad and well-balanced set of ***valued system characteristics*** that can be used to set the goals for development. Any party participating in this discussion should have at its disposal such a balanced set in order to strengthen its position. It will then be able to respond directly during the debate, if certain important characteristics have not been placed on the agenda. In other words, information-gathering by participants in the debate should ensure that all the issues are discussed. To that end, the object to be developed must be defined and known to all the participants in the debate.

IX.3.2 Defining the socio-environmental system

It will be clear from the *start-up* that the system to be defined and valued cannot be freely chosen by the party participating in the debate. The Rijnmond region is a given, due to governmental policy decisions in the past (Dutch Parliament, 1990). It might be attractive from the point of view of an organization to choose totally different regional boundaries, but this might mean being excluded from future debate. Since the application of the developed procedure is focused on the search for valuable characteristics, that option is not elaborated here.

System boundaries
The start-up documents provide a rough description of the *SEnS* at stake; the spatial boundaries are shown in figure IX.2. The objectives have been described as space for harbour and industrial activities, housing, recreation, and nature, and an increase in the accessibility of the region to the various types of land-use. These objectives can be used to select the subsystems and the related processes and structures for further analysis. From the point of view of a nature conservation organization, the conclusion could be that nature interests have been put on the agenda, offering perspectives for further discussion. But from an environmental point of view, there is more at stake: Figure IX.2 shows that while a large part of the area is reserved not for harbour facilities or housing, but for agricultural uses, agriculture is not fully recognized. The importance of the greenhouse area within the region is mentioned (the north part), but other farming activities do not feature among the objectives mentioned, although a large part of the area is still farmland (the south part). Thus the interests of farmers are not taken seriously, or have been deliberately ignored in order to avoid area claims later on in the debate. From the point of view of an environmental protection organization, this is a major flaw, since ignoring farmers' interests may mean losing potential allies, or being played off against farmers. This issue should be placed on the agenda in the debate even before goals are formulated (Figure IX.3). We will return to this point in Section IX.4.2. First, the *valuable system characteristics* must be identified, starting with the addition of agriculture to the list of major land-use activities. This does not alter the system boundaries, but it does affect the description of the system within those boundaries.

People involved
The description of land-use within the system boundaries results in a list of subsystems and their related land-use activities. In the search for a list of *valuable system characteristics* (step 3 in Figure IX.3), the people involved must be identified. Although the Development Plan does not explicitly mention the different groups, they can be identified by focusing on the major activities of the people involved within the subsystems. In urbanized areas, people may be involved as producers - those engaged in economic activities related to the harbour and the city - and as private citizens; in rural areas people may be involved as farmers, recreationists, or nature conservationists. All activities make use of the 'transport infrastructure', while this is also important for the system as a whole and for the higher-scale community, i.e., the Netherlands. Moreover, the overall contribution

of all activities and their environmental impacts must be identified.

IX.3.3 Selecting the valued system characteristics

Now that the categories of people involved and the contribution to the higher scale have been identified, the next question is what kind of needs and wants are at stake. On the basis of the objectives mentioned in the start-up, these will include economic, social, physical and environmental aspects (Section IX.3.1). In searching for *valuable system characteristics*, it is assumed that these aspects are relevant for each type of involvement. It is important to note that cultural aspects are not mentioned. As in case of agriculture, a party in the debate could take the initiative and place these aspects on the agenda of the debate.

On the basis of various types of involvement and the types of needs and wants, the valuable characteristics of the system must be identified by relating them to the framework (Chapter IV). In Table IX.1 the various types of involvement and the different types of needs and wants are used to specify the axes of the framework. The people and their types of involvement, with the exception of the higher-scale community, were regarded as a type of direct use (note that a person may be a member of different types of direct user groups); the main activities in the region are used to assess combined characteristics for direct use; the overall spatial and temporal qualities of the system are reflected in the combined characteristics for the total community. The contribution of the system to the higher-scale community is reflected in the characteristics of the various contributions of the system as a whole. In principle, each cell in the table stands for at least one potential valuable characteristic.

The actual *selection of valued system characteristics* can now be initiated. The list of issues presented in Table IX.1 was used to start the search for related valuable characteristics. The following steps were carried out:

1. The issues identified per user group, activity, etc. (Table IX.1) were fitted into the framework of Table IX.2; they were characterized by first two letters, for example income = in, area = ar. Since valued characteristics can be interpreted both structurally and functionally, each issue is described in functional and structural terms.
2. The objectives obviously refer not only to present use, but also to creating or improving the conditions for future use, such as infrastructure and potential building areas. This has been indicated by 'x' in Table IX.2.
3. Then the search turned to the characteristics relevant for a certain group which could also be relevant for other groups or for the higher-scale community: the association via the rows of the framework. Two groups of

Table IX.1. Selected types of needs and wants related to user groups, the activities, the higher-scale communities, and the aspects mentioned in ROM-Rijnmond (1992a) and summarized in the Rijnmond Development Plan (Dil *et al.*, 1993, pag i-viii). Letters between brackets - () - are used to fill in Table IX.2.

Characteristics related to direct users:

	economic	social	physical	environmental
Separate				
harbour/urban				
-citizen	. income (in)	. employment (em)	. area for housing (ar)	. impact on housing (im)
-producer	. production value/ volume(pv)	. employment (em)	. area for production (ar)	. emissions (es)
rural				
-recreationist	.	.	. area for recreation (ar)	. impact on recreation (im)
-nature conservation	.	.	. area for nature (ar)	. impact on nature (im)
Combined (Activities)				
harbour/urban				
-habitation	.	.	. housing facilities (fa)	.
-working	.	.	. working facilities (fa)	.
-transfer and production	.	.	. harbour and industry facilities (fa)	.
-transport	.	.	. transport infrastructure (fa)	.
rural				
-recreation	.	.	. recreation facilities (fa)	.
-nature conservation	.	.	. nature infra-structure (fa)	.

Characteristics related to higher scales

Combined: Activities				
-spatial quality	.	.	. infrastructure (if)	. public health (ph)
-temporal quality	.	.	. transport guidance (if)	. public health (ph)
Separate				
-community needs and wants	.	.	.	.
-higher-scale needs and wants	.	.	.	. emissions (es).

Table IX.2: Aspects of the Rijnmond Development Plan entered into the framework for identifying valuable characteristics: in= income, em= employment, ar = area, im = impact, pv = production value/volume, im= environmental impact, fa = facilities ph = public health, es = emissions, if = infrastructure and transport guidance. Characteristics: x = for future use or processing, derived from present characteristics, o = derived from the aims of the Plan, but not actually mentioned in the plan, . = not derived from the aims of the Plan, but from aspects mentioned in the Plan.

Characteristics related to direct users

	Processes				Structures				For future use				For future processing			
	ec	so	ph	ei	ec	so	ph	ei	ec	so	ph	ei	ec	so	ph	ei
Separate																
-citizen	in	em	ar	im	in	em	ar	im	x	x	x	x	x	x	ar	x
-producer	pv	em	ar	es	pv	em	ar	es	x	x	x	x	x	x	x	x
-recreationist	.	.	ar	im	.	.	ar	im	.	.	x	x	.	.	x	x
-farmer	o	o	o	o	o	o	o	o	o	o	o	o	o	o	o	o
-nature cons.	.	.	ar	im	.	.	ar	im	.	.	x	x	.	.	x	x
Combined:																
-habit./working	.	.	fa	.	.	.	fa	.	.	.	x	.	.	.	x	.
-production	.	.	fa	.	.	.	fa	.	.	.	x	.	.	.	x	.
-transport	.	.	fa	.	.	.	fa	.	.	.	x	.	.	.	x	.
-recreation	.	.	fa	.	.	.	fa	.	.	.	x	.	.	.	x	.
-farming	.	.	o	.	.	.	o	.	.	.	o	.	.	.	o	.
-conservation	.	.	fa	.	.	.	fa	.	.	.	x	.	.	.	x	.

Characteristics related to higher scale

	Processes				Structures				For future use				For future processing			
Combined																
-spatial	.	.	if	ph	.	.	if	ph	.	.	x	x	.	.	x	x
-temporal	.	.	if	ph	.	.	if	ph	.	.	x	x	.	.	x	x
Separate																
-community needs and wants	o	o	o	o	o	o	o	o	o	o	o	o	o	o	o	o
-higher-scale needs and wants	o	o	o	es	o	o	o	es	o	o	o	x	o	o	o	x

potentially relevant but missing characteristics were easily found. The first group consisted of farmers, following the addition of agriculture to the list of major activities within the region. The types of system characteristics valued for the industrial producers may also be important for farmers; for example, there is the role of labour, volume of production, area, etc. The second group comprises what can best be described as valuable characteristics on a system level, with special references to the needs of the higher-scale community for national planning. This includes the total production value of the region, the total production volume of the region, and the total employment within the region. These potential characteristics are identified by an 'o' in Table IX.2.

The search can be continued by focusing on other groups of people involved. For example, economic and social characteristics related to recreationists and nature conservationists, including a budget for the management of areas, are missing and may be at stake. In principle, for every type of characteristic identified, a search can be carried out for similar characteristics for other user groups. However, in view of the time and budgetary constraints, the search stopped here. The comparison with the results of the real-life process can now begin (Section IX.4), followed by a discussion (Section IX.5) to find improvements to the proposed procedure.

IX.4 Comparison with the current plan for the region

The application of the proposed procedure and its framework is analyzed by comparing the results of the search with the actual *valued system characteristics* selected in the real-life debate on the plan for the region. This debate resulted in the report 'Development Plan' (Dil *et al.*, 1993). For this comparison, the valued characteristics in the Development Plan must first be assessed and fitted into the framework.

IX.4.1 The valued aspects according to the Rijnmond Development Plan

Box IX.1 shows the major characteristics to be monitored, on the basis of the items mentioned in the Rijnmond Development Plan (Dil *et al.*, 1993). These characteristics are compared with the valued characteristics in Table IX.2 by inserting them into that table, which gives rise to Table IX.3. Although not explicitly mentioned in the plan, it is assumed that the items will be monitored and described in both functional and structural terms, or at least that the measurement will supply this information. For example, the percentage of 'people affected', which is mentioned (1 in Box X.1), is a structure variable, but it is assumed that the Development Plan also aims at describing the increase or decrease of the proportion of people affected, in other words, a functional characteristic. Furthermore, the items are examined to establish whether they are essential as conditions for future use as well (compare Chapter IV).

As expected, Table IX.3 shows that characteristics have been identified for producers and production activities, i.e., harbour-related activities (15, 16, 17, 18 & 19 in Table IX.3). The importance of space for activities leads to area-related

A. Environmental themes:		
1a	. dust	- % people affected by nuisance
1b		- % people seriously affected by nuisance
2a	. odour	- % people affected by nuisance
2b		- % people seriously affected by nuisance
3a	. safety	- risk contour of stationary sources
3b		- risk contour of mobile sources
4		- perception of safety
5		- number of houses within certain decibel contour
6a	. noise	- number of houses within certain decibel contour stationer sources
6b		- number of houses within contour mobil sources
7a		- % people affected by nuisance
7b		-% people seriously affected by nuisance
8		- ha sanctuary
9	. local air	- concentration NO2, benzene CO, BaP
10		- perception health
11a	. local water and soil quality	- heavy metals (cadmium, chrome, copper, zinc, lead, and mercury)
11b		- organic micropollutants (B(a)P, PCB, HCH)
11c		- fertilizers: N and P
11d		- oil: mineral oil
12a	. higher-scale emissions	- share CO_2 emission Rijnmond
12b		- kWatt linking of heat and power
12c		- kWatt other sources
B. Spatial quality themes:		
13	.nature	- ha planned, redesigned, rezoned
14	.recreation	- ha planned, redesigned, rezoned
C. Harbour themes:		
15	.harbour area issues	- ha ('containers, food, chemistry')
16	.reserve harbour area	- ha
17	.activity	- value added ('containers, food, chemistry'
18		- employment ('containers, food, chemistry')
19		- readiness to invest
D. Accessibility themes:		
20	.congestion	- % change of congestion
21	.modal split passenger/traffic	- kilometres passenger/traffic
22	.modal split goods/traffic	- ton km (road, water,rail, pipe)

Box IX.1 List of goals derived from the Rijnmond Development Plan (Dil *et al.*, 1993: pag. 7-81).

Table IX.3: Items mentioned in the Development Plan as goals or indicators for monitoring. Numbers refer to the items Box IX.1, __ = item that is assumed to give information on the characteristics of this cell. Other symbols as in Table IX.2. For discussion of this table, see text.

Characteristics related to direct users

	Processes				**Structures**				**For future use**				**For future processing**			
	ec	so	ph	ei	ec	so	ph	ei	ec	so	ph	ei	ec	so	ph	ei
Separate																
citizen	in	18	ar	5 6	in	18	ar	5 6	x	x	x	x	x	x	x	x
producer	17	18	15	im	17	18	15	im	x	x	16	x	19	x	x	x
recreat-ionist	.	.	14	8	.	.	14	8	.	.	14	x	.	.	x	x
farmer	o	o	o	o	o	o	o	o	o	o	o	o	o	o	o	o
nature	.	.	13	8	.	.	13	8	.	.	13	x	.	.	x	x
activities																
habitation working	.	.	21	.	.	.	21	.	.	.	21	.	.	.	x	.
producing	.	.	22	.	.	.	22	.	.	.	22	.	.	.	x	.
transport	.	.	20	.	.	.	20	.	.	.	20	.	.	.	x	.
recreation	.	.	14	.	.	.	14	.	.	.	14	.	.	.	x	.
farming	.	.	o	.	.	.	o	.	.	.	o		.	.	o	.
conser-ving	.	.	13	.	.	.	13	.	.	.	13	.	.	.	x	.

Characteristics related to higher scale

	ec	so	ph	ei	ec	so	ph	ei	ec	so	ph	ei	ec	so	ph	ei
combined																
spatial	.	.	if	3 9 11	.	4 10	21 22	3 9 11	.	.	x	x	.	.	x	x
temporal	.	.	if	1 2 7	.	4 10	20	1 2 7	.	.	x	x	.	.	x	x
Separate																
com-munity	17	18	14 13 15	5 6 8	17	18	14 13	5 6 8	o	o	16 14 13	o	19	o	o	o
higher-scale needs and wants	o	o	o	12	o	o	o	12	o	o	o	12	o	o	o	o

characteristics for the present and the future (15 & 16). In this sense, the Development Plan emphasizes harbour interests and the potential for expansion. The quality of the environment is also recognized as a *valued system characteristic*: attention is given to different kinds of emissions in spatial and temporal terms (risks/safety contours: 3, air: 9 and water qualities: 11). With respect to these characteristics, due attention is given to the impact on people in general (1, 2 & 7) and residents in particular (5 & 6). Even the perception of the quality of life is valued (safety: 4 and health: 10; cf. the quality of life indicators mentioned in

Chapter IV). The plan also considers important qualities of rural areas (sanctuary [8], and planned, redesigned, or rezoned areas: 13 & 14). On the level of the system as a whole, emissions (12) and transport (20, 21 & 22) are mentioned.

IX.4.2 Characteristics missing from the development plan

Despite the broad interpretation of the items mentioned in the plan, there are several types of characteristics that are valued in the Tables IX.2 and IX.3, but are not mentioned in the plan.

Starting with *empty rows and columns*, the table shows that characteristics related to certain user groups and conditions for future processing are missing. Because the activity 'farming' was not mentioned in the Development Plan, characteristics valuable for farmers and farming were not operationalized. Based on the defined system boundaries, the area for agriculture, at the very least, must be mentioned as a valued characteristic. Assuming that the immense economic power of the harbour will dictate the maintenance and expansion of the harbour area, the allocation of nature conservation concentrations will inevitably lead to conflicts with agriculture. Since these agricultural land-use interests are not recognized, this conflict is not on the agenda of the debate. If the plan is carried out "*as is*", nature conservationists will be played off against farmers, although in fact they may well have an important interest in common: restricting the expansion of the harbour, and promoting a more efficient use of the existing harbour area.

It is noteworthy that no attention is given to conditions for future processing. One would expect attention to be given to resources such as food, water, energy, and raw materials, and conditions that provide resources in the future, such as storage and distribution facilities. In general, this omission is understandable, because the supply of such resources is more or less guaranteed in the urbanized regions of developed countries, and it is not likely to frustrate the development of such regions. But one would expect some reference to vital resources, such as drinking water for the population and electricity for industry.

Table IX.3 shows that *within the columns and rows* there are three type of characteristics that are not in balance. Although explicitly mentioned as important when for the development of the region, no intention is expressed to monitor incomes and housing area ('in' and 'ar' in Table IX.3). This is unexpected, because these developments are essential from the standpoint of social stability and the cooperation of the citizens in development the region. In Table IX.3, 'reduce emissions' (12) is defined as a high-scale interest. However, there are no plans to measure the contribution of the individual groups. The type of emissions mentioned (organic micropollutants and oil, 11 in Box IX.1) suggests that the focus is primarily on industrial activities, but the list of *valued system characteristics* makes no mention of an explicit decision to measure only emissions of this type.

Similarly, the impact of emissions on housing is measured in terms of the number of noise-polluted houses (5 & 6 in Table IX.3) and the overall impact on people (1, 2 & 7 in Table IX.3). There are no variables which measure the impact of dust (1) and odours (2) on rural areas, such as recreation or nature conservation areas; the sanctuary area (8 in Table IX.3) is the only characteristic mentioned which is relevant for nature and recreation.

It is easy to forget that not only human beings suffer from the effects of dust and odours; the economic activities of certain groups of producers are also affected. In the present case, this would include producers involved in harbour-related activities, some of whom may them selves maintain high environmental standards for their activities ('im' in Table IX.3). Since the impact of emissions on producers is not designated as an important *valued system characteristic*, conflicts within the group of producers are masked. From the standpoint of environmental organizations, producers who suffer as a result of environmental impacts are potential allies, who can only be found if identified valuable characteristics are used as a starting point for further searches. This underscores the necessity to carry out in full the recommendation in Section IX.3.3 to verify whether a characteristic valued by one user group is also valued by another group.

IX.5 Discussion

The comparison shows that a rough application of the proposed procedure and the newly developed framework for identifying *valued system characteristics* could provide important information about missing activities and characteristics, which could be placed on the agenda of the debate. The focus might be on the interests of farmers, as well as on the conditions for future processing. This could lead to a proposal to monitor not only the environmental emissions of economic activities and traffic, but also those resulting from, say, housing and farming. And not only the environmental impact of economic activities on housing could be taken into account, but also the impact on the economic activities themselves. The study also shows that the framework can be used to analyze the completeness and consistency of lists of goals and variables included in the existing plans.

However, having at one's disposal the relevant information does not guarantee that it will be automatically accepted as relevant information in the debate. There is still a danger that the most powerful groups will use their influence to obstruct the participation of their opponents. In fact, this is what happened during the discussions on the Rijnmond Development Plan (Dil *et al.*,

1993). Industrial, business, and governmental organizations dominated the discussions to such an extent that the environmental groups ultimately withdrew from the planning process. Their demands that more attention should be paid to a number of serious issues were not recognized (Van de Klundert & Glasbergen, 1995). To avoid being identified with and held responsible for the decisions based on the discussions, they preferred to withdraw.

IX.6 Concluding remarks: improvements to the proposed procedure

On the basis of the simulation of the application of the proposed procedure, the study points up several ways to improve the guidelines for using of the newly developed framework.

First of all, the framework does not include the specific direct users, activities, or aspects to be recognized when selecting valuable characteristics. This means that cultural aspects, for example, are missing in the application. The axes of the basic framework had to be split up in order to make the framework suitable for the specific application at hand. This will, of course, be the case in any application.

Second, 'farming' was added to the list of activities, since it appeared on the map of the region. If the discussion on the definition of the system had not been based on a map of the area, the need to add agriculture might not have been recognised. Thus supplementary information is needed to complete the lists of direct users, activities, aspects, and higher-scale communities. Universal value lists derived from the literature (e.g., W.T. de Groot, 1992; Doyal & Cough, 1991; Max-Neef, 1992; Henderson, 1991) can be used to check the lists proposed in the existing plans or during the start-up of a procedure for sustainable development.

Third, it was stated that for every cell ('.' in Table IX.2) a characteristic could have been identified, but that this was not carried out because it was assumed that there was not sufficient time or money to carry out a full search for the needs and wants of people involved. If this had be done as recommended - by using identified characteristics to search for new ones and find the people who value them - it might easily have generated a huge list of potential but in the end non-relevant valuable characteristics. The best way of controlling this process would seems to be by cooperating with the people involved (or their representatives) in using the framework as an instrument for selection. As noted above (Chapter IV), the power of the framework is in the fact that it makes it possible to ask the appropriate questions when the people involved are consulted on what they regard as the valuable characteristics of the *SEnS* to be developed.

CHAPTER X

A Simulation of the Measurement of the Resource State of a Dune and Bulb-growing Region in the Netherlands.

CONTENTS

X.1 Introduction

Chapter VI stated that the proposed procedure for exploring the opportunities for a sustainable development should be useful not only for researchers or in planning procedures, but also as a means to help parties involved in the debate on sustainable development to develop their strategy. The focus in the previous chapter was on gaining insight into the efficiency of the developed procedure and its framework for identifying and selecting *valued system characteristics* in order to find shortcomings in the procedure. There the procedure was used to select the valuable characteristics of the Rijnmond region in the Netherlands. The next phase - the measurement of the available human, financial and biophysical resources in the region - was not carried out. In this chapter, the focus is on the efficiency of the proposed procedure to measure those resources - the *resource state* - in such a way that it provides insight into the development opportunities for a region (Van der Meij *et al.*, 1997b). Again, the procedure is applied in a situation designed to provide parties involved in a debate with additional information for strategy development. The procedure must be applicable when both time and money for gathering information are limited. Therefore, the measurement of the *resource state* should be possible within a short time, with little money, and be aimed at information of low, but adequate resolution. The idea is that this information of low resolution can be used by parties to participate in an early stage of the debate on regional development.

The application of the proposed procedure is simulated in order to examine whether the procedure would have been useful for identifying development opportunities by analyzing the *resource state* of the region. A comparison with existing plans may show up desired improvements to the proposed procedure.

As noted in Chapter IX, the Zuid-Hollandse MilieuFederatie (ZHMF) - a non-governmental environmental organization - was invited to cooperate in the simulation of an application designed to gain insight into the shortcomings of the proposed procedure. It was decided to focus on the development of a region with dunes and bulb-growing in the Netherlands known as the 'Duin- en Bollenstreek' (Figure X.1 & 2).

First, the method for application is presented (Section X.2), and then the proposed procedure is applied (Section X.3). In Section X.4 the application is compared with current plans for the development of the region and then discussed (Section X.5). Improvements to our procedure are proposed in Section X.6.

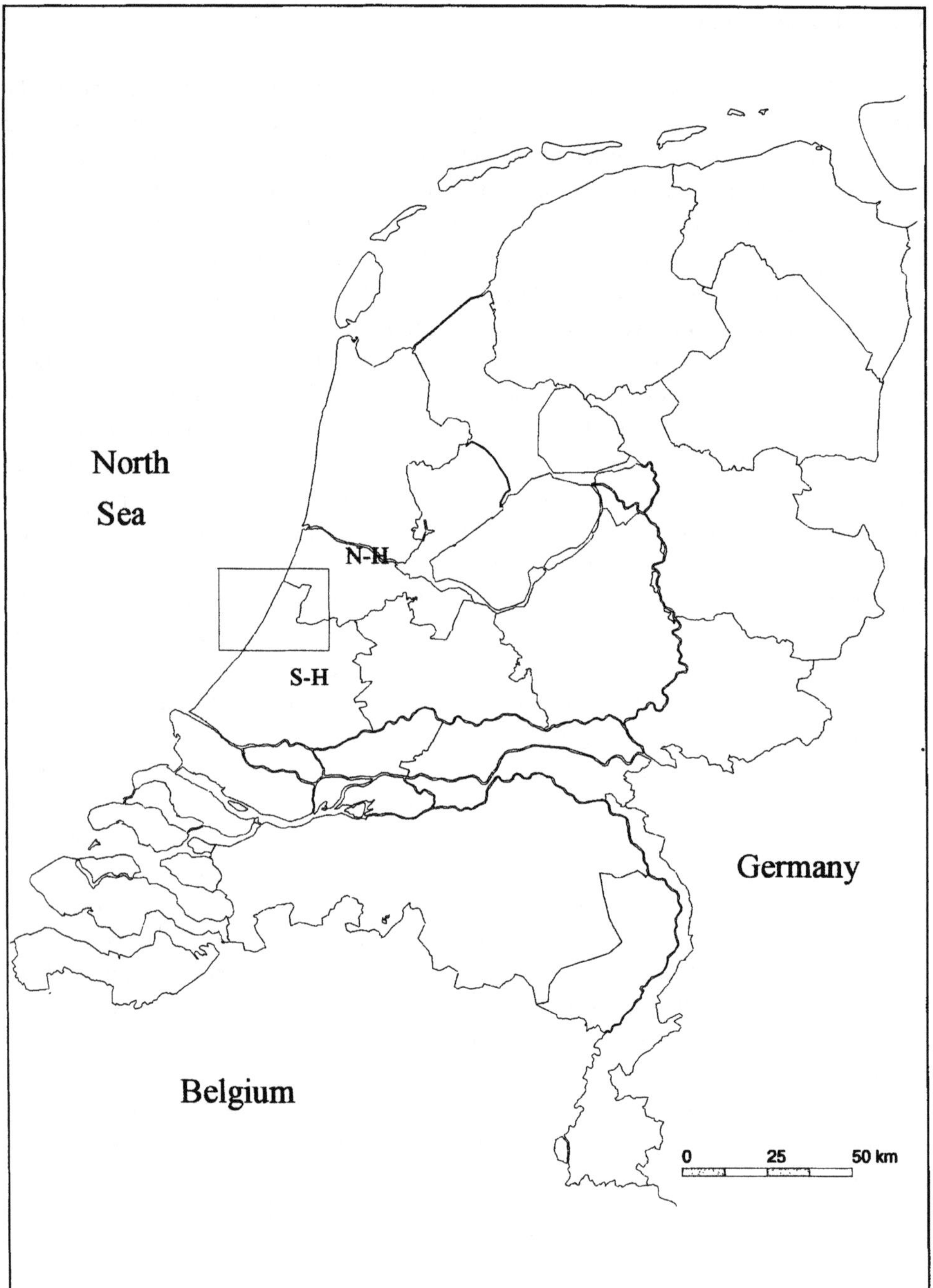

Figure X.1: Location of the dune and bulb-growing region within the Netherlands. S-H = the province of South Holland, and N-H = the province of North Holland

Figure X.2: The dune and bulb-growing region

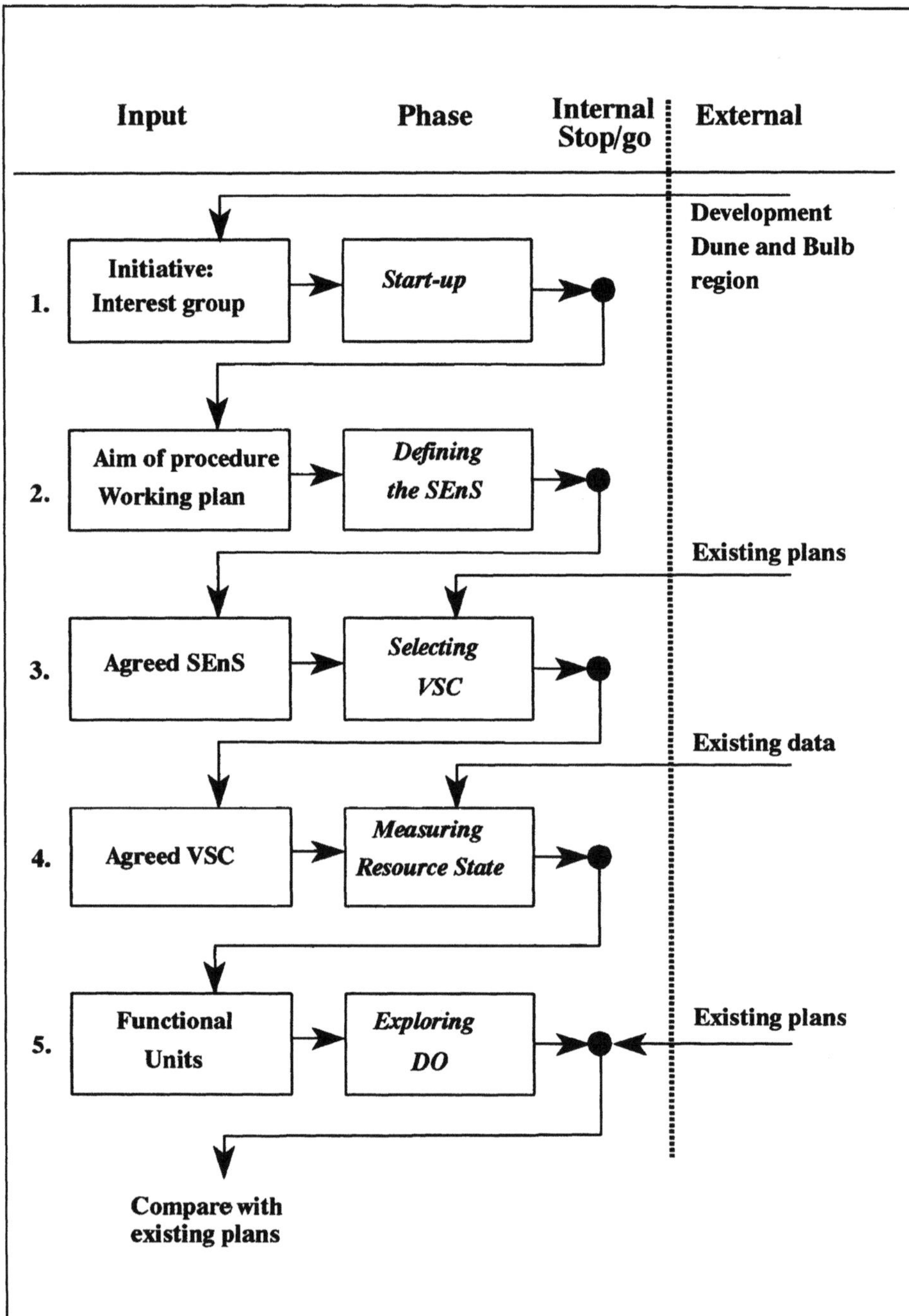

Figure X.3: The procedure for exploring the development opportunities of the dune and bulb-growing region

first, known as the 'Varianten Nota' of the Province of Zuid-Holland (Provincie Zuid-Holland, 1993), was a provincial report intended to review development opportunities for the region. It described four possible directions of development. The second, the 'Pact van Teylingen', was put together by a coalition of various groups in the region; It is a kind of development contract in which a number of development goals are specified (Projectgroep Toekomst Duin- en Bollenstreek, 1995). Data on needs and wants related to the region were collected from these two documents, and this information was supplemented by needs and wants derived from a reaction to the 'Varianten nota' by a coalition of environmental groups (Milieu Overleg Duin- en Bollenstreek, 1994; Projectgroep Toekomst Duin- en Bollenstreek, 1995).

As in the previous chapter, it was assumed here that information-gathering should be carried out within a few months by a small group of people, and that this period would not exceed the available time for either of the plans with which the results are to be compared.

Data were also required to apply the proposed procedure and measure the *resource state*. References are given in the table describing the resource characteristics (Table X.4). There was no sufficient time or financial resources to carry out a thorough consultation with the people involved, or an extensive collection of new data. This meant that the measurement of the *resource state* had to be based on a limited quantity of data. The measured *resource state* in the simulation was interpreted, in order to find development opportunities. The focus was on opportunities for three of the five major land-use activities within the dune and bulb-growing region: habitation, bulb-growing, and nature conservation. The results of this interpretation were compared with the goals of development proposed in the policy documents 'Varianten Nota' (Provincie Zuid-Holland, 1993) and 'Pact van Teylingen' (Projectgroep Toekomst Duin- en Bollenstreek, 1995).

X.3 Applying the proposed procedure to measure the spatial resource state of a dune and bulb-growing region

X.3.1 Start-up

The dune and bulb-growing region in South-Holland (Figure X.1 & 2) was named for two important landscape characteristics. Its western boundary is part of the Holland's North Sea dunes, which in turn are part of Europe's most important dune region. The mid and north-eastern part of the dune and bulb-growing region is where the famous Dutch tulips and other bulbs are cultivated, on the sandy coastal barrier deposits formed after the Ice Age. Lying within one of the most densely populated regions of the Netherlands, it is also part of the west wing of the chain of cities known as the 'Randstad' (Faludi & Van der Valk, 1994; see also Chapter IX).

The development of the region is complicated by a number of major problems (Provincie Zuid-Holland, 1987, 1990, & 1993; Milieu Overleg Duin- en Bollenstreek, 1994; Provincie Zuid-Holland, 1995b; Steekelenburg, 1995).

- The large number of bulb farms all face problems in expanding their farms or intensifying their business in order to remain economically sound.
- Bulb-growing is a pesticide-intensive form of agriculture and thus constitutes a major threat to the safety of the water reservoirs and the integrity of the nature reserves.
- There is huge pressure from encroaching urban areas to expand housing areas, but the possibilities are limited.
- Nature reserves must be expanded to ensure better protection for the dunes and to reduce their isolation by connecting them with inland nature reserves, but here, too, the possibilities are limited.

These are the major problems, but there are others. For example, the province of Zuid-Holland (1993) has recognized six problem areas, while the Projectgroep Toekomst Duin- en Bollenstreek (1995) has adopted seven themes related to regional problems (Box X.1).

A scheme which combines the expansion of bulb-growing and housing development with breathing space for nature would be a clear cut example of 'sustainable development' as defined in Chapter II. Is this possible? The major problem of the dune and bulb-growing region in this simulated application of the proposed procedure is a spatial one. Consequently, the following discussion will focus on this spatial problems, and the challenge of combining activities spatially. More specifically, the question is: Can the necessary information on the spatial opportunities to develop bulb-farming and housing, together with nature conservation be provided by measuring the *resource state* of the region?

X.3.2 Defining the socio-environmental system

Geographically, the dune and bulb-growing region is explicitly defined, but it should be noted that the boundaries correspond to the borders of the province of Zuid-Holland, and do not include the province of Noord-Holland (Figure X.1). The ten municipalities that make up the region are presently collaborating in developing the region. In that sense, it is an entity. Housing, agriculture, industry, recreation, and nature conservation are the major types of direct use that require space. There are also the substantial interests of the national government in the region, in the

1.	**Development in bulb farming**	A.	**Bulb-growing**
1.1	.bulb area	A1	.conservation of bulb-growing area
1.2	.bulb-growing	A2	.subsidiary bulb-growing area in the Haarlemmermeer
1.3	.bulb processing	A3	.no specialized greenhouses within the central bulb-growing area
1.4	.bulb trade	A4	.sustainable bulb-growing
1.5	.research	A5	.supporting policy and stimulation of large-scale bulb-growing farms
1.6	.tourism		
2.	**Development in horticulture**	B.	**Horticulture in greenhouses**
2.1	.open horticulture	B1	.increase in specialized greenhouse horticulture near Rijnsburg
2.2	.protected horticulture:	B1	.modernizing, consolidation, and revitalizing specialized greenhouses near Rijnsburg
2.2.1	.. subsidiary greenhouses		
2.2.2	.. large-scale greenhouses		
2.3	.auction hall		
3.	**Development in nature and landscape**	C.	**Employment**
3.1	.grassland	C1	.related to bulb-growing the highest priority in agricultural areas
3.2	.woods	C2	.not related to bulb-growing to be consolidated along the main road A4
3.3	.reserves		
3.4	.closed landscape on [coastwalls]	D.	**Habitation**
3.5	.open landscape in between [coastwalls]	D1	.house-building only for habitation needs of the inhabitants of the region
3.6	.ecological infrastructure	D2	.restrictive policy for house-building
3.7	.'green' areas within urbanised areas	D3	.agreement is needed on a location for house-building
		D4	.after the existing task for house building , no further house-building for non-inhabitants is allowed, except when the airport is approved as building location
4.	**Urbanisation**		
4.1	.number of houses		
4.2	.industrial areas		
		D5	.no 'bulb city'
5.	**Environment**	D6	.extension of 'green' areas in the southern, urbanized part of the region
5.1	.*supraregional*:		
5.1.1	.. mobility control		
5.1.2	.. energy use	E.	**Landscape, recreation and tourism**
5.2	.*regional*:		Conservation and reinforcement of the (open) landscape of the region:
5.2.1	.. noise		
5.2.2	.. air	E1	.in general
5.2.3	.. safety	E2	.regulation of subsidiary greenhouses
5.2.4	.. waste processing	E3	.regulation of non-land-use agriculture
5.2.5	.. soil cleaning&protection	E4	.planning rural area
5.2.6	.. water protection	E5	.reallocation of atypical enterprises
5.3	.*local*	E6	.supporting bulb tourism and other forms of recreation
5.3.1	.. pesticide use		
6.	**Mobility**	F.	**Nature and ecology**
6.1	.public transport	F1	.package of measures for conservation, reinforcement, and promotion
6.2	.east-west connections		
6.3	.opening-up of the area	G.	**Water and environment**
6.4	.cycle tracks	G1	.applying a water system approach

Box X.1: Left: the topics discussed in the 'Varianten nota' (Provincie Zuid-Holland, 1993; pag.: A2-A13). Right: the agreements reached in the 'Pact van Teijlingen' (Projectgroep Toekomst Duin- en Bollenstreek, 1995)

form of a military airport and a seawall. Other values at stake include water collection and the need to protect the quality of the dunes (Provincie Zuid-Holland, 1995b).

There are three major areas in the region: the dunes in the west; the bulb area in the centre and the north, and a mixed urban and agricultural area in the south (Figure X.2). The landscape is characterized by the presence of dunes running southwest to northeast, and 'coastal walls' , which are the remnants of coastal deposits. These coastal walls have sandy soils, with wet peaty soils in between. The landscape is open and used mainly for agriculture. The villages and main transport structures are located on the coastal walls. The old estates surrounded by woods are also located here. In the southern part of the region the river Oude Rijn runs east-west, cutting through the structure of the coastal walls. In the southeast there is a complex of small lakes known as the 'Kagerplassen', all part of the Old Rhine estuary.

In carrying out the proposed procedure, all information was collected on the basis of grid cells of 0.25 km^2. At this level, information would appear to be both manageable and relevant for exploring the overall development opportunities of the region. In 1994, the average size of a bulb farm was 15 ha (LEI & CBS, 1996).

The definition of the higher-scale system to which the region belongs is problematic. As stated, geographically and demographically, the region belongs to the west wing of the Randstad, an urbanized and densely populated area in the western part of the Netherlands. But the Randstad is not a governmental entity. The higher-scale governmental entity to which the region belongs is the province of South Holland, but it is also bordered extensively by the province of North Holland (Figure X.1). It was decided to consider only the province of South Holland as the higher-scale community and to collect information on that level.

X.3.3 Selecting the valued system characteristics

Selecting the valuable characteristics of a *SEnS* is a necessary step toward *exploring development opportunities* (Chapter VI). In this simulation, the topics and agreements included in the existing plans were used to execute this phase ('Varianten Nota' and 'Pact van Teijlingen': Box X.1). All of these are obviously related to a variety of system characteristics that may be important for measuring the *resource state*. In order to take the next step - *measuring the resource state* - the topics were interpreted as valuable characteristics and were placed in the framework for identifying *valuable system characteristics* (Chapter IV). The focus was on translating these relevant 'characteristics' into processes and structures related to the spatial use of the region, in order to find the relevant resources required and then to measure the *resource state*. The criteria laid down for fitting in the topics in the Boxes X.1 into the framework were:

- Topics that refer to products are placed in the first column.
- Structure features are placed in the second column.
- Topics that refer to the need for space are regarded as referring to a need for future land-use and are placed in the third column,
- All infrastructure features are regarded as referring to the concordance of features and thus are placed in the second or third row.

Applying these criteria resulted in Table X.1. The table was supplemented with topics suggested by the Milieu Overleg Duin- en Bollenstreek (1994) on the basis of the 'Varianten nota'.

X.3.4 Measuring the resource state

The *resource state* can only be quantified after the resources required for the processes and structures have been found. Chapter V identifies seven steps for measuring the *resource state*. In Figure X.4 these steps are described. The last step is the presentation of the *functional units*. These *functional units* show the potential to carry out different activities and combinations of activities which can be used to explore *development opportunities*.

The first two steps have already been taken; below the other steps are described in more detail.

Step 3: Processes and structures

In accordance with the proposed procedure, the third step is focused on identifying the processes and structures that determine which resources are needed to search for spatial development opportunities. This means a selection of the characteristics from Table X.1, resulting in Table X.2. The following procedure was followed:

- First, in Table X.1 a number of characteristics occur that are in effect constraints on development solutions, such as 'no greenhouses in bulb area' and 'open landscape between coastal walls'. Here these topics were skipped. But in view of the importance of greenhouses and landscape, 'landscape' was placed in Table X.2.
- Second, a number of characteristics related to nature conservation have not been well operationalized in Table X.1, e.g., 'nature and ecology'. All the characteristics related to nature conservation were translated into three specific types of processes and structures: (i) conservation of natural-historical features, (ii) conservation and promotion of biodiversity,

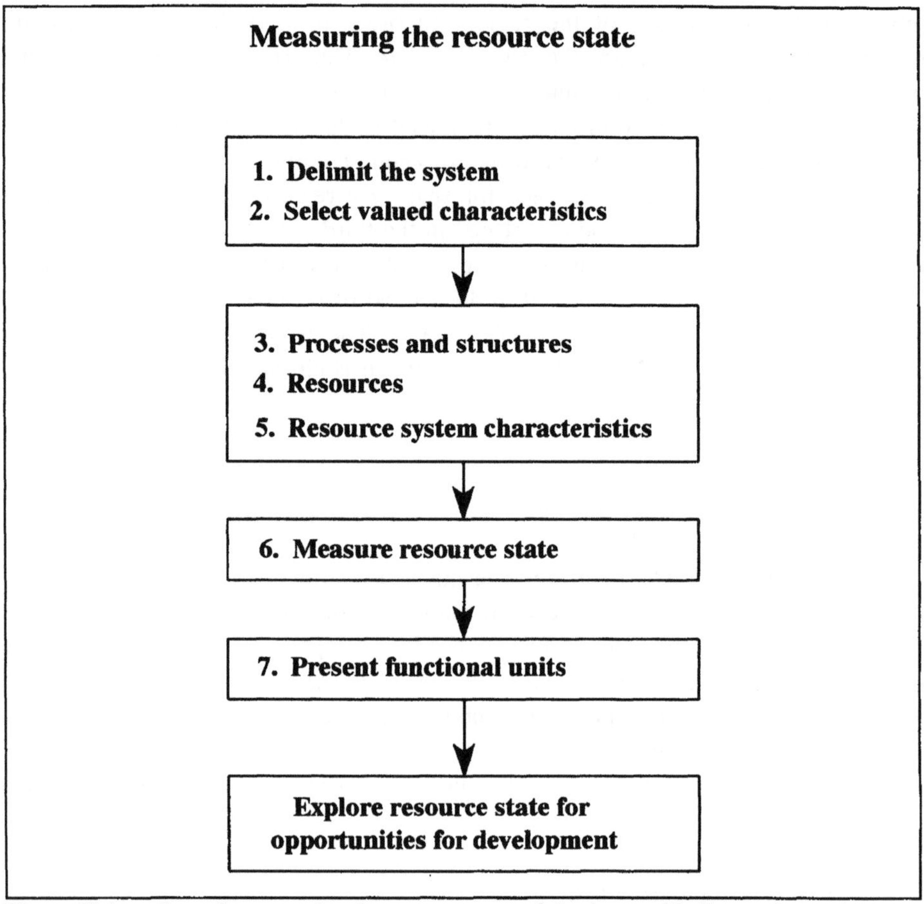

Figure X.4: Steps for measuring the resource state of the dune and bulb-growing region

including genes, species and ecosystems, and (iii) conservation and realization of ecological structure elements and corridors.

- Third, all construction was divided into two types: housing and all other buildings. The first type is located in 'habitation areas', the other type in 'industrial areas', which means that factories, greenhouses, auction-marts, etc. are all assumed to be located in 'industrial areas'.
- Fourth, in consultation with the ZHMF, a number of essential processes and structures that could not be derived from Table X.1 were added. In Table X.2, these activities can be identified by the fact that they do not refer to any of the documents used for selecting the processes and structures.

Table X.1: Valued characteristics in 'Varianten nota' (1.1, etc), and in 'Pact van Teijlingen' (A1, etc.). Supplemented by Milieu Overleg Duin- en Bollenstreek, 1994 (*). See Box X.1 for items and Section X.3.3 for criteria.

Characteristics related to direct users

	Processes	Structures	For future use	For future processing
Separate:				
-inhabitant	.house building (4.1; D1,2,4) .noise (5.2.1)	-	.habitation areas (D3)	-
-farmer	.bulb-growing (1.2) .bulb-processing (1.3) .bulb trade (1.4) .pesticide use (5.3.1)	.large enterprises (A5) .open horticulture (2.1) .protected horticulture (2.2; B2) .auction hall (2.3)	.bulb area (1.1, A1,2)	.research (1.5) .sustainable bulb farming (A4)
-industrial	-	-	.industrial areas (4.2)	-
-visitor	.tourism (1.6, E6)	-	-	-
-nature cons.	-	.grassland (3.1) .woods (3.2)	-	-
Combined:				
-living	-	-	.green areas in habitation areas (3.7; D6)	.public transport (6.1) .cycle tracks (6.4)
-farming	.bulb-related employment in agricultural area (C1)	.no greenhouses in bulb area (A3) .greenhouses near Rijnsburg (B1)	-	-
-industrial prod.	-	-	-	-
-visiting	-	-	-	-
-nature cons.	-	.woods as buffers (*)	-	.ecological infrastructure (3.6)

Characteristics Related to higher scale

Combined -spatial quality	.non-bulb-related employment near main road (C2)	.greenhouses in industrial areas (*) .closed landscape on coastal walls (3.4) .open landscape between coastal walls (3.5) .east-west connections (6.2)	.no 'bulb-city' (D5) .landscape (E1-E5)	.ecological structure with nature areas (*)
Separate -scale	.mobility control (5.1.1) .energy use (5.1.2) .waste processing (5.2.4)	.air (5.2.2) .safety (5.2.3)	.nature reserves (3.3) .nature and ecology (F1) .nature development areas (*)	.opening up of the area (6.3) .soil protection & 'cleaning' (5.2.5) .water protection (5.2.6; G1)

Table X.2: Processes and structures selected on the basis of Table X.1. Derived from 'Varianten nota' (1.1, etc.), from 'Pact van Teijlingen' (A1, etc.), from Milieu Overleg Duin- en Bollenstreek, 1994 (*). Items without reference were added by the ZHMF. See Section X.3.3 for criteria for selecting activities.

Characteristics related to direct users

	Processes	Structures	For future use	future processing
Separate	*Production of*: 1.bulbs (1.2) 2.dairy products (3.1) 3.crops (2.1; 2.2; B2) 4.raw material, drinking water, industrial products (4.2)	*Management of*: 6. bulb area (1.1, A1-A2) 7. habitation area (D3) 8. recreation area (1.6; E6) 9. dairy area (3.1) 10.arable land (2.1) 11.industrial area (4.2, 2.3) 12.area for storage of waste	*Realization of new*: 6. bulb area (1.1, A1-A2) 7. habitation area (D3) 8. recreation area (1.6; E6) 11.industrial area (4.2) 12.area for storage of waste	
Combined		*Management of*: 13.transport infra-structure (6.1; 6.4) 14.ecological infrastructure (3.6) 15.hydrological infrastructure		*Realization of*: 13.transport infrastructure (6.1; 6.4) 14.ecological infrastructure (3.6) 15.hydrological infrastructure

Characteristics related to higher scale

	Processes	Structures	For future use	future processing
Combined		*Management of*: 16.transport main structure (6.2; 6.3) 17.ecological main structure (*) 18.hydrological main structure 19.seawall 20.landscape (3.4; 3.5; E1-E5) 21.cultural history (1.6, E6) 22.natural history (1.6, E6)		*Realization of*: 17.transport main structure (6.2; 6.3) 18.ecological main structure (*) 19.hydrological main structure
Separate	*Processing* 5. waste (5.2.4)	*Management of*: 23.military terrain *Management of biodiversity*: 24.genes (3.3, F1,*) 25.species (3.3, F1,*) 26.ecosystems (3.3, F1, *)	*Strengthening biodiversity*: 24.genes (3.3, F1, *) 25.species (3.3, F1, *) 26.ecosystems (3.3, F1, *) Conservation of: 27.fresh water reserves (5.2.6; G1) 28.sandy layers	*Protection against*: 29.becoming brackish

Step 4: Resources

The processes and structures selected in Table X.2 are listed as the 29 items in the first column of Table X.3. The resources selected as essential for the processes and structures appear in the second column. They are based on current processes and structures and it is assumed that they will continue to be relevant in the future. Because of this assumption, it is not necessary to distinguish between present and future: both need the same type of resources. The related resources are described here in a straightforward manner: for 'waste processing' one needs space. Only those resources have been selected that have a direct relation to land-use. For that reason, resources such as 'energy', 'raw material', 'capital', 'skill', and 'education level' were not selected as resources for any of the activities in Table X.3. It was assumed that the availability of these resources is distributed evenly throughout the region.

Step 5: Resource system characteristics

The resource characteristics are provided in the third and fourth columns of Table X.3. A distinction is made between biophysical characteristics that are location-specific and those that define the relation of a location with its surroundings. For example, the location-specific resource characteristics related to bulb-farming are sandy, well-drained soil, no open water and no construction. The necessary relation with its surrounding is the presence of a link with the main transport structure. Together these two types of characteristics make a location suitable for bulb-growing. A total of 24 characteristics were found; these are indicated between brackets as a-x in Table X.3 and specified in Table X.4.

Selecting the *resource system characteristics* is not a simple, straightforward operation. Because the aim is to determine where combinations of activities that contribute to the goals of development may be located, the criterion for selecting a resource characteristic must be that either its presence or its absence is required in a location in order to make possible the desired activities in the region. Furthermore, to explore opportunities, the characteristics must be chosen in such a way that a distinction can be made between those that can be altered and those that cannot. And third, the characteristics chosen must be capable of demonstrating whether present or desired activities benefit or hinder one other (Chapter V). Features that are necessary for an activity, but as a rule are created or supplied when the activity is performed, should not to be selected. For example, houses can be built on flat land. Although this land must have a certain carrying capacity, here the 'carrying capacity of the ground' is not regarded as a *resource system characteristic*, because it is common practice in Holland to pile where the carrying capacity is too low. In Holland, a low carrying capacity is not sufficient reason to refrain from building, whereas the presence of health risks, such as noise

Table X.3: The processes and structures from Table X.2, their resources and the characteristics of that resources. The resource characteristics are operationalised in Table X.4. See Section X.3.4 for further explanation.

Processes/structures	Resources	Local characteristics	Relational characteristics
Production			
1. bulbs	'Bulb land'	Sandy soil (a) Good drainage (b) No open water (c) Unbuilt (d)	Link with transport main structure (s)
2. dairy products	Pasture land	No open water (c) Unbuilt (d)	Idem
3. crops	Arable land	No open water (c) Unbuilt (d)	Idem
4. industrial products	Built-up area	No open water (c)	Idem
Processing			
5. waste	Area	No open water (c)	Link with transport infrastructure (t)
Management and realization of:			
6. bulb farming area	Area	See 'bulb land'	Idem
7. habitation area	Area	No open water (c) No: - noise pollution (e) - health risks (f) - stench (g)	Link with transport infrastructure (t)
8. recreation area	Area	Idem, except 'no open water'	Idem
9. dairy farming area	Area	See pasture land	Link with transport main structure (s)
10.arable farming area	Area	See arable land	Idem
11.industrial area	Area	No open water (c)	Idem
12.area for storage of wastes	Area	-	Link with transport infrastructure (t)
13.transport infrastructure	Area	-	Link with transport main structure (s)
14.ecological infrastructure	Area	No:- over-manuring (h) - pollution (i) - disturbance (j) - drying-up (k)	Link with ecological main structure (u)
15.hydrological infrastructure	Area	Good drainage (b)	Link with hydrological main structure (v)
16.transport main structure	Area	-	Idem
17.ecological main structure	Area	No:- over-manuring (h) - pollution (i) - disturbance (j) - drying-up (k)	Link with ecological main structure (u)

Table X.3 (continued)

Processes/structures	Resources	Local characteristics	Relational characteristics
Management and realization of:			
18. hydrological main structure	Area	Good drainage (b)	Link with hydrological main structure (v)
19. seawall	Area	Elevation (l)	Link with other seawalls (x)
Management and reinforcement of:			
20. landscape	Landscape	Open space (m)	-
21. cultural history	Values concerning cultural history	Area with cultural history (n)	Link with transport infrastructure (t)
22. natural history	Values concerning natural history	Area with natural history (o)	Link with ecological infrastructure (w)
23. military terrain	Area	-	Link with transport main structure (s)
24. genes	Natural populations of animals and plants	Area with nature potential (p)	Link with ecological infrastructure (w)
25. species	Natural richness of species	Area with nature potential (p)	Link with ecological main structure (u)
26. ecosystems	Natural ecosystems	Area with nature potential (p)	Link with ecological main structure (u)
28. sandy layers	Sandy layers	Extractable (r)	-
29. Protection against becoming brackishing	Fresh water aquifers	-	-

and dust, is (Provincie Zuid-Holland, 1995a). Therefore, unlike carrying capacity, the absence of health risks is regarded as a necessary, but changeable characteristic of a habitation area.

A formalized selection scheme was not available to answer such questions as: What characteristics are essential for locating activities? And what essential characteristics are relevant for the level of scale of the simulation performed here? In the end, the *resource system characteristics* selected must be regarded as based on 'expert judgement'.

Table X.4: Definition of resource characteristics and source of data. (n.a.): not available; Bosatlas: Bosatlas (1988); OD205: Oosterbaan & Overbeek (1992); top. maps: 24H, 25C, 30E, 30F 1:25000 (TDN, 1988); hydr. maps: 24H, 25C, 30E, 30F 1:25000 (PWZH, 1980); VVV map: Provinciale VVV Zuid-Holland (*s.a.*); CML: Runhaar & Groen (1993); PZH: Province Zuid-Holland (1992); NBP map: Bakker *et al.* (1989).

		division	sources	criterion
a:	soil type	sand/no sand	Bosatlas, OD205	50% of the cell area
b:	drainage	good/bad for bulb farming	top. map, hydr. map	soils in high relief, surface water, or peaty soils below minus 1 Normal Amsterdam Level are poorly-drained; 50% of cell area
c:	open water	open water/mainland	top. map	50% of the cell area
d:	built	built/unbuilt	top. map, VVV map	urban areas (built-up areas including infrastructure) are built; 50% of the cell area
e:	noise (n.a.)	good/bad for living	- (equal to disturbance?)	on the basis of highways and air-traffic noise?
f:	health (n.a.)	better/worse for humans	- (equal to pollution?)	concentrations in soil/air/water?
g:	stench (n.a.)	stinking or not for humans	-	fish pocessing industry?
h:	over-manuring (n.a.)	over-manured or not	-	-
i:	pollution	clean/dirty (by pesticides)	CML	the bulb-growing region is assumed to be the only area polluted by pesticides; 50% of the cell area
j:	disturbance	disturbed/ non-disturbed	top. map, own obs.	built-up areas, main roads, or areas for intensive recreation with a border of 100 m around it; 50% of the cell area
k:	drying-up (n.a.)	dried-up/ not dried-up	-	-
l:	elevation	below/above Delta level	top. map	50% of the cell area ca 6 m above NAP
m:	open space	open/not open	top. map, own obs.	cells which are not or just marginally cut by main roads, lie (largely) outside the built-up area, and border at least one other open cell
n:	cultural history	valuable/not valuable	top. map, own obs.	cells containing estates or old town centre are assessed as valuable
o:	natural history	valuable/not valuable	top. map, PZH	cells containing nature areas, estates, or important nature values (e.g., meadowbirds) are assessed as valuable

Table X.4 (continued)

		division	sources	criterion
p:	nature potency (n.a.)	potentially valuable/not valuable	-	all cells potentially valuable?
q:	extractability of groundwater (n.a.)	extractable/not extractable	-	on the basis of elevation/soil pollution?
r:	extractability of sand (n.a.)	extractable/not extractable	-	indicate only sand collection areas?
s:	transport main structure	linked/non-linked	top. map, own obs.	cells which are linked directly or by a main road with good circulation (speed more than 80 km/h) with a driveway/exit of the A44 are assigned as 'linked'
t:	transport infrastructure	linked/non-linked	top. map	cells with a road (suitable for car traffic) are assigned as linked
u:	ecological main structure	linked/non-kinked	NBP map	over 50% of cell area in the EMS area are assigned as EMS
v:	hydrological main structure	linked/non-linked	top. map	cells with watercourses which are 10 m wide or more are assessed as linked with the hydrological main structure
w:	ecological infrastructure	linked/non-linked	top. map	cells part of the EMS (see u) or the hydrological main structure, or with more than 50% of the cell area within estates
x:	link with seawall	linked/non-linked	top. map	elevation above or below Delta level (exactly equal to l: elevation)

Step 6: Measuring the resource state

The *resource characteristics* (a-x in Table X.4) were transformed into digital variables to facilitate the decision on whether a grid cell has the characteristic or not. For this step, current quality standards for resources were assumed. Information from the literature and maps was collected to establish the characteristics of each grid cell. The source of this information and the criteria for deciding on the presence of each characteristic are given in Table X.4. The assessment of the characteristics per grid-cell resulted in an observation-set reflecting the distribution of resource characteristics within the region.

Since the set of resource characteristics present in each cell is now known, the possibilities for carrying out the activities defined can be analyzed. For the

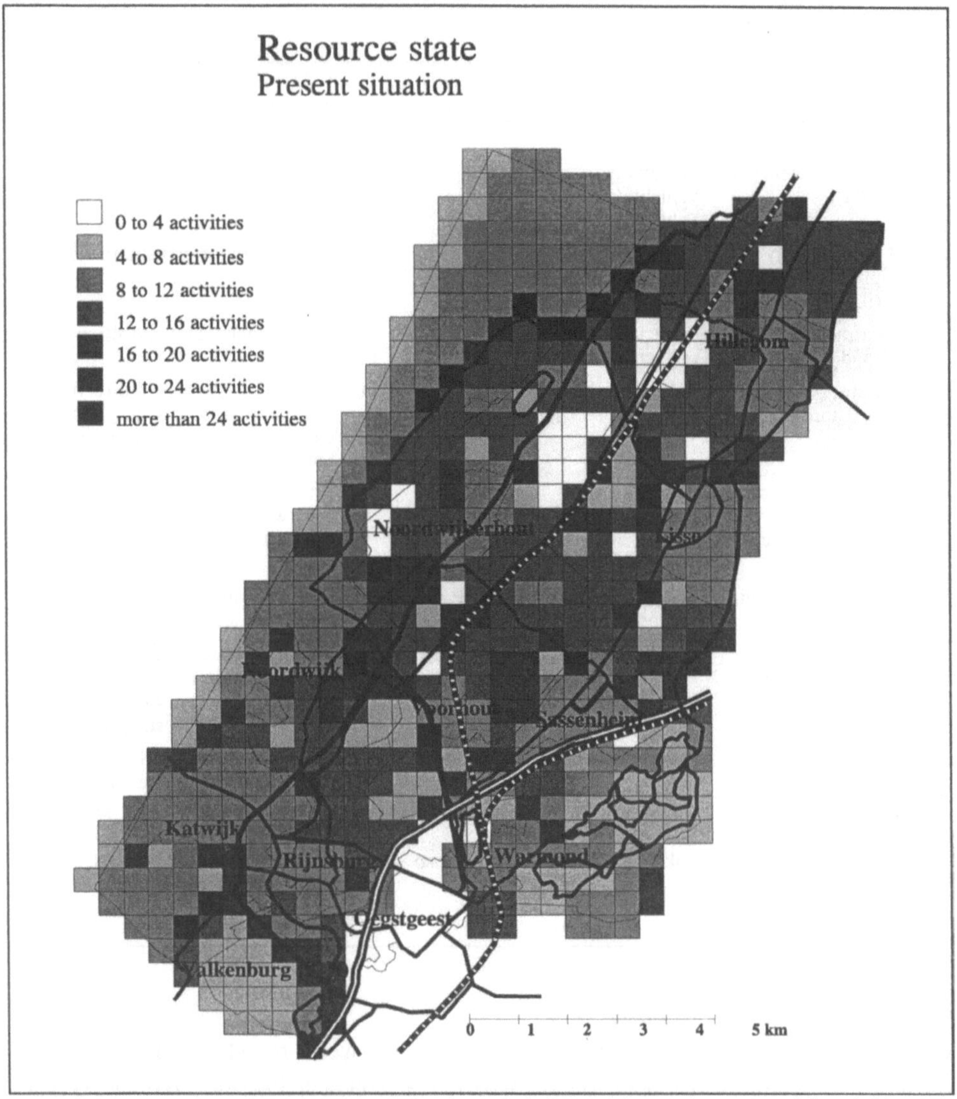

Figure X.5: Present resource state of the dune & bulb-growing region: potential activities per grid cell

analysis, a decision matrix is used in which the required states of the resource characteristics are described for each activity (Table X.5).

The data-set containing the resource characteristics in each grid cell may be seen as the available data on the *present resource state*. Figures X.5-7, which represent this state graphically in three different ways, will be discussed in the next section. It was not possible to find information on all the characteristics, and the

Table X.5: Decision matrix. All characteristics have to be present together to decide that the grid-cell has the potential for the activity (a-x see Table X.4)

Activities Char:	a	b	c	d	i	j	l	m	n	o	s	t	u	v	w	x
Production																
1. bulbs	1	1	0	0	.	.	.	.	.	.	.	1	.	.	.	.
2. dairy products	.		0	0	.	.	.	.	.	.	.	1	.	.	.	.
3. crops	.		0	0	.	.	.	.	.	.	.	1	.	.	.	.
4. industrial products	.		0	.	.	.	.	.	.	.	1	.	.	.	.	.
Processing																
5. waste	.		0	.	.	.	.	.	.	.	1	.	.	.	.	.
Management and realization																
6. bulb farming area	1	1	0	0	.	.	.	.	.	.	.	1	.	.	.	.
7. habitation area	.		0	.	1	.	.	.	.	.	.	1	.	.	.	.
8. recreation area	.		.	.	1	.	.	.	.	.	.	1	.	.	.	.
9. dairy farming area	.		0	0	.	.	.	.	.	.	.	1	.	.	.	.
10. arable farming area	.		0	0	.	.	.	.	.	.	.	1	.	.	.	.
11. industrial area	.		0	.	.	.	.	.	.	.	1	.	.	.	.	.
12. area for storage of waste	.		.	.	.	.	.	.	.	.	.	1	.	.	.	.
13. transport infrastructure	.		.	.	.	.	.	.	.	.	.	1	.	.	.	.
14. ecological infrastructure	.		.	.	1	1	.	.	.	.	.	.	.	.	1	.
15. hydrological infrastructure	.		.	.	.	.	.	.	.	.	.	.	.	.	.	.
16. transport main structure	.		.	0	.	.	.	.	.	.	1	.	.	.	.	.
17. ecological main structure	.		.	0	1	1	.	1	.	.	.	.	1	.	.	.
18. hydrological main structure	.		.	0	.	.	.	.	.	.	.	.	.	1	.	.
19. seawall	.		0	.	.	.	1	.	.	.	.	.	.	.	.	1
Management and reinforcement																
20. landscape	.		.	0	.	.	.	1	.	.	0	.	.	.	.	.
21. cultural history	.		.	.	1	.	.	.	1	.	.	1	.	.	.	.
22. natural history	.		.	.	1	.	.	.	.	1	0	1	.	.	1	.
23. military terrain	.		.	0	.	.	.	.	.	.	1	.	.	.	.	.
24. genes	.	.	.	.	1	.	.	.	.	.	.	.	.	.	.	.
25. species	.	.	.	.	1	.	.	.	.	.	.	.	1	.	1	.
26. ecosystems	.	.	.	0	1	1	.	1	.	1	0	.	1	.	1	.
27. fresh water reserves	.	.	.	.	.	.	.	.	.	.	.	.	.	.	.	.
28. sandy layers	.	.	.	0	.	.	.	.	.	.	.	1	.	.	.	.
29. protection against brackish	.	.	.	.	.	.	1	.	.	.	.	.	.	.	.	.

1= must be present, 0 = must not be present

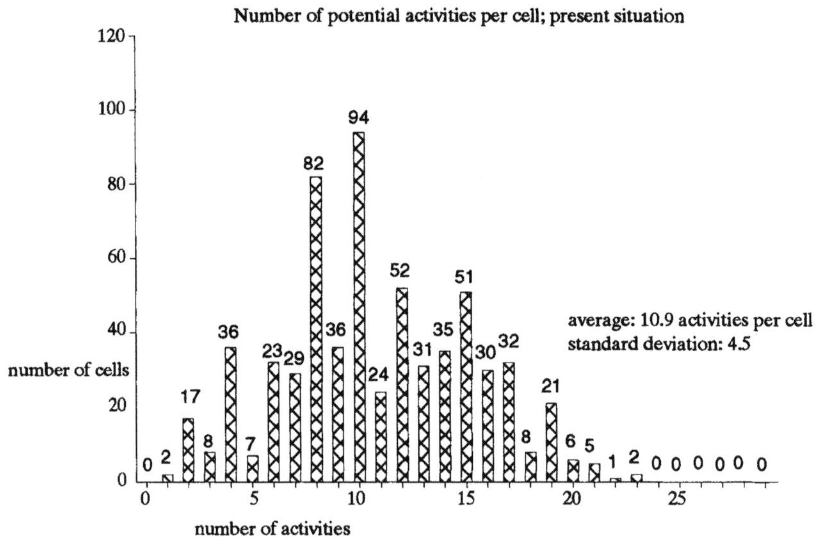

Figure X.6: Present resource state of the dune & bulb-growing region: distribution of number of potential activities per cell

results presented here cannot be regarded as a totally adequate description of the region's *resource state.*

Step 7: Functional units

The proposed procedure prescribes that the *resource state* is to be presented as a table of *functional units*, i.e., a table containing all the combinations of resources occurring in the region, their area, and the potential for carrying out activities. However, no such table of *functional units* is provided here because it was clear that such a table would have been too long. The total of all possible combinations of activities is 2^{29}. Obviously, only a small fraction of all these possible combinations will actually occur in the dune and bulb-growing region: that number can never be exceed 614, the total number of grid cells. And on the basis of Figure X.6 it can be concluded that the number is equal to or higher than 23. In any case, a complete table of *functional units* would be almost unmanageable.

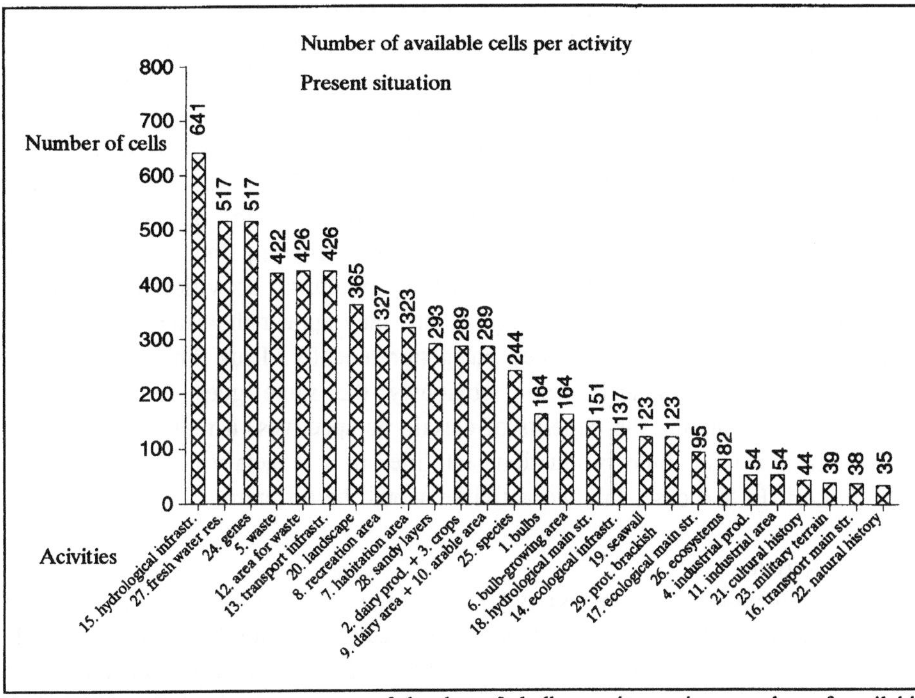

Figure X.7: Present resource state of the dune & bulb-growing region: number of available grid-cells per activity

X.3.5 Exploring development opportunities

In the following several ways of *exploring the resource state* to find development opportunities will be discussed.

Potential activities

Figure X.5 shows the number of activities actually possible per grid cell; the present possibilities for activities are not distributed equally within the dune and bulb-growing region. Some areas have a high potential, notably the inner dune border, the border of Valkenburg, and, to a lesser extent, a large area in the southern part of the region that has Voorhout as its centre. Considering the large number of activities that are possible, many different actors may lay claim to these areas, making competition for space in these areas extremely intense. In contrast, there are other areas with conspicuously few potential activities, such as the area in the northern part of the region along the railway line. This is due to the fact that many activities demand a good link with surrounding areas, which is lacking here.

In Figure X.6 the frequency distribution of the potential activities per grid cell is presented, in order to provide a reference for the impact of changing resource characteristics (see also 'Changing present resource state'). The figure shows that the number of activities actually possible per grid cell varies from 1 to 23, with an average of 10.9 activities per cell.

Figure X.7 shows the number of grid-cells in which specific activities can take place at present. It shows that 323 of the 641 (50%) cells can be used for habitation, 164 (26%) for bulb-growing, and 158 (25%) for the conservation of nature. The figure shows two similar bars for bulb-growing: 'bulb production' and 'bulb area'. The same is true for other agricultural activities and industry. Although the two activities are not equivalent, they appear the same in this figure, because the *resource state* is worked out geographically. Thus 'bulb production' is expressed in terms of area while obviously the same goes for 'bulb area'. However, in calculating the benefits of certain perspectives, it may be necessary to express the potential 'bulb production' activities in a monetary currency and 'bulb area' in km^2. In that case, the two activities would be different.

Scale of activities

In regional planning, it is necessary to assess not only the activities needed in the region, but also the extent to which those activities can be carried out. The *resource state* offers information that can be used in a debate on the quantitative aspects of the activities. From Figure X.7, for instance, it can be concluded that there are few good possibilities for industry in the region, because it requires a link with major transport facilities. This raises the question of whether industrial development in the region should be pursued. There is also little space available for, say, military area. Thus the exploration of development opportunities could start by excluding certain impossible or improbable activities, even if those activities have been previously accepted.

The next step would be to answer the question To what extent should the remaining activities be carried out?. Should, for instance, bulb-growing be expanded spatially? Is nature conservation so important that it is necessary to develop nature corridors between the dunes and the southeastern area which exclude both habitation and bulb-growing? How many housing facilities should the dune and bulb-growing region offer? These questions cannot be answered independently of each other. Representing the *resource state* on maps gives us the chance to both identify and locate the conflict points where those choices must be made.

Opportunities for combining activities

Although the Figures X.5, 6 & 7 illustrate the actual potentials for locating activities, they tell us little about how to combine different activities within a cell.

For the assessment of this potential, more detailed information is needed on the level of separate grid-cells. Not only the number of possible activities must be determined, but the possibilities of certain combinations of activities must also be analyzed, as defined in Chapter V. This means that the *functional units* within the region must be assessed. However, no complete table of *functional units* is provided, because it would have been much too long. Nevertheless, such a table would be needed if it were necessary to describe all possible combinations of activities, and the scale at which these combinations occur. Here the analysis of potential combinations will be elaborated only where bulb-growing, nature conservation, and habitation are concerned. This means that only the *functional units* 'bulb-growing/habitation', 'bulb-growing/nature conservation', and 'habitation /nature conservation' will be described and discussed.

Figure X.8 shows that bulb-growing is currently feasible in the north central part of the region. By contrast, bulb-growing is either not possible or not permitted in the grid cells near the western border and in the southern part of the region. The potential for habitation exists largely in the southern half of the region and along the eastern border. In the areas bordering the north central part of the region both activities are feasible, but only in isolation. Generally speaking, it is not possible to combine the two activities 'bulb-growing' and 'habitation'. Because 'habitation' stands for houses in residential areas, bulbs cannot be grown in places which are inhabited. Conversely, the intensive use of pesticides in bulb-growing renders the surrounding ground, water and air unsuitable for human habitation ('health risk').

This contrast suggests that habitation should be expanded in areas where it, rather than bulb-growing, is feasible. Similarly, bulb-growing should be expanded in areas where it is feasible, and habitation is not. In that way, expanding neither bulb-growing nor habitation hinders the other activity. The question remains, however, whether this results in enough potential for expansion to accommodate both habitation and bulb-growing. The answer would appear to be 'yes'. Habitation occupies only part of its potential area. Moreover, bulb-growing utilizes also only part of its current available space, although its potential for expansion is smaller than that of habitation (compare Figures X.8 & 9).

In about 3% of the grid cells it is possible to combine 'bulb-growing' with 'nature conservation' (Figure X.10). In most of the grid cells only one of the two activities is possible. The reasoning is similar to that in the case of habitation and bulb-growing. In areas with intense concentrations of pesticides in the ground, water and air are incompatible with the resource demands of nature conservation in general, and 'species conservation' in particular.

However, there is obvious potential for combining 'habitation' and 'nature conservation' activities (Figure X.11). Here 'nature conservation' means 'species

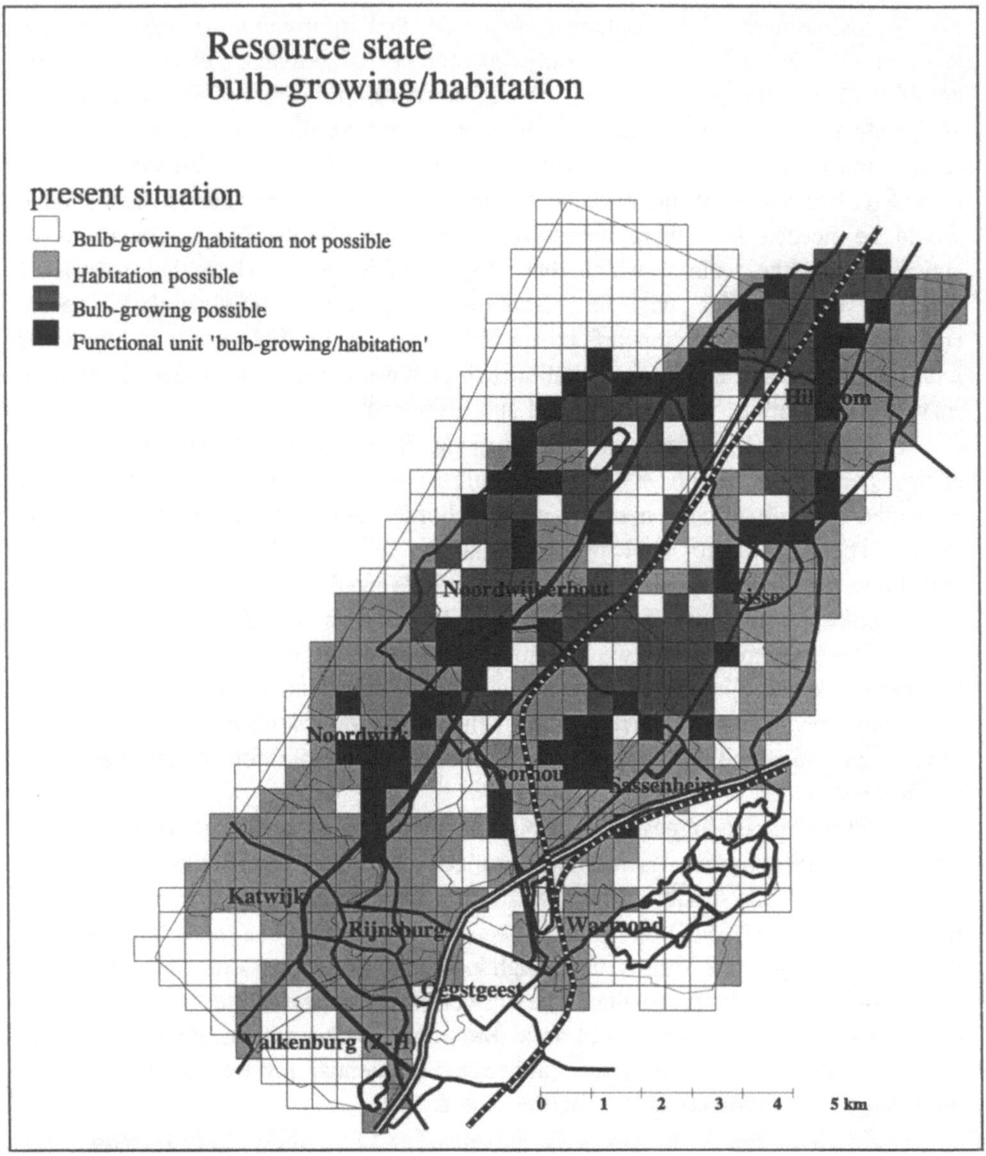

Figure X.8: Present resource state of the dune & bulb-growing region: potentials for bulb farming and habitation

conservation'. In almost 15% of the grid cells both activities are possible.

Moreover, in the context of the present situation it seems to be possible to develop one nature corridor between the dunes and the southeastern part of the region, and another between the southeastern and northeastern border. However, these corridors would run through areas that are also suitable for habitation, and

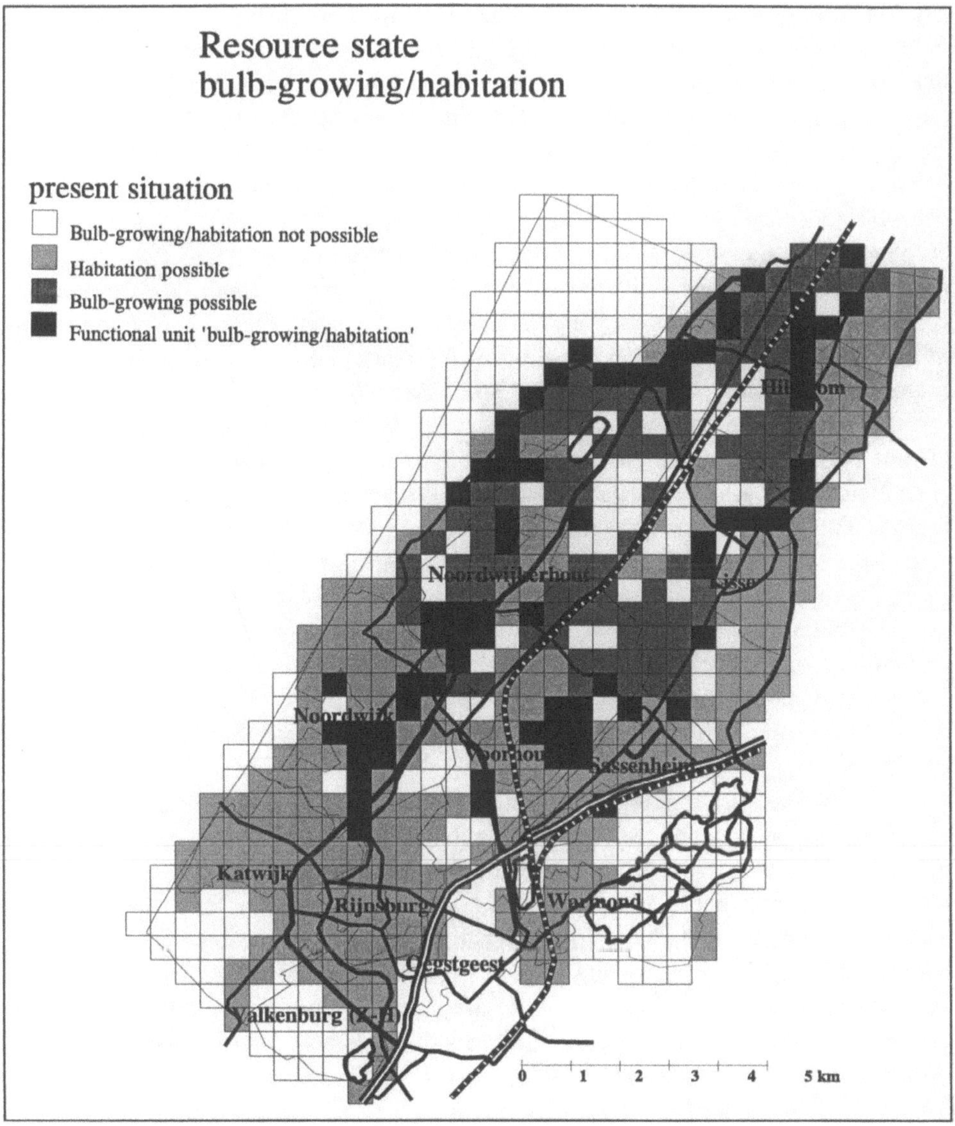

Figure X.9: Present resource state of the dune & bulb-growing region: potentials for expanding bulb-growing

this creates a dilemma. Actualizing these possibilities would render the areas unsuitable for habitation while building houses would make them unsuitable as corridors. This is the result of our assumption that habitation is accompanied by considerable human disturbance which conflicts with the ecological function of the corridors, i.e., with 'nature conservation' in general.

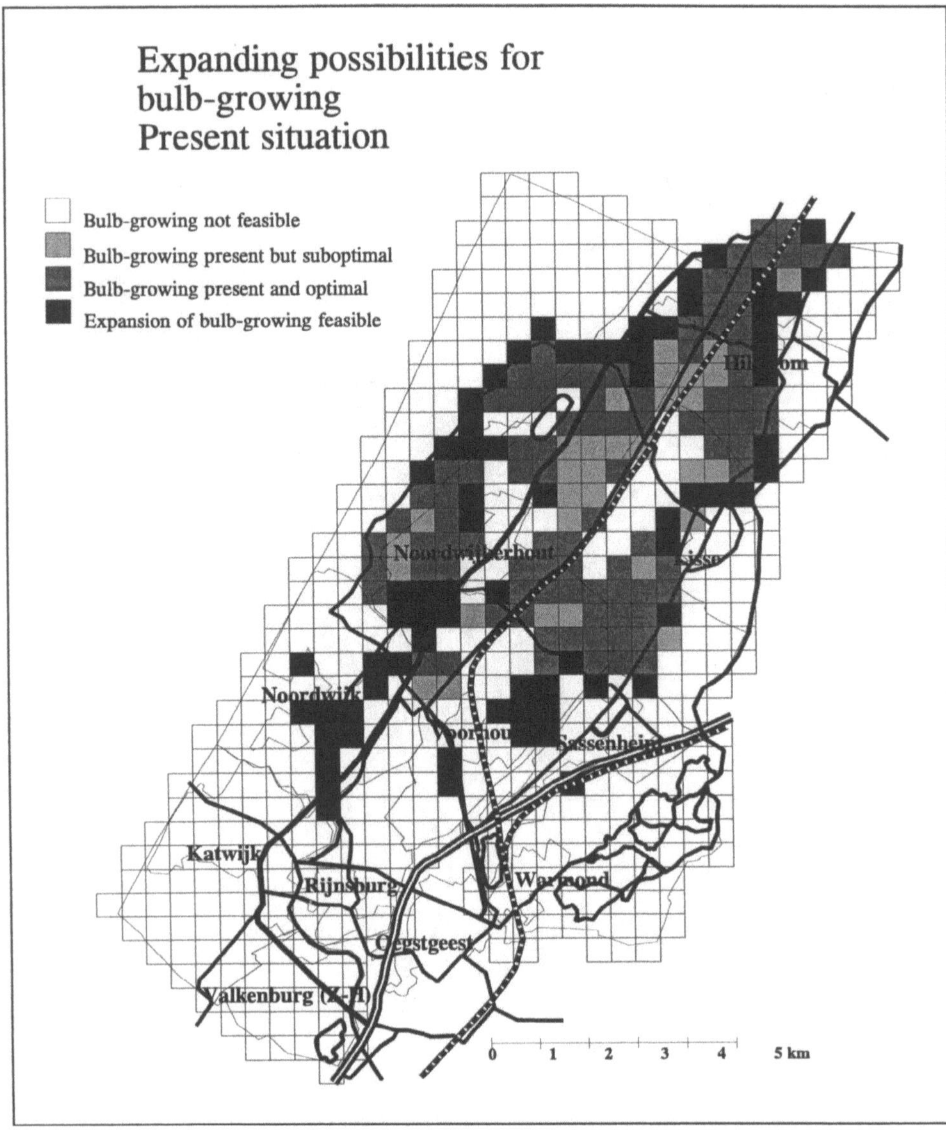

Figure X.10: Present resource state of the dune & bulb-growing region: potentials for bulb farming and nature conservation

The same maps that were used to analyze the conflicts between habitation, bulb-growing, and nature conservation can also be used more creatively, to look for new solutions. If it is decided that the nature corridors are necessary and also that the suitable area is also highly attractive for current modes of habitation, why not look for an alternative type of housing that is compatible with such a corridor?

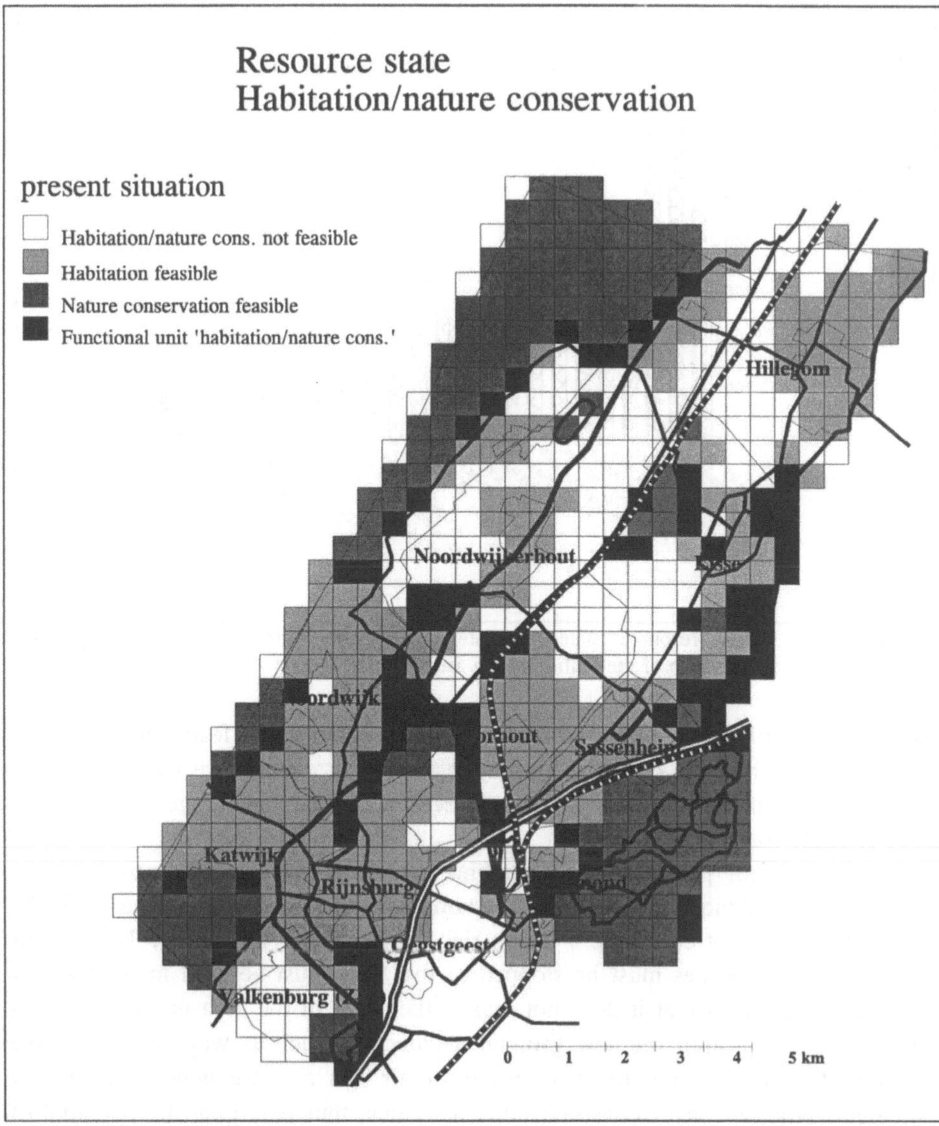

Figure X.11: Present resource state of the dune & bulb-growing region: potentials for habitation and nature conservation

The green parts of suburbs and cities are already rich in flora and fauna (Melchers & Timmermans, 1991; Denters *et al.*, 1994; Rebele, 1994; Melchers & Daalder, 1996). If the challenge of living in a nature corridor is taken seriously throughout the design and construction of new residential areas, combining habitation with ecological corridors might well become feasible (see also Tjallingii's 'strategy of

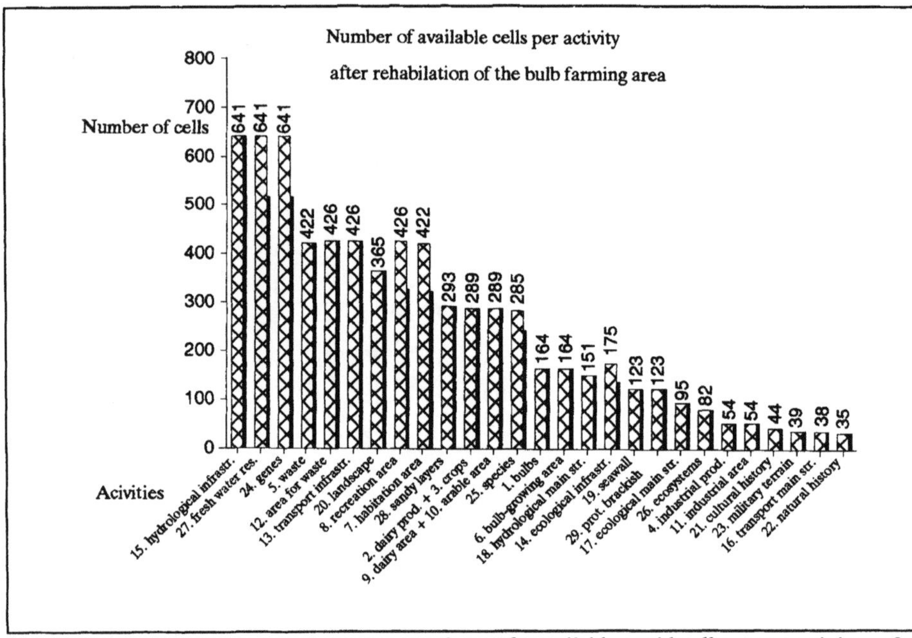

Figure X.12: Future resource state: number of available grid-cells per activity after rehabilation of the bulb-growing area

two networks', 1995 & 1996). For example, green zones could featuring ditches or dense vegetation barriers could be employed to prevent disturbance by residents or their pets, and the 'guardians' of these areas could be encouraged to practice nature-friendly management.

Looking for possibilities to combine activities will also bring to light the need for new techniques. An obvious example is the need for a type of bulb-growing that does not constitute a health risk for people or for nature. This means that the use of pesticides must be stopped or that they must be used in such a way and in such quantities that it does not make other uses of the area unfeasible. This will intensify the search for new growing techniques, such as ways to reduce the emission of pesticides into the soil, water or air. As a consequence, the use of pesticides would no longer constrain other activities, thus enlarging the potential of the region.

New opportunities: Changing the present resource state

Tables and maps make it possible to display the *present resource state* in terms of potential activities. The same is true of the *future resource state* that will become manifest when a set of measures is taken which increases the resource potential. The importance of different type of measures can be demonstrated by assuming the modifications that are opted for in the distribution of the resource characteristics.

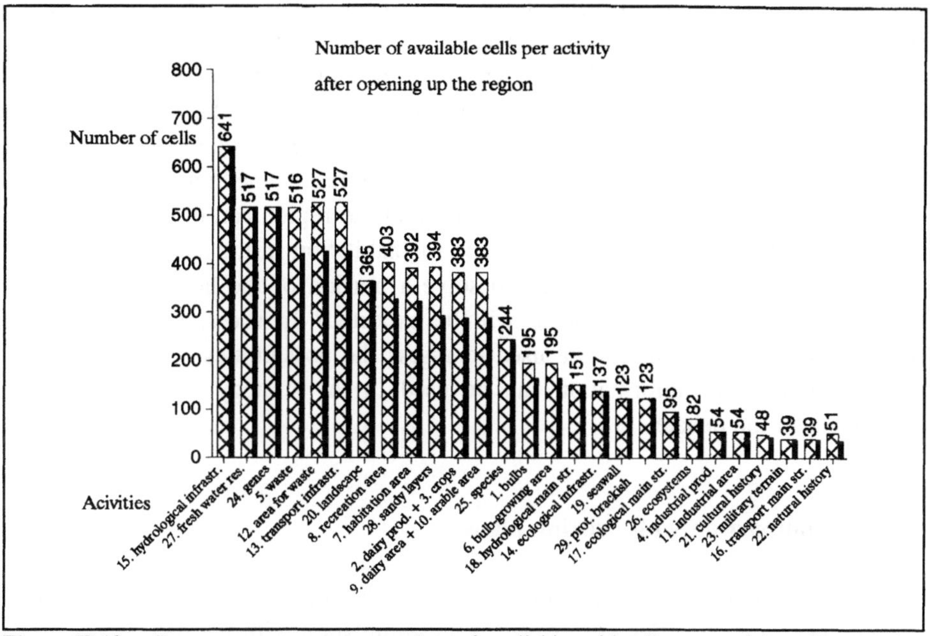

Figure X.13: Future resource state: number of available grid-cells per activity after opening up the dune & bulb-growing region

For example, new technology in the bulb-growing will result in zero grid-cells with pesticide pollution. Or, it can be assumed that all cells will have links to the transport infrastructure. Figures X.12 & 13 show the consequences of these two measures for changing the *resource state*: the rehabilitation of the bulb-growing area, and opening up the region by improving the infrastructure. As expected, each measure has a markedly different impact on the number of potential activities. Rehabilitation by means of new technology (Figure X.12) enlarges the potential for nature conservation, habitation, recreation, and the conservation of freshwater reserves. Opening up the region by making new connections (Figure X.13) enlarges the potential for transport, for agriculture and the production of raw materials, for recreation and habitation, for waste processing and storage, and for the conservation of natural-historical features. The average number of potential activities per cell increases from 10.9 to 11.7 after rehabilitation, and from 10.9 to 12.3 after the region is opened up.

Potential exchange of activities with other systems

Up until now, little attention has been given to the relationship between the dune and bulb-growing region, and other regions. The reason for this omission is that only the *resource state* of this region has been measured, as opposed to that of the west wing of the Randstad or the provinces of South Holland and North Holland.

On the basis of this information alone, we cannot discuss what its role could or should be in the context of the Randstad or those two provinces.

Some of the region's problems can be solved by exporting them, for example, by transferring bulb-growing elsewhere. Will this be acceptable to the inhabitants of those areas? Would acceptance be easier if there was a trade-off, i.e., if the dune and bulb-growing region offered something in return? The *resource state* can also be used to examine what the region can offer. For example, it has been shown that there is still space for habitation in the southern part of the region. Perhaps this could be used to solve a portion of the housing problems of the region to which bulb-growing will be exported? In short, knowledge of the *resource state* of a *SEnS* can be used to assess the possibilities for trade-offs with other regions. This information can be used in the strategy development of an participant in the debate

X.4 Comparison with existing plans

It is not possible to fully compare the results of the exploration of development opportunities performed in this simulation with the development goals proposed in the existing plans. The exploration was not complete and did not culminate in a set of alternative directions of development. The exploration described here is more like a working method than an outcome that can be compared with the results of other procedures. Thus existing plans would be needed that include a description of the way the results were established. However, one of the shortcomings of the documents 'Varianten Nota' and 'Pact van Teylingen' on the development of the region (Section X.3.1.) is that no explicit attention has been given to the properties of the region in relation to the activities to be performed there (Van der Meij *et al.*, 1997). This type of information is especially important for a sound comparison. Nonetheless, in a number of respects the results of the simulation can be compared with existing plans.

- First, the *measured resource state* confirms the description of the problems in the region. It shows that enlarging of bulb-growing area conflicts with habitation, and that in the southern and eastern parts of the area, as well as along the border of the dunes, both nature conservation and habitation are possible. This can also be concluded from the two plans. The *measured resource state* apparently represents a realistic picture of the region, even if it is incomplete. Moreover, the *measured resource state* also reveals some less expected features: bulb-growing can still be expanded within the region, and few areas seem to exist that could be claimed by both nature conservation and bulb-growing to enlarge their area.

- Second, by analysing *functional units* the focus is on combining activities. This calls for an exploration of the possibilities to develop habitation areas within ecological corridors. In the 'Pact van Teijlingen' there is agreement on locating 'green areas' within the urbanized southern part of the region, both to increase the quality of the habitation areas and for 'ecology'. Obviously, one does not have to measure the *resource state* to identify such potential opportunities, but the *measured resource state* shows that the necessary biophysical space does in fact exist and where it is located.
- Third, it is easy to estimate the impact of measures aimed at changing resource characteristics on the *resource state*. Thus, it is possible - though it is not demonstrated in this simulation - to study the *future resource state* by assuming a coherent set of measures and new activities (Chapter XI). This considerably enlarges the possibilities to explore development opportunities, making it possible to answer such questions as How would new bulb-growing techniques change the potential of a region? The existing plans do not answer these type of questions. They result in overall pictures of possible layouts of the region ('Varianten nota') or a set of agreed measures ('Pact van Teijlingen'), but are not keyed to future potential.

X.5 Discussion

The aim of the study was to gain insight into the possibilities to apply the proposed procedure - especially the measurement of a *resource state* - in a real-life situation, in order to find weakness in our procedure. The study simulated a situation in which time and money are constraints. The *present resource state* was measured on the basis of easily available information. A number of practical problems were encountered while this simulation was being carried out.

In *defining the system*, an adequate higher-scale system to represent the interests of the people indirectly involved was not available. The system that first seemed appropriate, the Randstad, is not a governmental entity. The province of South Holland was chosen as the higher-scale entity, because it is the next governmental level that includes the region. But the province of North Holland is also involved in the development of the dune and bulb-growing region, because it has a longer common border with the region than South Holland (the dune and bulb-growing region excluded). This troublesome definition of the higher-scale system could be a common problem in comparable real-life situations. Defining such a higher-scale entity is essential in order to avoid burdening areas that are not part of the *SEnS* with serious side-effects of the development of the *SEnS*. In this case, problems could easily be exported to North Holland, without becoming visible in the analysis of trade-offs. The importance of this problem is demonstrated by the fact that both the 'Varianten nota' and in the 'Pact van Teylingen' suggest extending bulb-growing into the Haarlemmermeer, which is part of North Holland (Provincie Zuid-Holland, 1993; Projectgroep Toekomst Duin- en Bollenstreek, 1995).

In this simulation the *selection of the valued system characteristics* was not carried out in its entirety. Rather, topics from existing plans were fitted into the framework for identifying *valuable characteristics*, and the relevant processes and structures were selected on the basis of these topics. The selected processes and structures were discussed with the ZHMF, and no problems were encountered.

The *measurement of the resource state* was based on a number of steps, starting with the list of processes and structures. The major resources needed for these processes and structures were assessed and the required qualities - the resource characteristics - described. In each of these steps, assumptions and choices were made. This part of the proposed procedure proved to be the most difficult to perform. As no method for selecting resource characteristics is yet available, expert judgement was needed (see for a discussion of this problem in land evaluation Van Diepen *et al.*, 1991). Thus it seemed advisable to seek the support of experts and the people involved. Because all the steps are explicitly documented in the Tables X.3-5, the assumptions and choices can be discussed and their influence on the *measured resource state* can in principle be estimated. For example, Figure X.12 can be also regarded as the representation of the *measured resource state*, if the assumption that the use of pesticides in bulb-growing constrains certain activities is left out.

Since no information was available on a number of variables measuring the *resource characteristics* (Table X.4), the measurement of the *resource state* is incomplete. Consequently, while the measurement is sufficiently worked out for illustrative purposes, the number of potential activities per grid cell may be overestimated. This is probably the case for habitation and nature conservation, but not for bulb-growing, since information was available on all resource characteristics of bulb-growing. More time and money would have solved the problem of incomplete data, while improved availability of information would have increased the quality of the exploration of the development opportunities, although it is unclear to what extent.

A complete table of *functional units* was not generated. The analysis focused on the feasible combinations of bulb-growing, habitation, and nature conservation, disregarding all other possible combinations. In general, the measurement of the *resource state* delivers a large data-set, viz. the resource characteristics of all grid cells (Musters *et al.*, 1994 & 1995). This data-set can be represented in figures and tables in a number of different ways, each showing only a part of the information captured. Which of this information is relevant and which is not depends on the aim of the representation and ultimately on the aims of the proposed procedure laid down during the *start-up*. Choosing the right representation may be crucial for the exploration of development opportunities. On the other hand, the different types of representation can easily be generated from the data-set, thus making possible a kind of 'trial and error' approach.

Since none of the problems encountered is theoretically unsolvable, this study suggests that the *resource state* of any given area can, in principle, be measured in order to provide information for strategy development, although support by experts may be needed. Where resources are limited, the available information may be constraining, but even with limited financial support, it is possible to provide useful information about development opportunities.

Because for this application the spatial *resource state* was measured, the results bear considerable resemblance to conventional land evaluation (e.g., FAO, 1983; Fresco *et al.*, 1995; Chapter V). However, there are a number of differences: (i) land evaluation does not incorporate activities intended to benefit other activities, such as the realization of infrastructure; (ii) land evaluation does not include the prevention of certain activities by other activities, such as nature conservation by bulb-growing; (iii) the possibilities to examine the results of modifying resources are limited in the case of land evaluation, such as the rehabilitation of the bulb area; and (iv) land evaluation centres on optimal land productivity under biophysical, environmental, and social constraints, while the *resource state* focuses on the potential of the region to provide combined resources for the present and the future, and on the potential increase in resources available for new opportunities for development.

The *measured resource state*, a data-set that requires manipulation and interpretation if it is to provide information on opportunities for development, does not in itself appear to be a suitable input for a debate on sustainable development. In any case, the results will only be fruitful if they are used to design a comprehensive set of *development opportunities* for debate and subsequent decision making. Chapters XII and XIII describe applications in which such *development opportunities* are indeed designed.

X.6 Concluding remarks: improvements to the proposed procedure

It is possible to provide useful information about opportunities for sustainable development, even with limited financial support and availability of information. However, during the execution of this simulation, the selection of resource characteristics proved to be a step that includes many choices. A formalized selection scheme was not available to answer such questions as What characteristics are essential for locating activities? And what essential characteristics are relevant for the level of scale of the simulation performed here? In the end, the *resource system characteristics* selected had to be regarded as based on 'expert judgement'. A formalized selection scheme should be designed (based on experiences in land evaluation, Van Diepen *et al.*, 1991) and added to the proposed procedure.

Although certain *functional units* could fruitfully be used for exploring development opportunities, a table of *functional units* could not be presented, due to excessive length. The prescription that the *resource state* should be presented as a table of *functional units* will have to be dropped from the proposed procedure. It may be replaced by recommendations on the use of different possibilities for presenting 'partial information' derived from the *resource state*, as illustrated in the figures.

PART THREE

The Application of the Proposed Procedure within a Decision Making Process

Part Three focuses on a real-life application of the entire information-gathering procedure embedded in a decision making process as proposed in Part One. In the application the desired modifications discussed in Part Two have been incorporated. To monitor the quality of the procedure, a planning project for sustainable development had to be selected into which the information-gathering procedure could be incorporated. At the beginning of the nineties such an opportunity arose. Part Three describes this application, which is aimed at supporting the development of new, multiple-land-use activities A regional approach was chosen, to show the potential benefits of the new land-use activities in practice. Information on the potential benefits is provided, to stimulate the development of such activities and the new technology and organizational structures required.

Applying the procedure to real-life decision making has major consequences for the way methodological problems encountered during the application are dealt with. In a real-life application such problems cannot immediately be supported by extensive additional research, as that would hamper the decision making process. Problems must be solved by expert judgment, and only afterwards can the consequences for the consistency of the results of the procedure be analyzed and discussed. This will indicate which shortcomings in the information-gathering procedure hamper the use of the procedure when external decision making is used to steer the information-gathering. This may give rise to new methodological research aimed at improving and supplementing the methodological basis of the procedure.

CHAPTER XI

Introduction: Opportunities for Multiple Land-Use in Rural Areas

CONTENTS

XI.1 Introduction

In the previous chapters, the information-gathering procedure for sustainable development was applied as an instrument to re-analyze projects (Chapter VII & VIII) and to help parties in the debate on sustainable development to develop their strategy (Chapters IX & X). The studies were performed to find improvements to the proposed procedure. All decisions were made by researchers. For this reason, the decisions on the information-gathering procedure were not embedded in an external decision making process.

The following chapters discuss the execution of the procedure for information-gathering as described in Chapter VI, within the framework of the project 'Sustainable Land-Use' of the interdepartmental programme for Sustainable Technological Development (STD, 1991; Weterings & Opschoor, 1992; STD, 1992; De Kuijer, 1995; Musters & De Graaf, 1996; De Graaf, & Musters, 1997; De Kuijer *et al.*, 1997a & b).

The project presented here is not aimed at producing a regional design for land-use that will actually be implemented. Options for land-use are presented, not as blueprints for realization, but to show the potential benefits of new land-use activities. This is supported to stimulate the development of such activities, while identifying the new technologies and organizational structures required. In that sense, this project is not a land-use planning project, but rather a development-stimulation project.

Since the information-gathering procedure described in Chapter VI is a theoretical one, embedded here in practical, external decision making, there may be some tensions between the need to carry out the information-gathering as prescribed, and the directives formulated by the decision makers within the process. This raises the major question to be answered in Part Three: Can the phases proposed in Chapter VI be carried out as described, and is the procedure effective and efficient in a context where gathering support and stimulating development are the overall aims, and where the decision making process constrains the available time and money for information-gathering? This question can be specified still further: Can a list of *valued system characteristics* be derived from interviews with the people involved? Can the *resource state* be measured? Can *development opportunities* be explored? Can information about *valued system characteristics*, the *resource state* and its use, and *development opportunities* be presented in a form suitable for decision making? Is the proposed procedure sufficient and flexible enough to respond to the decisions made?

XI.2 Method for analyzing the application

A real-life application cannot be compared with the results of an alternative procedure within the same decision making context, because such an alternative procedure cannot be carried out simultaneously without mutual influence, even assuming that this is financially feasible.

Furthermore, choosing a real-life application has major consequences for the manner in which methodological problems are solved. They cannot be solved by extensive additional research, because that would stagnate the decision making process. Any problems encountered must be solved immediately by expert judgment; later on, the consequences for the soundness of the procedure can be analyzed and discussed, in order to identify any shortcomings in the information-gathering procedure. The conclusions may give rise to new methodological research designed to improve the methodological basis of the proposed procedure. It is such an analysis of the quality of the proposed procedure within an external decision making process which is performed here.

A project for sustainable development had to be found in which the information-gathering procedure proposed Chapter VI could be applied. In the early nineties, we ourselves became involved in such project. The Dutch government started an interdepartmental programme for Sustainable Technological Development (the STD programme) which consisted of five subprogrammes: (1) food, (2) chemistry, (3) transport, (4) housing, and (5) water (De Kuijer, 1995; STD, 1997).

When the participants in the subprogramme 'Food' decided to focus one of the projects on sustainable multiple land-use, the department of Environmental Biology of Leiden University was invited to participate in the process of defining 'sustainable multiple land-use' and, later, a spatial exploration of the extent to which multiple land-use was possible after the present problems involved in combining activities had been solved (Chapter XII). We also took part in an exploration of the potential benefits of multiple land-use on a regional level (Chapter XIII). Other researchers were invited to carry out these steps simultaneously on the enterprise level (Aarts *et al.*, 1996; Aarts & De Kuijer 1997).

The participation in project 'Sustainable Multiple Land-use' of the subprogramme 'Food' was of particular interest, because in principle it provided the possibility to apply the procedure proposed in Chapter VI. The actual application of the proposed procedure within the project 'Sustainable Multiple Land-use' provided the material by which the quality of the procedure was studied. The application itself and the experience with the execution of the application are described in Chapter XII & XIII. Special attention will be given to the influence of external decision making on the application of the procedure. The working hypothesis is that the proposed procedure can be applied within a decision making process that constrains both time and resources, and that it will produce results that facilitate the subsequent stages of that process. Since these results cannot be compared with those of an alternative procedure within the same decision making process, simply because no alternative results are available, the quality of the results cannot be assessed. However, the analysis and discussion of the application may point up necessary improvements to the proposed procedure (Chapter XIII).

In the present chapter, the project will be introduced (Section XI.3), and the research strategy presented (Section XI.4).

XI.3 Background: the programme 'Sustainable Technological Development'

XI.3.1 The need for new techniques

Initiated by the conclusions drawn in the reports of the World Commission on Environment and Development (WCED, 1987), the Dutch Government embraced the concept of sustainable development and formulated a policy aimed at operationalizing this concept. It was recognized that sustainable technological development is indispensable in solving or preventing economic, sociocultural, or environmental problems. Opschoor & Weterings (1992) illustrate this when they estimate that if the environmental impact is to be halved during the next 50 year, techniques will be needed that make the production and consumption of goods and services possible while causing twenty times less environmental impacts per unit of production or consumption than todays. This estimate used very crude simplifications, which were based on Commoner's scaleless and timeless formula, I=PAT, i.e., 'Impact' (environmental strain) equals 'Population size' times 'Affluence' (average per unit of prosperity per person) times 'Technology' (environmental strain per unit of prosperity). It demonstrates that even under assumptions of very mild population growth, limited welfare-growth in developed countries, and of a sound global distribution of prosperity, significant and wide-reaching technological changes are needed to prevent negative environmental impacts from escalation.

XI.3.2 Governmental policy

The Dutch government has recognized this need for new techniques and in 1993 initiated the Dutch interdepartmental programme Sustainable Technological

Development (STD-programme, 1996 & 1997). It was set up by the Ministries of Economic Affairs, Agriculture, Nature Management and Fisheries; Education, Culture and Science; Public Works and Water Management; and Housing, Spatial Planning and Environment (STD, 1991). They joined forces and started an overall programme that consists of five subprogrammes: (1) food, (2) chemistry, (3) transport, (4) housing, and (5) water. The subprogramme 'Food' was further subdivided into five projects: (i)'Sustainable Land Use'; (ii) 'High-tech Agroproduction'; (iii) 'Whole Crop Utilization'; (iv) 'Novel Protein Food'; and (v) 'Sensor Technology'. (STD, 1996 & 1997). The 'Sustainable Land Use' project is not aimed at developing technology only for producing food, but was aimed at new land-use activities that produce more than food alone: a 'sustainable multiple land-use'.

XI.3.3 Development process

In view of the technology to be developed, the questions were not only 'what is possible' but also 'what is desirable and acceptable' (STD, 1992). Thus sustainable development is regarded not as a purely technical issue, and the technology development for sustainable development is not primarily an activity carried out by research institutes (STD, 1997), but a process in which from the beginning parties that have interests in development should take part. The process starts with (i) an inventory of the environmental problems, followed by (ii) a rough design of a sustainable future by the people involved, i.e., they first determine the goals of a specific development: the products desired and the problems to be solved. Then the process aims at (iii) joining forces and strengthening the cooperation between the different people involved to analyze the opportunities to achieve these goals in a backcasting process, and (iv) carry out the necessary activities to realize these opportunities in a process in which it is demonstrated that new production techniques can be developed that reduce the impact of production techniques by a factor 10-20. If this process shows that goals cannot be achieved because of lack of technology, and inadequate organization structures, the technology needed must be developed and structures changed.

XI.4 The project 'Sustainable Multiple Land-use'

The subprogramme 'Food' of the STD programme (De Kuijer, 1995; STD, 1997; De Kuijer *et al.* 1997a) began in the early nineties with an analysis of the environmental problems related to the current food supply, and a description of future food demands (step 1 in Box XI.1). Organizations involved in food research, production, and consumption were consulted during the drafting of a qualitative design for a sustainable future (step 2 in Box XI.1). During the next step these organizations suggested focusing one of the projects of the STD programme on 'Sustainable Multiple Land-use' (step 3 in Box XI.1). The subsequent steps were aimed at identifying the opportunities for new land-use activities, in order to reach consensus on the actual development of such activities.

Step 1: *Strategic analysis of present problems*
- analysis of present food supply, and its environmental impact
- expected development in terms of food demands
- assessment of present environmental impact on separate types of food products

Step 2 *Qualitative design of a sustainable future*
- consultation of governmental and private commercial and non-profit organizations
- assessment of future food supply demands
- inventory of directions for research into new technology

Step 3 *Backcasting*
- inventory of building stones for a sustainable food supply
- selection of building stones:
  - (i) **Sustainable Multiple Land-Use**
  - (ii) High-tech Agroproduction
  - (iii) Whole Crop Utilization
  - (iv) Novel Protein Food
  - (v) Sensor Technology

Step 4 *Definition of building-blocks, using the example of Sustainable Multiple Land-Use*
- goals and constraints
- description of scale level of sustainable land-use
- identification of technologies
- selection of regions

Step 5 *Illustration process*

a *Exploration of the potential to combine land-use processes*
- in the rural area at regional and enterprise level and plots

b *Exploration of the potential benefits of sustainable multiple land-use*
- design new land-use activities on the basis of assumed new technology
- determine potential contribution to a spectrum of products
- exploration on regional and company level

Step 6 *Start of demonstration and development projects aimed at sustainable multiple land-use*

Box XI.2: The process strategy of the research programme Sustainable Technological Development applied to the subprogramme 'Food' and elaborated for the project 'Sustainable Multiple Land-Use' (De Kuijer, 1997).

XI.4.1 Premises

One of the fundamental premises of the 'Sustainable Multiple Land-use' project is that rural activities with more optimal use of all available resources - which is why they are called 'multiple' land-use activities - use fewer resources and have a smaller environmental impact per unit of overall production than a set of conventional land-use activities that together produce the same package of products.

The project 'Sustainable Multiple Land-use' was not to be aimed at developing new techniques for producing food, but rather at developing new land-use activities that 'produce' more than food alone. In other words, 'sustainable multiple land-use' is intended to provide a broad spectrum of products and values which are desired by people involved in the development of rural areas. This is particular relevant in the case of densely populated areas where space is limited, and where a broad spectrum of products and values are supplied from relatively small rural areas, as in the Netherlands (Johnson, 1995; Brinkman, 1995).

The second fundamental premise of the project 'Sustainable Multiple Land-use' is that it is not sufficient to know that new multiple-land-use activities are technically possible. When such activities are available, they must actually be applied within enterprises. Because they are new, it may be necessary to change current production structures and develop new structures, creating new forms of mutual cooperation between producers. Therefore, the project was aimed not only at the development of new activities, but also at the development of new production and cooperation structures, tailored to the need to combine the different activities within new enterprises. To be able to steer developments, monitoring was regarded as of crucial importance.

On the basis of these premises, the concept of 'sustainable multiple land-use' was further elaborated (Step 4 in Box XI.1). The subprogramme 'Multiple Land-use' of the STD programme centred on the development of new land-use activities that have a good chance of replacing conventional activities. It must be shown that all the costs of new land-use activities will be more than recouped by the benefits gained from the replacement of the existing land-use activities by new ones. Furthermore, in order to stand a chance of replacing existing land-use activities, the new ones must be socially acceptable in the sense that they enable land-users to make a decent living. Therefore, the new activities must be economically feasible as well as efficient. In addition to being economically viable and socially acceptable, the new activities must also respect the cultural and historical characteristics of a given rural area. These characteristics may be regarded as 'products' of land-use activities, not only because they are often highly valued side-effects of traditional activities, but also because they have an economic potential as attractive recreational features.

All the various demands on the new land-use activities indicate that a 'sustainable development' approach is needed. The challenge of the 'Sustainable Multiple Land-use' project is: How to develop new land-use activities that deliver a broad combination of products and values, and are at the same time ecologically, economically and socially sound, contribute to solving the current environmental problems, and do not have negative side-effects?

XI.4.2 Definition of sustainable multiple land-use

Land in rural areas is often used to produce only one kind of product at a time, e.g., food, energy, drinking water, or wildlife. For this purpose, only those characteristics of the land are used that maximize production. It has often been argued that if other characteristics of the same land are used to produce many different products at the same time, the environment would be used more efficiently and the overall productivity would increase per unit of environmental impact (e.g., WRR, 1980, 1983; Folke *et al.*, 1996; Giampiettro, 1997; Vandermeer *et al.*, 1998). In the agricultural literature, this alternative form of land-use has been called 'multifunctional agriculture' (e.g. Jurgens, 1993; Vereijken, 1997a), 'multi-purpose agriculture' (e.g. Van Mansveld & Mulder, 1993), or 'multi-objective farming' (Vereijken, 1997b). Here we will employ the term 'multiple land-use' to stress the fact that the concept refers to the optimal use of all available resources and that it is not necessarily concerned with agriculture (e.g., Johnson, 1995).

Multiple land-use can be defined at different levels of scale (Giampietro, 1997; Vereijken, 1997a; Vandermeer *et al.*, 1998). At the regional level, it is the optimal use of regional resources and the sound coordination of the inputs and outputs of subregions or enterprises. At this level, subregions or enterprises may be monofunctional, but the use of the region as a whole is multiple when all resources are used optimally and losses are minimalized. However, the focus can also be on a lower level of scale, for example, the enterprise level. In that case it may be defined as fine-tuning land-use within a enterprise to small-scale differences in land characteristics, such as soil, availability of water, and exposure to sunlight. It may lead to a small-scale mosaic of monofunctional or multiple-use plots, interconnected and dependent upon each other. Multiple land-use can also be attained at an even lower level of scale, for example, when each square meter of rural area is used to provide food, to offer space for wildlife, and to collect drinking water.

It will be clear, that at each level of scale, multiple land-use demands

different techniques and production structures. However, for decades development in agriculture, nature conservation, resource management, and industry has been aimed at specialization and segregation, resulting in an almost exclusive focus on technological development for purpose of monofunctional land-use (Giampietro, 1997; Vandermeer *et al.*, 1998). Modern techniques for multiple land-use are few and far between (Giampietro, 1997). In order to develop these techniques, new forms of land-use activities have to be designed that can make more efficient use of the available resources - space, air, soil, water, energy, and raw materials - while delivering the products and values desired by the people. Once a concrete picture of these new land-use activities is available, the techniques that are needed to realize them may be developed. Creating that picture of new, multiple-land-use activities is the aim of the studies described.

XI.4.3 Selection of a region

To minimize the risk that such a process would result in purely theoretical solutions, it was seen as essential to select a specific rural area for which the new land-use activities could be developed (the start of Step 5) In other words, an illustrative process was proposed for the development of new land-use activities.

The search in the project 'Sustainable Multiple Land-use' turned to regions where new land-use activities could actually be developed. The search focused on different regions, because they are likely to offer a different potential to realize multiple land-use on the different levels of scale, depending on such characteristics as soil diversity, water supply, infrastructure, etc.

Two types of regions were characterized as being suitable for selection. The first consisted of regions where it was possible to focus on the fine-tuning of the inputs and output of separate enterprises, thus reducing resource use and losses at a regional level of scale. The second type included regions in which the emphasis can be placed on realizing multiple land-use on a lower level of scale, i.e., on using all the resources available within enterprises or even within plots.

In selecting regions for a development project, the following criteria were used:

- The region is facing serious environmental problems.
- The people involved are more or less organized, in order to minimize the risk of starting an unmanageable process.
- There is a real chance that the people involved will participate in the process of developing new land-use activities.
- The new land-use activities to be developed are also applicable in other regions, to justify investments in the development.

Two possible regions were selected: a polder area in the middle of the Netherlands and a sandy-soil area in the eastern part of the country. Both regions met the above criteria, but the choice fell on an area around Winterswijk, a small town in the province of Gelderland, within a sandy region in the eastern part of the country (Figure XI.1).

Winterswijk was considered a good study area to illustrate the process aimed at the development of new multiple-land-use activities at the level of enterprises and plots. This decision was based on the following considerations.

- First, the region was facing serious problems. As in many other sandy areas in the Netherlands, there are major socioeconomic and environmental problems (Hetsen, 1993; WCL Winterswijk, 1996).
- Second, the region also had real potential for multiple land-use, because of the diversity of its small-scale landscape elements and its biophysical and hydrological characteristics. In addition to agriculture, it had great potential for collecting drinking water, creating recreational locations, preserving nature, and stimulating recreational activities based on its attractive cultural and historical landscape features (Grondmij/Biologisch Station Zwillbrock, 1994; Provincie Gelderland, 1996a; Veen & Jorna, 1995; Zandvoort Kappelhoff Advies, 1994).
- Third, an institutional structure already existed that could be used to identify and consult the people involved or their representatives, i.e., the stakeholders. The region had been designated a 'Valuable Man-made Landscape' (WCL-Winterswijk) by the Dutch government (LNV, 1992), and many stakeholders were already working together within a foundation to explore, stimulate and exploit the potential of the region (Box XII.2; WCL-Winterswijk, 1996).
- Fourth, the stakeholders consulted - such as agricultural, nature and governmental organizations - had demonstrated an interest and could be expected to cooperate in the search for new technology.
- Fifth, the solutions - new multiple-land-use activities - would be of value not only for this particular region, but also for many other sandy regions.

XI.5 Survey

Chapter XII describes the study aimed at assessing opportunities for multiple land-use (Step 5a in Box XI.1). The decision making at the end of that study led to the start of a second study designed to explore the potential benefits of new land-use activities (Step 5b in Box XI.1). The results of that study are described in Chapter XIII.

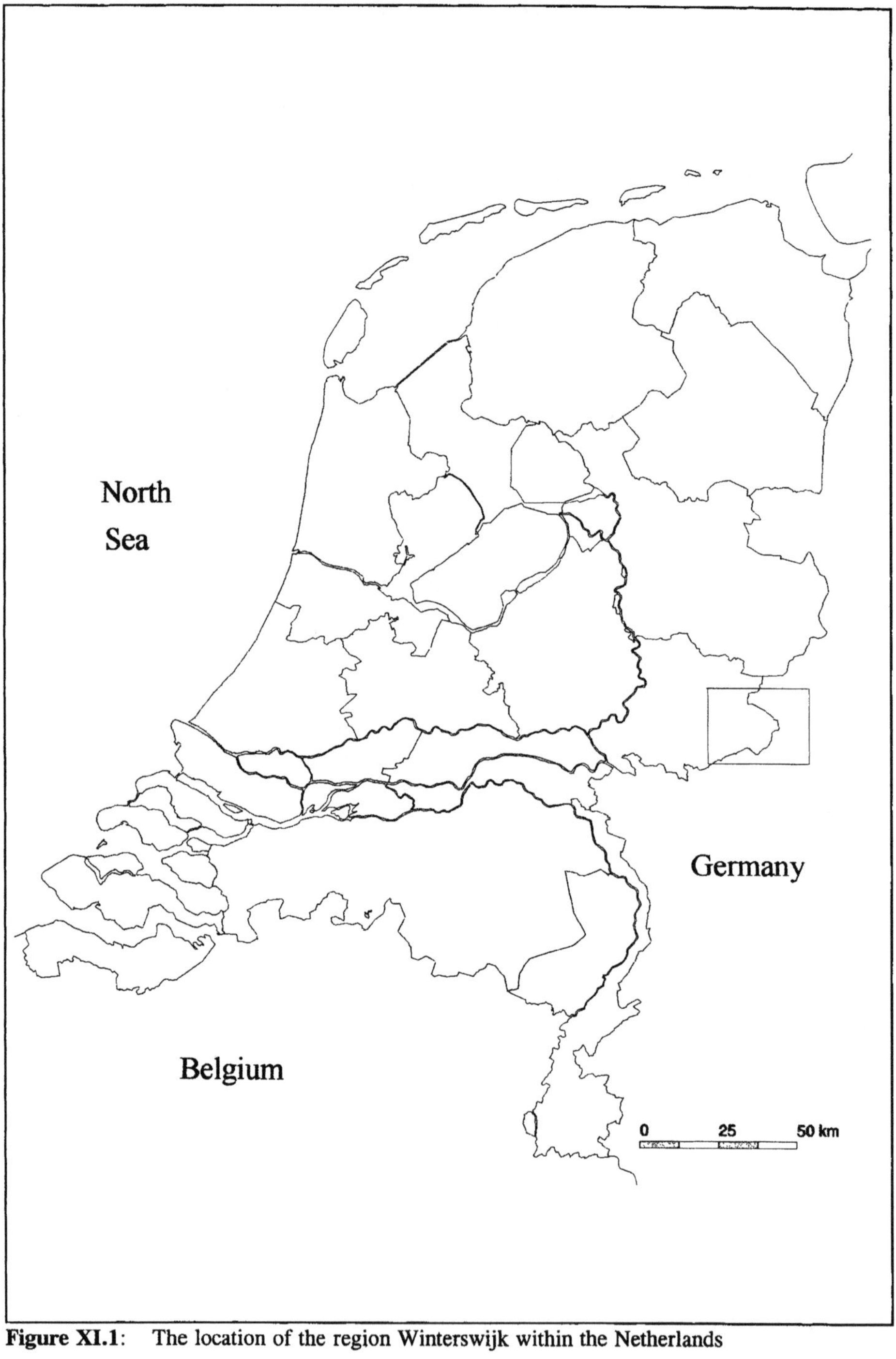

Figure XI.1: The location of the region Winterswijk within the Netherlands

Organizations united in the foundation Valuable Man-made Landscape Winterswijk (WCL-Winterswijk)

- municipality of Winterswijk
- municipality of Eibergen
- municipality of Lichtenvoorde
- municipality of Aalten
- local agricultural organizations
- local branches of rural women's organizations
- Platform for Nature and Landscape of Winterswijk
- Society for the Preservation of Nature
- Foundation Het Gelders Landschap
- study-group NSW Estates of Winterswijk
- local recreation entrepreneurs
- Association for Tourist Information Winterswijk
- organization of recreation boards

Advising members

- Ministry of Housing, Spatial Planning and Environment
- Ministry of Agriculture, Nature Management and Fishery
- Waterboards
- Recreation Board Oost-Gelderland
- Province of Gelderland

Box XI.3: Organizations collaborating in the Foundation Valuable Man-made Landscape Winterswijk (the 'Stichting WCL Winterswijk'), to explore, stimulate and exploit the potential of the area (WCL Winterswijk, 1996)

Each of the two subsequent steps (5a & b) may be regarded as a case study for the application of the procedure proposed in Chapter VI; they are both presented as part of the discussion on the quality of the proposed procedure within a decision making process characterized by temporal and financial constraints. Consequently, the results of all phases of the procedure will be presented in both studies, i.e., the *start-up* phase, the *definition of the socioenvironmental system*, the *selection of valuable system characteristics*, the *measurement of the resource state* and its use, and the *exploration of development opportunities*.

In these two applications the major challenge was to formulate, analyze, and present the information needed. The information required for each step and the method used to gather information will be described. Several organizations collaborated with us in collecting and processing the data.

CHAPTER XII

The Exploration of the Potential to Combine Land-use Processes

CONTENTS

XII.1 Introduction

The present chapter discusses the execution of the procedure to explore the potential to combine land-use processes, within the framework of the project 'Sustainable Land-Use' of the interdepartmental programme for Sustainable Technological Development (Musters & De Graaf, 1996). As noted in Chapter XI, the project presented here is aimed at options for land-use, not as blueprints for realization. It is not a land-use planning project, but rather a development-stimulation project that shows the opportunities of new land-use activities.

The following phases of the procedure for information-gathering will be described.

1 *Start-up*
2 *Defining the socio-environmental system*
3 *Selecting the valued system characteristics*
4 *Measuring the resource state*
5 *Exploring the development opportunities*

XII.2 Start-up

The idea of starting the development of sustainable multiple land-use in Winterswijk was discussed with the foundation WCL-Winterswijk, who agreed to participate in the organization in the following steps of the project.

The aim of the present study within the project was formulated as: (i) to reach consensus among the stakeholders on the necessity to demonstrate the potential benefits of new land-use activities, (ii) to stimulate the readiness of stakeholders to take responsibility for, and to cooperate in, the development of new land-use activities, and (iii) to raise funds to make this development possible.

XII.2.1 Research strategy

As stated in Chapter XI, one of the major guiding principles of the 'Sustainable Multiple Land-use' project is the assumption that new forms of land-use can only be successfully developed with the cooperation of the people who will have to invest in those activities. This strategy will increase the likelihood that the new land-use activities will be tailored to the potential users, and that they will actually be carried out. However, the readiness of the stakeholders to invest money, skills, and time will depend on the benefits they expect to gain from their investment. But how can this information on the opportunities of new land-use activities be provided, when the activities themselves are not yet available, and may be constrained by present problems? To avoid an impasse, the technique of backcasting was applied, in which a picture of future land-use is presented which entails the solving of present problems and shows the spatial potential of multiple land-use. This is designed to motivate people to start the actual development of new land-use activities. Looking back from that picture of the future to the present, the necessary changes can be determined and the new land-use activities developed.

XII.2.2 Working plan

The programme agency of the STD programme and the representatives of the WCL decided to start the process by studying the spatial potential of multiple land-use (Phase 1 in Figure XII.1). The STD steering group and regional stakeholders would then decide on a follow-up: an assessment of the potential benefits of new land-use activities on the enterprise and region level. Because of the nature of these aims, the management of the project was entrusted to a project team which included not only researchers and the STD programme management, but also the local organization for the improvement of the man-made landscape WCL (Box XII.1). The project team would continue to define the *SEnS* (Phase 2 in Figure XII.1; Aarts *et al.*, 1996; Musters & De Graaf, 1996). In order to accomplish the aims of the study, it was considered crucial to consult with various organizations on the direction of change, and the goals, means, and activities they desired (Phase 3 in Figure XII.1). For consensus-building a strategy was followed in which the stakeholders were informed about the progress of the study, and were regularly consulted, so that their needs and wants could be taken into account, and to stimulate the change in attitude with regard to developments. At the end of the study, they were both informed and consulted about the opportunities for multiple land-use in the region (external decision making in Figure XII.1). A picture of future land-use that entails solving the present problems and demonstrating the spatial potential of multiple land-use had to be generated and presented, based on exploration of the *resources* available in the surroundings of Winterswijk (phase 4 and 5 in Figure XII.1). At the same time, a broad description was to be presented of the type of land-use systems to be elaborated.

XII.3 Defining the rural area of Winterswijk as a socioenvironmental system

After formulating the research strategy and the working plan, the region Winterswijk had to be defined. The Dutch municipality of Winterswijk consists of

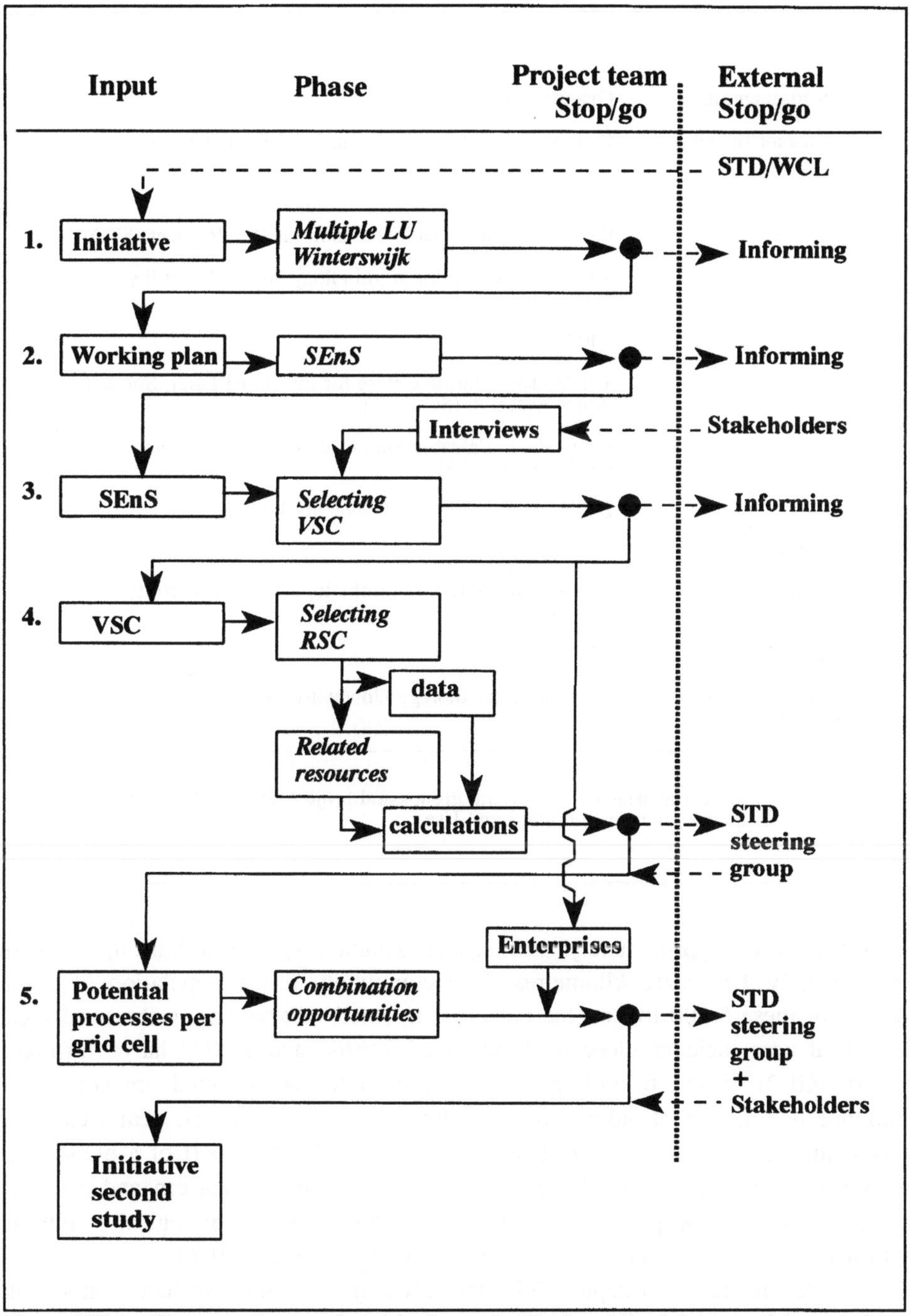

Figure XII.1: The procedure employed in the study 'Exploration of the potential to combine land-use processes'.

Project management	**Organization**
project leader secretary	Programme Sustainable Technolocal Development
Project team	Programma Sustainable Technolocal Development (STD)
	Research Institute for Agrobiology and Soil Fertility (AB-DLO)
	ARCADIS
	The Winand Staring Centre for Integrated Land, Soil and Water Research (SC-DLO).
	Foundation Valuable Man-made Landscape Winterswijk (WCL-Winterwijk)
	Environmental Biology University Leiden
Working groups	
- Technology	Research Institute for Agrobiology and Soil Fertility (AB-DLO)
- Communication	ARCADIS
- Development and monitoring	Environmental Biology University Leiden

Box XII.1: Project management, working groups and project team of the STD, Subprogramme Multiple Land-Use.

a rural area of approximately 140 square kilometres, and a built-up area of approximately 10 square kilometres. The community has a population of about 30,000; of these 5,500 live on the outskirts, in small villages and in the rural area. This rural area includes close to 2000 ha of forests, and 11,000 ha of farmland (Figure XII.2). Some 6 million m^3 of groundwater is collected annually. The landscape is a mosaic of old and recent cultivation, country estates, peat areas, and woods, and is criss-crossed with hedgerows and brooks. It is part of a pleistocene sandy-soil area that includes large portions of the country's eastern and southern districts. The landscape is considered valuable from a biological, cultural, historical, and recreational point of view (WCL-Winterswijk, 1996).

As noted in Chapter XI, the area has serious socioeconomic and environmental problems. The topsoil layer is high in nitrates and phosphorus, too much ammonia is deposited, and large parts of the area are overly dry (Provincie

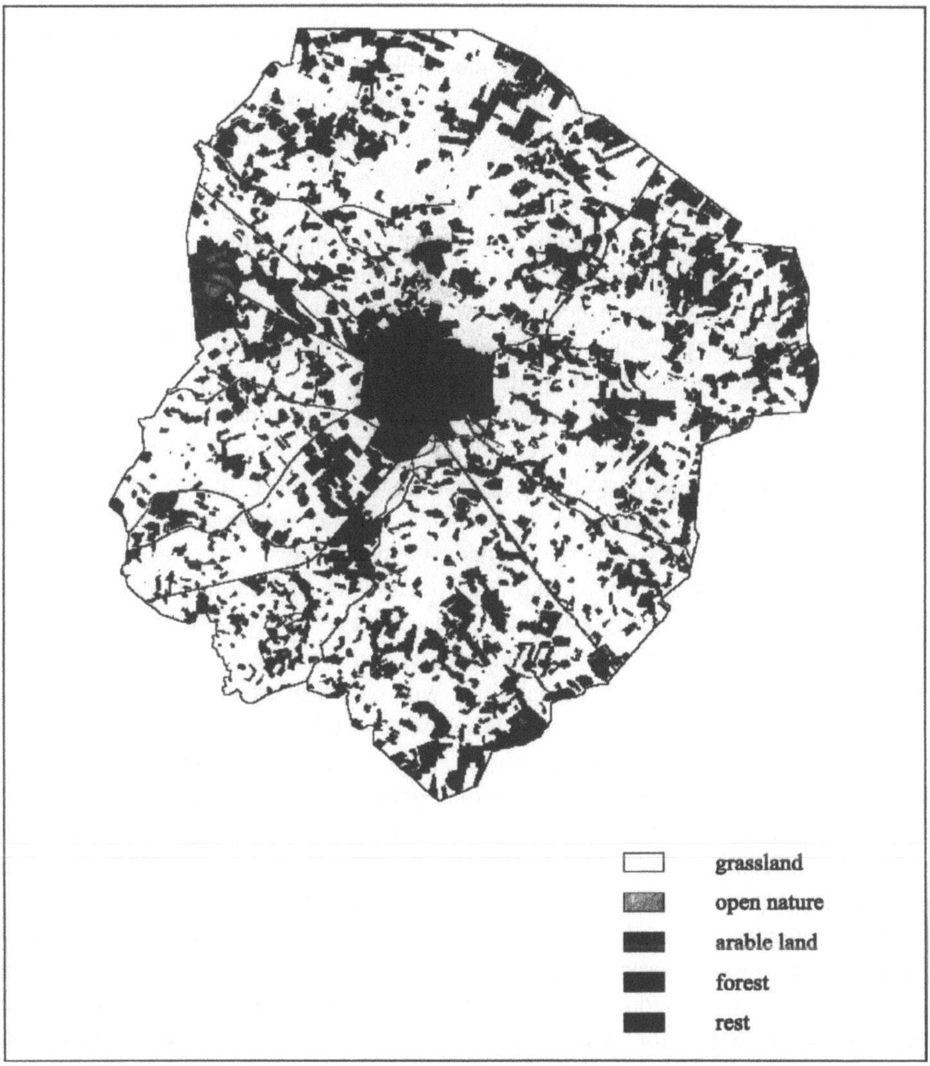

Figure XII.2: Present land cover in the region Winterswijk

Gelderland, 1996a, b & c). Together these factors form a threat to the valued landscape and to its attractiveness for recreation. There has been no agricultural development, and within the next 10 years the number of farms is expected to decline by 15-30% (WCL-Winterswijk, 1996). These and other factors endanger the liveability of the region, and to some extent its economic potential, for example, as regards drinking-water collection (WCL-Winterswijk, 1996).

Chapter III stated that the *system boundaries* must be chosen in such a way that the system can be steered. For that reason, and on the basis of practical considerations - such as the availability of data and financial resources - it was decided to focus only on part of the WCL area. The boundaries chosen were the boundaries of the municipality of Winterswijk, covering about 14,000 ha of the WCL area. Since the project aims refer only to rural land-use activities, the built-up area of the town was excluded and designated as part of the context of the rural area.

The *resolution or grain* of information on the system's potential for multiple land-use should be adequate to identify the opportunities for new land-use activities within the region. For a broad analysis - as preferred by the decision makers at this stage of the project - a low resolution was recommended, in order to keep the data-gathering manageable and provide the desired information within budget and time constraints. In analyzing a region of 140 km^2, a resolution could be chosen that describes the potential for multiple land-use in units of, say, a square kilometre. However, two reasons, a higher resolution was preferable. First, with regard to the major activities presently taking place in the region, such a resolution would have been too low, as there are many woods in the region that are smaller than a square kilometre. Second, in intensively used rural areas such as those in the Netherlands, most individual landowner or land-user manage a much smaller area than one square kilometre. Since changes to land-use cannot be carried out without the cooperation of the landowners and land-users and their enterprises, replacing conventional land-use activities will take also place on a smaller scale. For this reason, a grain had to be chosen which at least approximates the size of existing land-use enterprises in the region. A grain size of 0,25 km^2 was considered sufficient for the assessment of the region's potential for multiple land-use.

A *time scale* had to be chosen, in order to make explicit the extent to which modifications to activities and the availability of resources could be assumed. The STD project team considered a time scale of several decades most feasible. Since no attempt was made to predict the future, selecting such a time scale simply indicates the possible assumptions about the types of new technology available and the kind of environmental problems to be solved. It does not indicate a time horizon for scenarios for land-use planning in which all changes are described for the coming decades.

Within the chosen time scale, two situations were distinguished. One shows what opportunities are offered by new techniques where there are no social, economic, or environmental constraints, and the other shows what opportunities are offered given the current environmental problems, but assuming that no social or economic constraints exist.

The *people involved* were regarded to be represented by the stakeholders. These stakeholders were defined by the project team as those groups who are involved in present land-use and who are confronted by the present regional problems. They include present land-users and landowners, suppliers of means of production and technology, users of products, and representatives of higher-scale, governmental organizations· responsible for socioeconomic and environmental policies.

XII.4 Selecting valued system characteristics

The selection of the *valuable system characteristics* of the rural area around Winterswijk was based on interviews with the stakeholders. Since the focus is on the opportunities for multiple land-use, the aim of this phase was to select the major products and values of the region - in order to find the relevant processes and structures, and from that the needed resources - and to identify the major environmental problems that currently hamper multiple land-use.

XII.4.1 Method for information-gathering

At the end of the preliminary study (Chapter XI), an initial consultation with the stakeholders was organized by the project team; the main intention was to generate support for a study focusing on the Winterswijk region. Therefore, a second round of interviews was required to provide sufficient information to carry out the selection of *valuable system characteristics*. The interviews also provided insight into the kind of land-use activities desired. Moreover, the interviews were used to motivate stakeholders to participate in the development of new land-use activities. The interviews were conducted by the working group 'Communication'.

Because of the different objectives of the interviews, the information-gathering procedures presented in Chapter IV could not be followed. The project team felt it would not be fruitful to employ the framework to search for *valuable system characteristics*. The interviews were in the form of open talks, or rather strategy discussions, which did not zoom in on a direct identification of desired characteristics, but usually ended up by recording a comprehensive list of topics. These included goals of development, means to achieve them, constraints on activities, wishes, and development aims (Box XII.2). The investigators themselves then selected the *valuable system characteristics* from the list of topics.

A. Agricultural organizations Winterswijk:

.High-value-added animal products
.Region-specific products
.Products for consumption
.Income from producing clean water
.land-use employment > 500 places
.land-use income: from recreation and from nature conservation
.Income farmers: agriculture 70%, nature 20%; recreation 10%

D. Dairy industry: Coberco:

.Keeping cows in the pasture during winter: other cow breeds
.Soft butter: using less green maize
.Effective use of proteins from milk
.Change milk quality
.Deliver of milk without water to factories
.Less energy use in production
.Reduce losses of raw material

E. Energy and water: NUON:

.High-tech agroproduction
.Use of solar energy
.Less energy use in farms: effective farm buildings
.Electric tractors
.Solar collectors on roofs
.Small-scale windmills
.Crops for energy and food
.Waterpower from groundwater and/or rainfall

G Province Gelderland:

.Quiet, passive forms of recreation (day & stay)
.Payment technologies: tourist taxes
.High-tech agriculture
.Divers food production
.Extraction of phosphates and nitrates out of water
.Domestic windmills

H: Higher scale: national

Facilities and infrastructure for:
.education
.economy
.hydrology
.recycling
.transport
Environmental claims:
.net energy, land, water, mineral, raw materials use
.global: CO_2
Regional environmental impacts:
.soil: nitrate and phosphate, pesticides
.water: nitrate, phosphate pesticides
.air: NH_3, pesticides
Overall:
.composition of population
.liveability rural area

N. Nature conservation: 'Natuurmonumenten':

.Segregation and integration
.One cow per hectare
.Support for nature development (1000 ha)
.Design of new farming models
.Realization of landscape elements: hedges, pastures on limestone, wet haylands, moorland, heathlands, brook valleys, heights.
.No drying out (-95%)
.No emissions (-95%)
.Realization of ecological infrastructures
.Restoration cultural-historical landscape by agriculture
.Maintenance of old farms
.Maximization of biodiversity in agriculture
.Small-scale recreation

P. Purification Board Oost-Gelderland

.Equilibrium fertilization
.New crop-protection measurements
.Other crops
.Mechanical weeding
.Fields of reed for purifying water
.No low water levels
.Nature development along watercourses
.Prevention of planological side-effects

Box XII.2: Consulted stakeholders and stakeholder's ideas for reaching development goals

XII.4.2 Results of the interviews

In order to identify the characteristics of the Winterswijk region valued by the stakeholders, the topics mentioned in the interviews were rearranged according to the framework for *valuable system characteristics*. Each description of the topics mentioned was compared with the types of characteristics presented in the framework. Four checks were carried out.

- First, did the topics mentioned refer to needs or wants of a direct user-group or to an amount, without specifying the user group? In the interviews, for example, the demand for at least 500 incomes from farming was mentioned. Other topics obviously described desired amounts or qualities without specifying the user-group, such as 'reduce losses of raw material'. These latter topics were considered to be desired characteristics of the system as a whole. When fitting topics, the focus was on the 'needs or wants' described in the topic. Therefore 'less energy use in farms' is regarded as a higher-scale topic, i.e., it is related to the need for effective resource use, not to an explicit need on the part of farmers.
- Second, did the topics refer to direct use or conditions for use in the future? For example, the maintenance of employment was mentioned, but also the creation of new nature areas (1000 ha mentioned by the nature conservation stakeholder).
- Third, did the topics refer to processes or to structures? A distinction was made between fluxes, such as food, energy, and ammonia emission, and structure elements such as desired area or maintenance of landscape features.
- Fourth, did the topics refer to a valuation of separate or combined system characteristics? For example, a distinction was made between 'effective sprinkling' and 'total production of greenhouse gasses'.

The results are given in Table XII.1. Capitals between brackets refer to the organizations mentioned in Box XII.2. It is noteworthy that very few present structure characteristics occur. This is understandable, since the interviews were more or less strategic discussions, in which the present lay-out of the region was not really an issue. Reserves for future use for the higher-scale community are also missing. This is less understandable, considering that the region has, for example, important aquifers.

Table XII.1: Topics based on the discussions with stakeholders as described in Box XII.2.

Characteristics related to direct users

	Processes	Structures	For future use	For future processing
Separate	. high-valued-added animal products (A) . region-specific products (A) . end products for consumption (A) . income from nature conservation (A) . income from recreation (A) . income from collecting clean water (A) . use of solar energy (E) . crops for energy and food (E) . quiet, passive forms of recreation (G) . water catchment (W)	. Employment > 500 land-users (A)	. keeping cows in the pasture during the winter: other cow breeds (D) . payment technologies: tourists taxes (G) . Facilities for water catchment (H) . Realization of landscape elements (N) . design of new farming models (N) . Nature area 1000 ha (N) . Nature development along water courses (P) . Field of reed for purifying water (P) . other crops (P)	. soil: nitrate and phosphate, pesticides (H) . water: nitrate ha>25mg/l, phosphate pesticides (H) . high-tech agroproduction (E,G) . no drying out -95% (N) . no emissions -95% (N) . new crop-protection measurements (P) . no low water levels (P) . reduce number of dams in brooks (P) . mechanical weeding (P) . water level enhancement (W)
combined	. income from agriculture 70%, nature 20%, recreation 10 % (A) . soft butter using less green maize (D) . small-scale recreation (N) . change milk quality (D) . effective use proteins from milk (D) . delivery of milk without water to factories (D) . waterpower from ground-water and or rainfall (E) . divers food production (G)	. infrastructure for: education economy hydrology transport (H)	. land-use (H)	. tractors on electricity (E) . solar collectors on roofs (E) . small-scale windmills (E) . domestic wind mills (G) . facilities and infrastructure for education (H) recycling(H) . one cow per ha (N) . segregation and integration (N) . realization of ecological infrastructures (N)

Table XII.1 (continued)

Charactreristics related to higher scale

	Processes	Structures	For future use	For future processing
C o m b i n e d	. reduce losses of raw materials (D) . less energy use in production (D) . less energy use in farms: effective farm buildings (E) . winning phosphates and nitrates out of water (G) . effective sprinkling (P) . equilibrium fertilization (P)	. restore cultural-historical landscape by agriculture (N)	. liveability rural area(H) . maximize biodiversity in agriculture (N)	. Composition of population (H)
S e p a r a t e	. energy use (H) . water use (H) . minerals (H) . raw materials (H)	. maintenance of old farms (N)	-	. water: nitrate ha>25mg/l, phosphate pesticides (H) . air: NH_3 ha>600mol, pesticides (H) . global: CO_2 (H)

Left out:
. support for nature development (N)
. prevent planological side-effects (P)
. research on impact of water level (W)
. more responsibility of urban areas for water use (W)
. government enlargement of waterboards (W)

XII.4.3 Valuable system characteristics

It was clear that the complete list of topics could not immediately be treated as characteristics. Some of the them are barely measurable, while others were seen as not specific enough to be included in the decision making. Therefore, the original list was translated into the workable set of *valued system characteristics* given in Table XII.2.

In some cases, the topics mentioned by the stakeholders were not aimed at the right level of scale (e.g., 'restoration of specific hedges'). These goals had to be translated into a type of characteristic that matched the detail level of regional

Table XII.2: Valued system characteristics derived from interviews with stakeholders and from governmental documents; * major products and constraints; - sub-divisions; ! added characteristics.

Characteristics related to direct users:

	Processes	Structures	For future use	For future processing
Separate	*products*: * food products * drinking water * energy * electricity ! heat ! wood products ! fodder ! fertilizer *income*: * agriculture * other products * recreation	*employment*: * agricultural land use meat, etc. services *land use*: * grassland * arable land * forest area * divers growing * nature area: * - general - special - hedges * housing area *facilities for*: * day recreation	*land-use potential*: * grassland * arable land * forests area * divers growing * nature area * housing area	*soil*: * ammonia * pesticides *water*: * water quality * nitrate
Combined	*diversity of*: * products * sources of income	*infrastructure*: * transport ecological * hydrologic * education financial	*diversity of*: ! land use potentials	*new infrastructure*: * transport ecological * hydrologic * education * financial * recycling

Characteristics related to higher scale:

Combined	! efficiency resource use	* landscape *mainstructures*: ! transport ! ecological ! hydrologic ! financial]	* liveability of rural area	*new mainstructure*: ! transport ! ecological ! hydrological ! financial * composition of population
Separate	*external*: * water * fodder area *recycling*: ! manure ! waste water	! overall employment * cultural-historical features	! biodiversity ! aquifers	*environment*: * fossil energy * greenhouse gases * water quality * air quality

description - for example, 'presence of ecological infrastructures' which stands for all the specific structure elements mentioned, while the separate elements were used for calculations. For the same reason 'animal products', 'milk', and 'butter' are translated into 'food products'. It also became apparent that some important development topics were not brought up in the course of the interviews by any of the stakeholders. For example, the production of wood was not mentioned, even though about 15% of the area is forested. To prevent discussions about 'forgotten' valuable characteristics later on the procedure, 'missing' items were added by the investigators. These are indicated by an '!' in Table XII.2. By adding characteristics, at least one characteristic in each cell of the framework was taken into consideration during the project.

XII.5 Measuring the resource state

In exploring of the potential of multiple land-use, one first has to *measure the resource state* for this new type of use (phase 4 in Figure XII.1 and Chapter V). This means measuring the units suitable for combining processes that deliver products and values based on the different resource characteristics of the land. These units are called '*functional units*' (Chapter V). Since a backcasting approach is applied, these *functional units* should be based on the assumed availability when there are no economic, social, environmental, or technological problems. In fact, the potential of multiple land-use will be made visible by assuming that the existing problems have been solved.

XII.5.1 Method of information-gathering

To measure the availability of resources, five steps must be taken after the *selection of the valued system characteristics* (Steps 3-7 from Chapter V): *(a) selection of the processes and related structures*, *(b) selection of the required resources*, *(c) selection of resource system characteristics*, *(d) measurement of the resource state*, and *(e) representation of functional units*.

XII.5.2 Selection of the processes and related structures to be combined

Table XII.2 was presented to the project team responsible for selecting the major land-use processes. Thirteen processes were selected, including the production, management or development of *general* or *specific* 'nature values', 'recreational facilities', 'food (for livestock and humans)', 'drinking water', 'energy', 'landscape and cultural and historical features', and 'recycling'.

A *general* type of process is one that can be carried out almost anywhere in the region; a *specific* type applies to specific biophysical conditions in particular parts of the region. This makes it possible to do justice to the potential of specific production circumstances that may provide products with a higher added value than can be expected of the general production and management processes. These specific types of processes are not by definition region-specific, and may also be found in other regions.

For sake of simplicity the project team decided to refrain from selecting all the processes to supply the characteristics in Table XII.2. For example, no process was selected for the construction of infrastructures or main structures, or managing or creating distribution facilities, on the assumption that it was not necessary to open up the region. Other valued characteristics, such as emissions and resource efficiency, were considered to be relevant only for the way the new land-use activities will be carried out. These must be taken into consideration when analyzing the benefits of new land-use activities, but need not be reflected in distinct processes.

The selection resulted in the following list of 13 major processes to be used for assessing resource characteristics (Table XII.3). No distinction was made between general and special recycling, since recycling is always possible when one of the other processes occurs. The emphasis is on the local recycling of organic waste and waste water. This list shows the basic processes which, in varying combinations, will be used to define new land-use activities.

XII.5.3 Resources and resource characteristics

The next step was the selection of the required *resources* and *resource characteristics* which must be derived from the processes that deliver the *valued system characteristics* (Phase 4). Since the focus is on the opportunities in a situation where there are no socioeconomic constraints, the resources sought refer to biophysical resources. The major biophysical resources of the 13 processes are listed in column 2 of Table XII.3. The third column lists the required resource characteristics.

The next step was the assessment of the distribution of *resource characteristics* within the region. Here it was necessary to describe the criteria for the measurement of each characteristic. This step results in Table XII.4, which is used to verify the presence or absence of the required characteristics, on the level

Table XII.3: The selected processes, their resources and required characteristics (G: general; S: related to specific characteristics)

	Processes	Resource	Required characteristics
1.	Nature conservation (GN)	undisturbed space	rural area (a) not overmanured (b) not dried-out (c) not polluted (d) not acidified (e) not acidification (f)
2.	Nature conservation of specific wildlife (SN)	undisturbed space with specific characteristics	rural area (a) not overmanured (b) not dried-out (c) not polluted (d) not acidified (e) not acidification (f) specific soils (g) diversity of soils (h) specific hydrology (i) diversity of hydrology (j) diversity in elevation (k)
3.	Recreation outdoors (GR)	area with lodging facilities	rural area (a) no health risk (l) no noise pollution (m) no stench (n)
4.	Recreation connected with specific landscape features (SR)	area with attractive features	no health risk (l) no noise pollution (m) no stench (n) recreation features (o)
5	Agriculture (GA)	arable land	arable land (p) irrigation water (q)
6	Agriculture on land with steep gradients or producing special products (SA)	arable land in specific places	diversity of soils (h) diversity of hydrology(j) diversity in elevation (k) arable land (p) irrigation water (q) landscape features (r)
7	Energy production through biomass production (GE)	biomass	agriculture (s)
8	Energy production from flow resources such wind or water(SE)	streaming water wind	drop in water level (t) open areas (u) elevation (v)
9	Drinking water collection, 'whole' area local (GW)	surface water of drinking quality	rural area (a) not overmanured (b) not polluted (d) not acidified (e) not acidification (f)
10	Drinking water collection, large-scale, but specifically located groundwater collection (SW)	ground water of drinking quality	not overmanured (b) not polluted (d) not acidified (e) not acidification (f) sand layer (w)
11	Conservation of cultural-historical landscape features (GH)	cultural-historical landscape	landscape type (x) no excavation (y)
12	Conservation of cultural-historical elements (SH)	cultural-historical features	no acidification (f) cultural-historical features (z)
13	Recycling waste (SO)	organic material & waste water	agriculture (s)

Table XII.4: Resource System Characteristics; a-z refer to the required characteristics in Table XII.3; n.a. = data not available; top. map: topographic map 1:25,000 41E (TDN, 1994); dig. map: digital map in GIS of ARCADIS; Y/N: the characterisitc is present or not; >50%: the characteristic covers at least 50% of the cell area.

	Resource system characteristics	Description	Sources	Criterion
a	rural area	non building area	top. map	>50%
b	over-manured	agricultural area, groundwater protection areas excluded	dig. map land cover	>50%
c	dried-out	too dry for nature	dig. map	>50%
d	polluted	arable farming, groundwater protection areas excluded	dig. map land cover	>50%
e	acidified	whole area except peat and built up area	dig. map land cover	>50%
f	acidific-ation	cells with pig farms with more than 2500 kg NH3 emission/year	Landbouworganisaties Winterswijk (1995)	Y/N
g	specific soils	podzols	dig. map land cover	>50%
h	diversity of soils	cells with more than 3 different soil classes	dig. map land cover	Y/N
i	specific hydrology	presence of seepage	Biologisch Station Zwillbrock (1995)	Y/N
j	diversity of hydrology	cells with more than 4 groundwater tables	dig. map land cover	Y/N
k	diversity in elevation	standard deviation of elevation within cell higher than 1.2	dig. map land cover	Y/N
l	health risk	-	n.a.	-
m	noise	area within 200 m from highways or railroad	top. map	>50%
n	stench	stench due to presence of pig breeding farm (2500 kg nh3/year)	Landbouworganisaties Winterswijk (1995)	Y/N
o	recreation features	1.cultural-historical values 2.cultural-historical points 3.geographic objects 4.springs 5.nature protection area 6.recreation: campings 7.recreation: open water 8.golfcourse 9.stone quarry	-Grondmij/Zwillbrock (1994) -Biologisch Station Zwillborck (1995) -Lörzing & Keijsers (1996)	Y/N Y/N Y/N Y/N Y/N Y/N Y/N Y/N Y/N

Table XI.4 (continued)

	Resource system characteristics	Description	Sources	Criterion
p	arable land	all cells except peat, water, built up area, brooks, stone quarry, large forests, recreation terrain.	dig. map land cover ARCADIS	>50%
q	water of irrigation quality	all cells		
r	landscape features	1.steepedges 2.ashes	Biologisch Station Zwillborck (1995)	Y/N Y/N
s	agriculture	agricultural activity	dig. map land cover ARCADIS	
t	drop in water level	cells with 'large' fall in groundwater levels	dig. map contour lines groundwater levels ARCADIS	Y/N
u	open areas	cells which are not, or just marginally, cut by main roads, lie outside the built-up area, and border at least one other open cell	top. map	>50%
v	elevations	elevations	top. map	>50%
w	sand layer	sandy layers at depth of more than 20 meter depth	Biologisch Station Zwillborck (1995)	>50%
x	landscape	all cells		
y	excavation	-	n.a.	-
z	cultural-historic features	system characteristic o 1-5 and r 1,2 (see above)		idem o and r

of the chosen resolution, i.e., for each grid-cell of 0,25 km². Table XII.5 is a decision-matrix that describes the conditional relationship between characteristics and processes possible within a cell. On the basis of the available data, the potential processes per grid were calculated (Figure XII.1). Thus the *resource state* is described as the potential combination of processes per cell which measures the *present resource state* of the region.

As noted before, the *present resource state* does not show those opportunities for multiple land-use that are restricted by present environmental problems. However, the characteristics that determine the *present resource state* cannot be regarded as unchangeable, since they may be simply the result of current inefficient and polluting activities. The techniques used in new land-use activities could mitigate or eliminate such inefficiencies and pollution, thus altering the resource characteristics within a grid. By assuming that all those environmental

Table XII.5: The conditional relations between the system characteristics and the potential processes (b-w see Table XII.4)

		b	c	d	e	f	g	h	i	j	k	m	n	o 1	o 2	o 3	o 4	o 5	o 6	o 7	o 8	o 9	p	r 1	r 2	s	t	u	v	w
	Changeable	*	*	*	*	*						*	*																	
	Processes and structures																													
1-G	Nature	0	0	0	0	0	.	.	.	.	.	.	.	.	.	.	.	.	.	.	.	.	.	.	.	.	.	.	.	.
2-S	Nature	0	0	0	0	0	1	1	1	1	1	.	.	.	.	.	1	.	.	.	.	.	.	1	.	.	.	.	.	.
3-G	Recreation	.	.	.	.	.	.	.	.	.	.	0	0	.	.	.	.	.	.	.	.	.	.	.	.	.	.	.	.	.
4-S	Recreation	.	.	.	.	.	.	.	.	.	.	0	0	1	1	1	1	1	1	1	1	1	.	.	.	.	.	.	.	.
5-G	Agriculture	.	.	.	.	.	.	.	.	.	.	.	.	.	.	.	.	.	.	.	.	.	1	.	.	.	.	.	.	.
6-S	Agriculture	.	.	.	.	.	.	1	.	1	1	.	.	.	.	.	.	.	.	.	.	.	1	1	1	.	.	.	.	.
7-G	Energy	.	.	.	.	.	.	.	.	.	.	.	.	.	.	.	.	.	.	.	.	.	.	.	.	1	.	.	.	.
8-S	Energy	.	.	.	.	.	.	.	.	.	.	.	.	.	.	.	.	.	.	.	.	.	.	.	.	.	1	1	1	.
9-G	Water Collection	0	.	0	.	0	.	.	.	.	.	.	.	.	.	.	.	.	.	.	.	.	.	.	.	.	.	.	.	.
10-S	Water Collection	0	.	0	.	0	.	.	1	.	.	.	.	.	.	.	.	.	.	.	.	.	.	.	.	.	.	.	.	1
11-G	Cult-hist Features	.	.	.	.	.	.	.	.	.	.	.	.	.	.	.	.	.	.	.	.	.	.	.	.	.	.	.	.	.
12-S	Cult-hist Features	.	.	.	.	0	.	.	.	.	.	.	.	1	1	1	1	1	.	.	.	.	.	1	1	.	.	.	.	.
13-G	Recycling Wastes	.	.	.	.	.	.	.	.	.	.	.	.	.	.	.	.	.	.	.	.	.	.	.	.	1	.	.	.	.

G= general: whole region
S= specific: specific characteristics

0: quality prevents activity and
1: quality enables activity

*: quality has disappeared thanks to new techniques
a + x = whole area
l + y = n.a
z via column o and r

problems have actually been solved, the *future resource state* can be estimated, which can be used to show in greater detail the opportunities for multiple land-use. The options for changing resource characteristics, however, were limited beforehand. Large-scale operations that would change the geographically, hydrologically, or historically valued characteristics of the area were unacceptable, because they would obviously conflict with various *valued characteristics* described in Table XII.2, such as the maintenance of landscape. Seven resource characteristics remain, the distribution of which can be changed by new techniques: overmanured, dried-out, polluted, acidified, acidification, noise, and stench ('*' in Table XII.5).

XII.5.4 Results: present and future Resource State

The results were recorded in tables of *functional units* for further analysis. For external communication the results were presented graphically as maps of the *resource states* expressed in potential activities (Figures XII.3 and 4).

Figure XII.3 shows the future *resource state* expressed in terms of the number of potential processes, assuming that the present environmental problems have been solved. In that case most of the area has a considerable potential for multiple land-use. But the question remains what potential is available in the near future, when environmental problems have not yet been solved? This potential is visualized in Figure XII.4 as the number of processes possible per grid-cell. It shows that even with the present environmental problems, there is a potential for combining processes, thus pointing up the challenge to use current knowledge and techniques to combine processes in multiple land-use.

In contrast to what is prescribed in Chapter VI, these results were not checked by comparing the valuable characteristics estimated from the *present resource state* with the *present value state*. Because here the focus was on the potential for combining processes, the check was not carried out here, but rather in the next part of the project (Chapter XIII).

XII.6 Exploring development opportunities

The aim of this phase is to explore the *resource space* in terms of the potential to combine processes. For purposes of this exploration, the possible combinations of processes, now and in the future, will be analyzed.

The Figures XII.3 and 4 specify the *resource state* in terms of the number of processes that can be combined, but they do not specify what those combinations

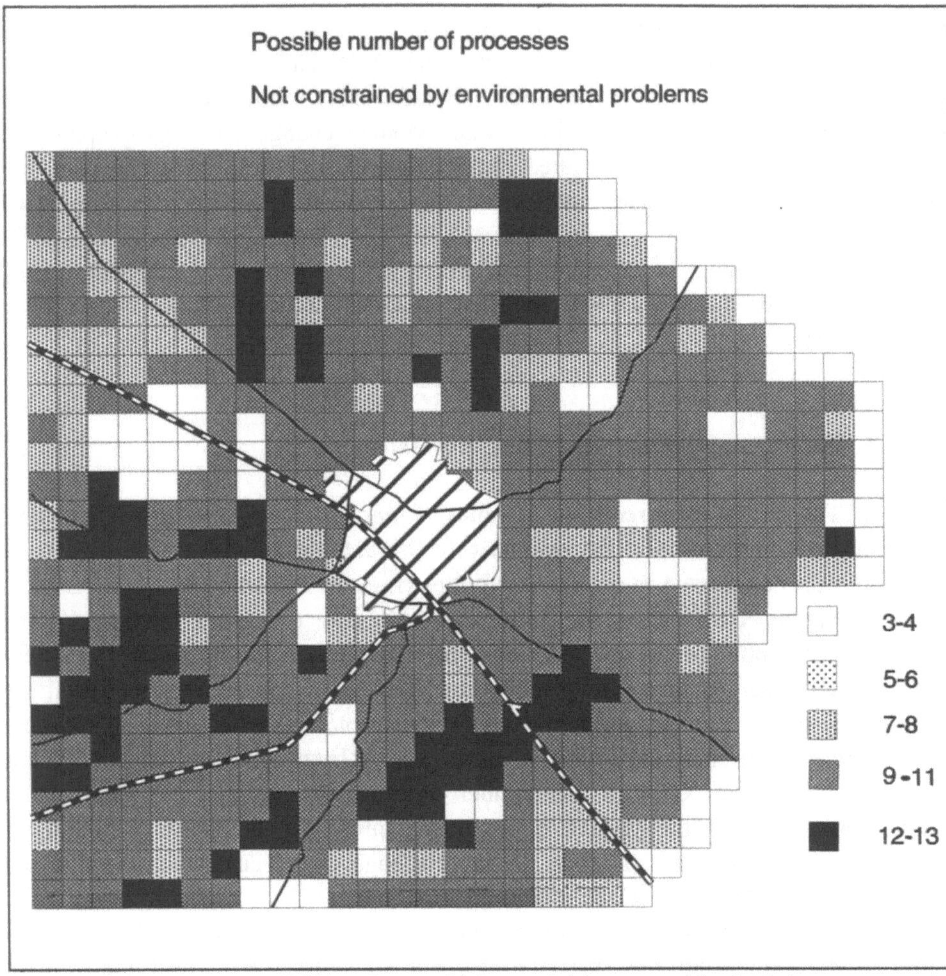

Figure XII.3: Future possibilities for combining activities per grid of 25 hectares, on the assumption that new techniques will be found and applied

are. For example, if three processes are possible in a given grid cell, this might be a combination of 'nature conservation', 'agriculture', and 'recreation', but it could also be a combination of 'agriculture', 'recreation' and 'water collection'. In this phase, the aim was to determine the extent to which certain combination are possible in the region, in order to be able to select those combinations which cover large parts of the region. This was important in order to avoid working out only very general combinations which have no specific value for the stakeholders. At the same time this will ensure that the focus is not on combinations that may have a high potential for multiplicity in land-use, but are unsuitable to replace the major current land-use activities. The aim was to find those specific combinations of

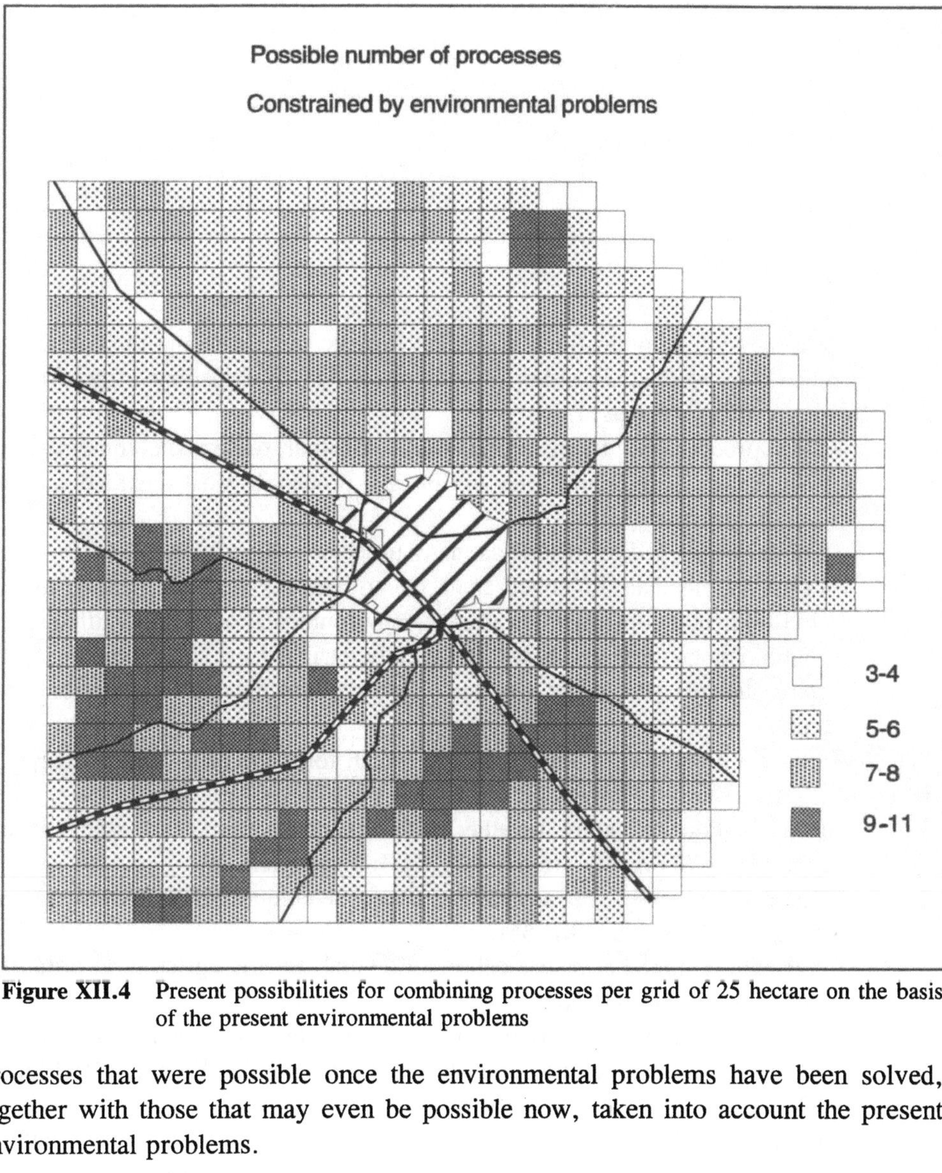

Figure XII.4 Present possibilities for combining processes per grid of 25 hectare on the basis of the present environmental problems

processes that were possible once the environmental problems have been solved, together with those that may even be possible now, taken into account the present environmental problems.

XII.6.1 Method of information-gathering

The analysis now centred on the number of grid-cells in which certain combinations of processes can occur. Because the general types of processes were defined as capable of taking place almost anywhere in the region - with the exception of nature conservation areas and forests - they appear in each combination in each grid cell. Therefore, the search was restricted to combinations of the six processes which use specific characteristics present only in a particular part of the region. Again, these processes are not regional specific - they may also occur outside the region - but they cannot take place in the entire Winterswijk area.

XII.6.2 Results

The possible combinations of the six specific processes and the frequency of occurrence were studied. In order to select promising combinations of processes, the following criteria were used:

- Each process had to be represented. For that reason six combinations were selected in which one of the six processes was regarded as the 'major' process.
- Combinations had to include as many processes as possible.
- Combinations had to cover as large an area as possible, either under environmental constraints or after the environmental constraints lifted.
- Combinations had to be technologically challenging, i.e., combinations were selected that could be made more promising with the aid of new techniques.

Six combinations of processes were selected, which form the basis for defining six types of multiple-land-use activities.

1. Conservation of nature together with the production of drinking water (combining SN & SW from Table XII.3: Figure XII.5a).
2. Recreation focused on agriculture within a cultural or historical setting (SR, SA, SH: Figure XII.5b).
3 Agriculture aimed at nature conservation and specific agricultural products (SN, SA: Figure XII.5c).
4. Water collection, recreation, and the production of specific agricultural products within a cultural or historical setting (SW, SR, SA, SH: Figure XII.5d).
5. Production of energy, recreation, and specific agricultural products within a cultural or historical setting (SE, SR, SA, SH: Figure XII.5e).
6. Production of specific agricultural products within a cultural or historical setting (SA, SH: Figure XII.5f).

Figure XII.5g illustrates which part of the region is covered by at least one of these six types. It appears that these six combinations cover 60% of the area of the region, indicating that they provide a basis for the development of new activities throughout a large part of the region.

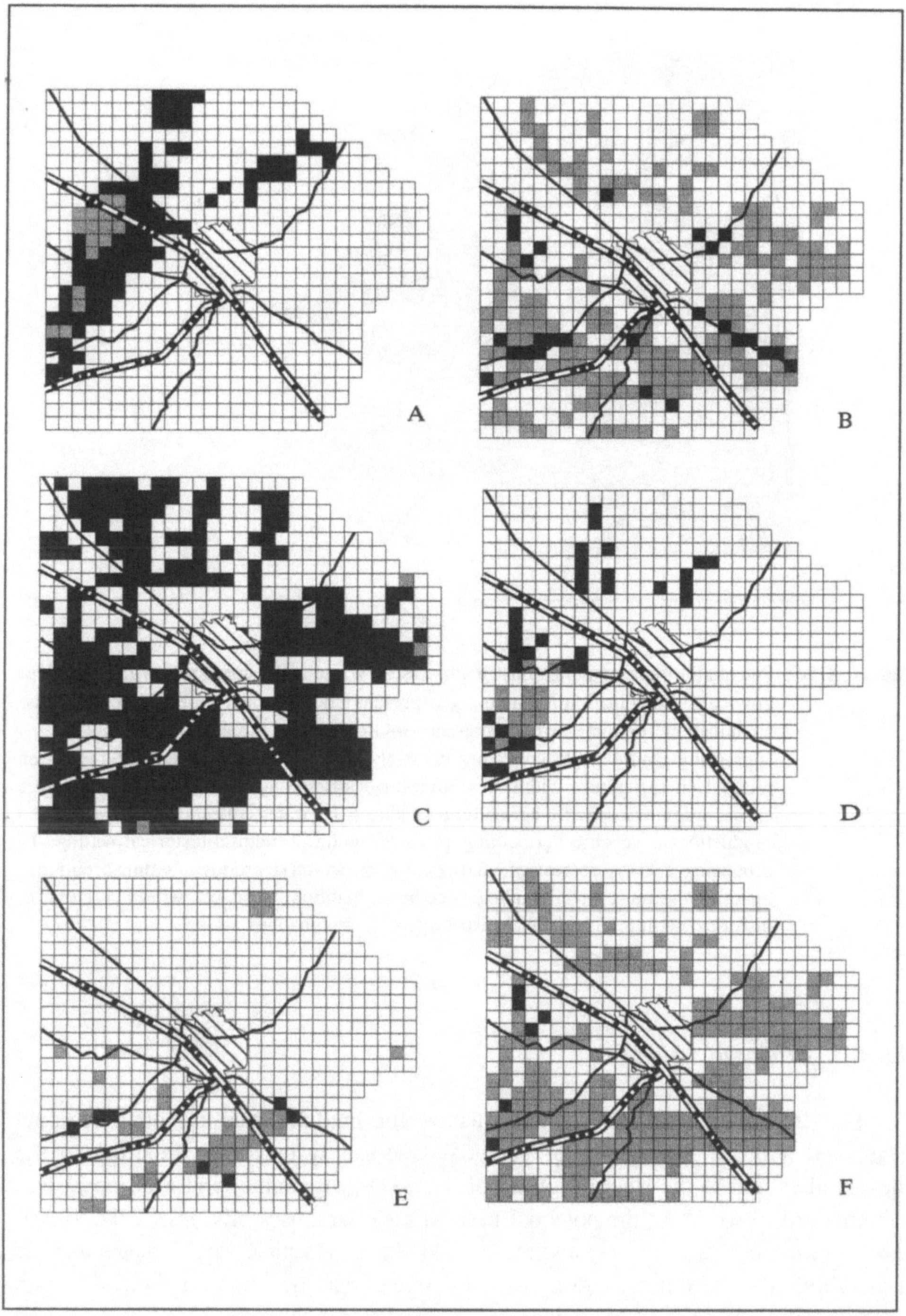
A
B
C
D
E
F

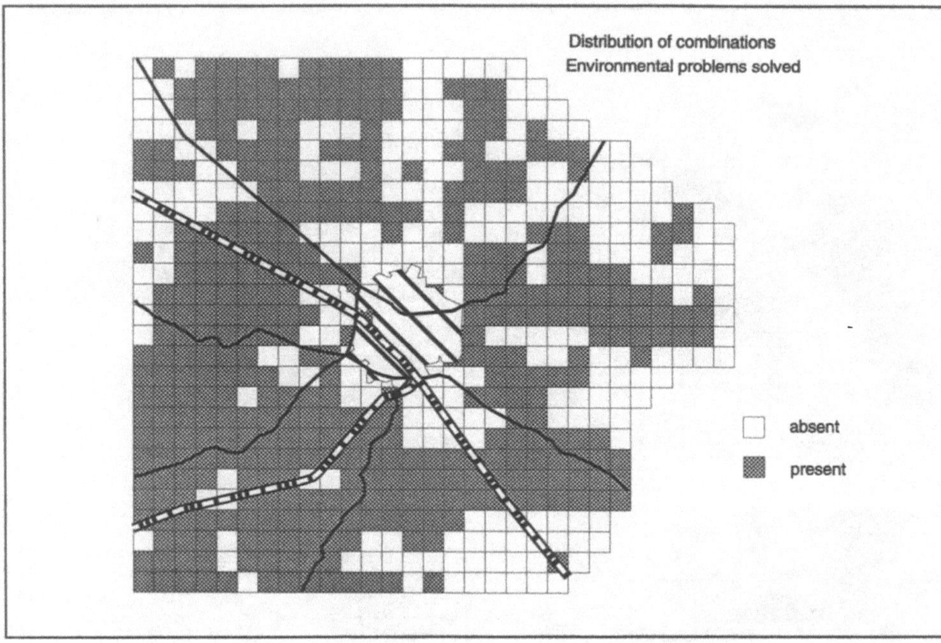

Figure XII.5: Potentials of six combinations of processes. White cells: combination not possible; grey cells: combination possible under present environmental problems; black cells: combination possible when present environmental problems are solved; A: conserving nature while collecting drinking water from deep layers; B: recreation focused on agriculture within a cultural-historical setting; C: agriculture for nature conservation and specific agricultural products; D: water collection, recreation, and production of specific agricultural products within a cultural-historical setting; E: producing energy, recreation, and specific agricultural products within a cultural-historical setting; F: producing specific agricultural products within a cultural-historical setting; G: potential distribution of combinations a-f.

XII.6.3 Proposal

Information on the spatial potential for multiple land-use on a regional level and a broad description of potential land-use systems was reported to the stakeholders and to the steering group of the STD programme, and a proposal was put forward to work out the potential new multiple-land-use activities (STD, 1996). To ensure that the new activities would have practical and socioeconomic feasibility, the recommendation was to work out the new activities within economically viable land-use enterprises. The project team proposed that several projects be set up to describe such enterprises assuming that there are no organizational or technological constraints. In this way insight could be gained into the technological development required, and the necessary changes to the organizational structures.

XII.7 Discussion: problems encountered by applying the proposed procedure

Defining the *socioenvironmental system* did not present major obstacles. The region had been selected in a preliminary study. Because the aim of the project was not to steer land-use planning, but to steer the design of new land-use activities, the definition of the system did not become an issue in the external debates.

The *valued system characteristics* were chosen by going from the topics mentioned in the interviews to a list of *valued system characteristics*; several items not mentioned by the stakeholders were added. The proposed framework in Chapter IV was used to systemize the topics and find items the stakeholders had omitted. However, it was not used during the interviews with the stakeholders. The list of *valued system characteristics* was used to select the major processes that supply these characteristics; this choice was made externally, by the project team. This greatly simplified the *measurement of the resource state* by assuming only thirteen major processes; however, this may result in an underestimation or overestimation of opportunities. For example, by disregarding the need for 'infrastructure', the need to distribute products is likewise disregarded. In the project, this is partially overcome by focusing on the local use of products. But if energy is to be generated, than it must also be made available to the consumers, and the costs of distribution should be taken into account. A strong point in favour of the procedure is that such simplifications are made explicit and subject to discussions.

On the basis of the assumed processes and the available information on resource characteristics, the *present resource state* was measured. Certain necessary information was not available, while other information was not available on the correct level of scale. While the consequences of this situation are difficult to estimate, it may have interfered with the accuracy of the analysis of possibilities offered by combining new land-use processes.

A more fundamental problem involved in the application of this phase is that no social, economic, or cultural resource characteristics were recognized in the study. This seemed justified by the backcasting strategy in which the options for the future were identified by assuming that certain problems have already been solved. In this project, it was assumed that the social, economic, and cultural problems involved in the application of new land-use activities would be solved. However, by not recognizing social, economic, or cultural resources, no insight is gained into what these problems are and how they can be solved. This contrasts with the environmental problems, which where explicitly defined as delimiting resource characteristics. This problem can be partially solved by describing the new land-use activities in such a way that they are part of economically viable enterprises. But at the regional level of scale, the presence of education facilities, say, or payment technology is not made visible, even though it may play a crucial role. Obviously, the fact that these characteristics were valued by the stakeholders does not mean that they will also be taken into consideration in an external decision making context.

CHAPTER XIII

Exploration of the Potential Regional Benefits of Multiple Land-use

CONTENTS

XIII.1 Introduction

Chapter XII described the study aimed at assessing opportunities for multiple land-use (Step 5a in Box XI.1; De Graaf & Musters, 1997)). The decision making at the end of that study led to the start of a second study designed to explore the potential benefits of new land-use activities (Step 5b in Box XI.1). The results of that study are described in the present chapter. Again the following phases of the procedure for information-gathering will be described.

1 *Start-up*
2 *Defining the socio-environmental system*
3 *Selecting the valued system characteristics*
4 *Measuring the resource state*
5 *Exploring the development opportunities*

XIII.2 Start-up

On the basis of the results of the first study and on the description of examples of new entreprises (Chapter XII, Aarts *et al.*, 1996), the STD programme management decided to continue the project, and to explore the potential benefits of multiple-land-use activities, in preparation for the decision at the end of the study about the actual development of those activities. Again, the question was How can this information on the potential benefits be provided, when the performance of those activities is not yet known? Again, the technique of backcasting was applied. The aim was to show the quantitative benefits of new land-use activities, assuming that these new activities are available, i.e., that the techniques and organizational structures are available that make optimal use of the available resources and even contribute to an increase in those resources. This entails the assumption that present problems on the level of the region and enterprises that could hamper new activities have been solved, resulting in an increased *resource and use space*. Returning to the present, this information can be used to select the new land-use activities and the enterprises to be worked out in demonstration and development projects.

XIII.2.1 Organization of decision making

In order to ensure that the interests of the stakeholders are taken into account in the development of new land-use activities, representatives of organizations interested in the development of new land-use activities were asked to join a steering group. The participation of the stakeholders should also promote a change in attitudes and provide a basis for starting demonstration and development projects later on in the project. The steering group included people from research institutes, utility enterprises, commercial business, nature and environmental conservation groups, and agricultural organizations. The group was responsible for decisions to be made during the study, including:

- the selection of the valued characteristics to be expressed in the variables to be measured
- the choice of the type of design for the region and its enterprises in order to demonstrate the overall benefits of new land-use activities (not for land-use planning but for gaining insight into the direction of technological development)
- the selection of new land-use activities that will be demonstrated and the new technologies that must be developed in follow-up projects.

XIII.2.2 Organization of management

The management of the project was entrusted to the project team involved in the previous study (in modified form), which had the support of several working groups. This team prepared the proposals for decision making and was responsible for consensus-building within the steering group. The following activities were coordinated:

- the presentation of the results of information-gathering to the steering group and the larger community
- the description of new enterprises, together with new activities and new techniques
- the assessment of the potential benefits of new land-use activities on a regional level
- the development of an instrument for monitoring the development of new technology and new land-use activities and the impact of their implementation on the region.

The results discussed here are the result of the information-gathering, the analyses of that information, and the exploration of opportunities for regional development. The group made use of information about new land-use activities provided by other researchers.

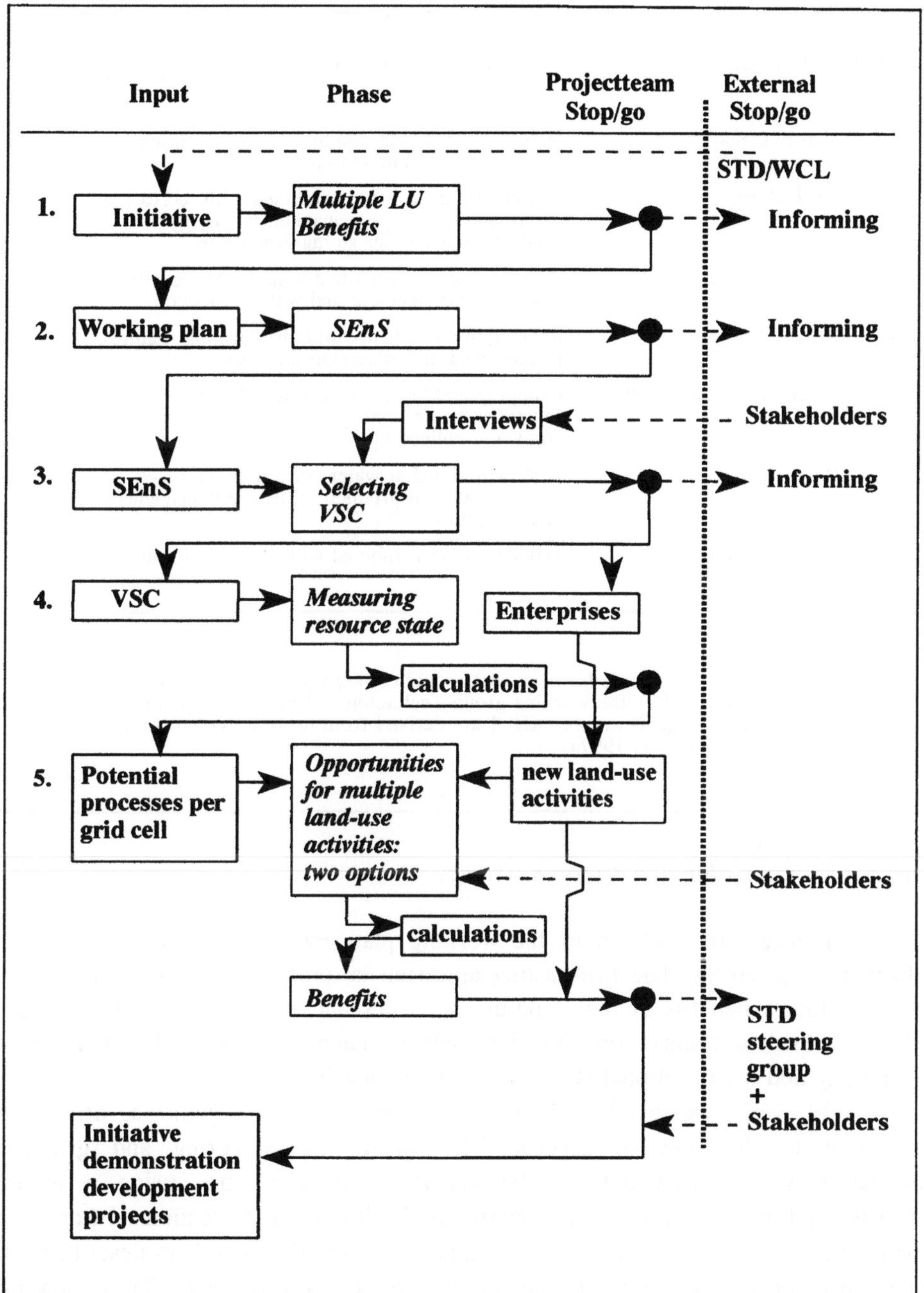

Figure XIII.1: The procedure employed in the study 'Exploration of the potential benefits of multiple land use'

Multiple-land-use system		Combinations of activities
1.	Nature management	Nature management combined with water and biomass production and recreation
2.	Agriculture	Agriculture production combined with recreation, nature and landscape management, energy production and water storage/catchment
3.	Pork production	Pork production combined with waste recycling, landscape management and water collection
4.	Recreation	Recreation combined with dairy farming, nature management and education facilities
5.	Cropping and horticulture	Cropping and horticulture with subsidiary functions: nature management, waste recycling and water collection
6.	Water collection	A system aimed at water collection combined with agricultural production, nature management and recreation
7.	Dairy farming	Dairy farming combined with processing milk

Box XIII.1: Type of multiple-land-use system and combinations of activities elaborated by the working group Technology. Data for contribution to selected goals in table XIII.1 are derived from these activities (Aarts & De Kuijer, 1997a)

XIII.2.3 Working plan and research strategy

Figure XIII.1 illustrates the working plan and research strategy for the information-gathering. The backcasting approach is aimed at a comparison of the potential future benefits of new land-use activities and the benefits of the current activities. Such a comparison will show which changes are profitable and which technology and organizational structures must be developed.

We were responsible for carrying out the information-gathering procedure pertaining to the regional aspects. The spatial data-gathering and grid-cell calculation was carried out by the Winand Staring Centre, using a model constructed for this purpose (PROcedure for Multicriteria Allocation of Land-use PROMAL, Jansen, 1997). At the same time, a group of researchers described the quantitative benefits of new land-use activities on a enterprise level. They drew up a technical description of these activities based on assumed new technology and production structures (see Box XIII.1; Aarts & De Kuijer, 1997a & b). For the description of the enterprises see: Aarts & De Kuijer (1997a). The quantitative information on the activities was used in the regional quantification of the benefits of the new land-use activities.

In order to assess the benefits of new land-use activities, more detailed information was required, than that collected in the first study. As a consequence, the whole proposed procedure for gathering information described in Chapter VI had to be repeated, and the same steps taken as in the previous study (Phase 3 & 4 in Figure XIII.1). To quantify the potential benefits of the new land-use activities, various types of information were consolidated: (a) regional resource characteristics, (b) the characteristics required by the new land-use activities, and (c) the demands of the stakeholders in the steering group (Phase 5 in Figure XIII.1.). Two options for future land-use and their potential benefits were presented.

In the following sections each phase of the proposed procedure will be described and results presented. The results of the project as a whole, up to the summer of 1997, were described by the STD (STD, 1995; De Kuijer & De Bruijne, 1996; De Kuijer *et al.*, 1997a).

XIII.3 Defining the system

To quantify the potential benefits of new land-use activities, it is necessary to know their assumed products and other outputs, the input needed, and the quantitative relationship between inputs and outputs. A model was required in which these activities could be related to each other and to the resource characteristics of the region (PROMAL, Jansen, 1997). For this purpose, the rural area of Winterswijk is assumed to encompass a number of activities which are characterized by their major products. It is assumed that the new land-use activities will replace these activities in the future. The major processes selected in the previous study were used for modelling. However, to quantify the regional benefits, information was needed on a smaller scale, in order to do justice to the need for product diversity, income, and land-use potentials formulated in Table XII.2. Thus information of a higher resolution is needed on such characteristics as soil differences and water availability. Again, data management restricted the choice of the grain. While the smallest workable unit for *measuring the resource state* was a grid-cell size of 25m^2, such resolution was not available for all the data.

XIII.4 Selecting valued system characteristics

The aim of this phase was to select the list of products, inputs and conditions for processing on which the benefits of the new activities were to be assessed. The focus was on the selection of variables to express the contributions to the goals. In principle, all the characteristics from Table XII.2 - the *valued system characteristics* in Table XIII.1 - should be taken into account.

Not all *valued system characteristics* were operationalized into variables to calculate the benefits of the new land-use activities: those which were not operationalized are indicated between brackets ([...]) in Table XIII.1. Of the characteristics related to direct users, only those referring to infrastructure were omitted. As noted in Chapter XII, this was done because the project team did not believe it would be necessary to open up the region. Of the characteristics related to higher scales, not only the main structures were omitted, but also sociocultural characteristics and a number of the environmental and ecological characteristics. Sometimes this was due to a lack of knowledge about the quantitative relationship between those characteristics and the new land-use activities, as in the case of 'biodiversity'. The operationalization of 'biodiversity' is an important issue in current research, but as yet there are no workable solutions for incorporating it into regional models (Hawksworth, 1995; Gaston, 1996; Musters *et al.*, 1997). In other cases, a characteristic was not operationalized because important aspects of that characteristic were already incorporated. For example: 'liveability' is determined by 'income', 'employment', 'landscape', and 'diversity of sources of income'. Whether these aspects are sufficient to cover 'liveability' is open to discussion.

XIII.5 Measuring the resource state

To increase the resolution of the measurement of available resource characteristics, new criteria for measurement were formulated. Since these steps follow the procedure described in Chapter XII, they are not discussed extensively here. The major differences of the newly measured characteristics with Table XII.4 are the absence of the characteristics pertaining to noise and stench. More specific norms for nitrate and pesticides were added. Instead of counting water tables in each grid cell, the current table was measured. The present land cover was also measured in more detail, to facilitate the modelling of land-use activities. In addition, a new decision matrix was made, resulting in a more detailed map of the present *resource state* (Figure XIII.2).

In order to check the model, the *present value state* of the region was calculated, on the basis of the *present resource state*, the location of current land-

Table XIII.1: Valued system characteristics derived from interviews with stakeholders and from governmental documents; * major products and constraints; - sub-divisions; [] not operationalized.

Characteristics related to direct users:

	Processes	Structures	For future use	For future processing
Separate	*products*: * food products * drinking water * energy * electricity * heat * wood products * fodder * fertilizer *income*: * agriculture * other products * recreation	*employment*: * agricultural land use meat, etc. services land use: * grassland * arable land * forest area * divers growing * nature area: * - general - special - hedges * housing area facilities for: * day recreation	*land-use potential*: * grassland * arable land * forests area * divers growing * nature area * housing area	*soil*: * ammonia * pesticides *water quality*: * water quality * - nitrate
Combined	*diversity of*: * products * sources of income	[*infrastructure*: * transport ecological * hydrologic * education financial]	*diversity of*: * land use potentials	[*new infrastructure*: * transport ecological * hydrologic * education * financial * recycling]

Characteristics related to higher scale:

	Processes	Structures	For future use	For future processing
Combined	* efficiency resource use	* landscape [* *mainstructures*: transport ecological hydrologic financial]	[* liveability of rural area]	[*new mainstructure*: * transport * ecological * hydrological * financial] [* composition of population]
Separate	*external*: * water * fodder area *recycling*: * manure * waste water	* overall employment [* cultural-historical features]	[* biodiversity] [* aquifers]	*environment*: * fossil energy * greenhouse gases [* water quality] [* air quality]

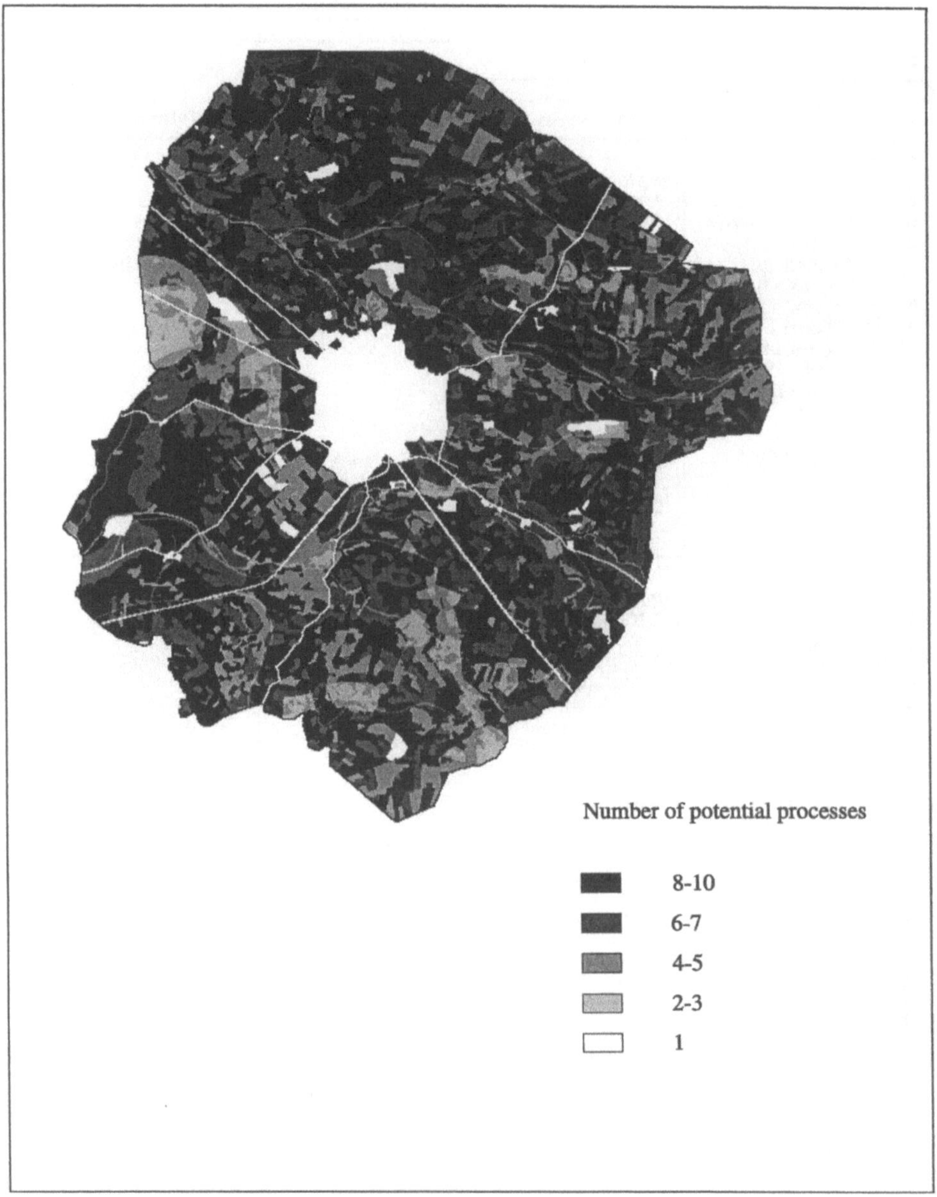

Figure XIII.2: Present resource state expressed in number of potential processes; high resolution

use activities, and the quantitative relationships between the inputs and outputs of current land-use activities (Aarts & De Kuijer, 1997a). Furthermore, the assumed autonomic development of land-use within the region and current environmental and nature conservation policies were used to calculate the expected benefits of that development (Vreke, 1997). This estimate was used as a reference to compare the benefits of the new land-use activities. The results of these two calculations are discussed in the next section, together with the results of the two options for replacing current land-use activities by new ones.

XIII.6 Exploring the benefits of multiple land-use

XIII.6.1 Method for information-gathering

The enterprises

In order to identify the new land-use activities, the enterprises were described as specifically as possible by several technology working groups (the types of entreprises are listed in Box XIII.1). They first described seven enterprises and these descriptions were presented to the steering group. It was decided that three of them would be worked out, and elements of the other enterprises incorporated into these three. These three enterprises are (i) a nature-conservation system for collecting water, (ii) a country estate with dairy farming, arable farming, and forestry, and (iii) a pig-rearing system using fodders and waste from the region to produce meat and usable metabolic warmth (Aarts & De Kuijer, 1997a).

Potential land-use

The *present resource state* and the descriptions of the new land-use activities can be used to calculate the maximum area available for each new land-use activity once the environmental constraints have been lifted. These are given in Table XIII.2, together with the products they can deliver. This table also shows that all the activities supply more than one product. However, these maxima cannot be used to estimate the benefits of the actual application of the new land-use activities, because only one new multiple-land-use activity is possible at a given moment in a given location. Here a certain distribution of new land-use activities must be assumed. Two options for distributing the new activities are used to explore the benefits of the new land-use activities.

The exploration

After the new land-use activities have been described, their benefits can be estimated, by assuming the replacement of old, monofunctional land-use activities

Table XIII.2: Maximum available area for each multiple land use activity and maximum production per multiple-land-use activity

Multiple Land-use	pasture	hay land	arable land	specialized agriculture	forest	open nature	water
Areas (ha)	10342	10806	6893	5228	5821	10806	1268
Cows (1000x)	16	0	0	3	0	0	1
Production							
Energy total (1000 GJ/yr)	36	0	4	6	0	0	2
Biomass fodder (1000 ton/yr)	-64	46	48	14	0	48	-6
Water (1000 m^3/yr)	0	0	0	0	0	0	19025
Wood (1000 m^3/yr)	6	4	13	78	36	0	0
Nature, general (ha)	1034	10390	689	0	5821	0	0
Nature, specific (ha)	600	418	774	210	210	11496	1478
Recreation (1000 plcs/yr)	1117	292	502	2855	10624	1920	0
Recreation paths (km)	410	499	639	0	611	1081	0
Hedges (km)	780	416	1128	0	0	540	0

by new, multiple-land-use activities. (Jansen, 1997; De Graaf & Musters, 1997). Of the many possible options for replacement, only two could be worked out, given the time and money constraints of the project. These two options have not yet been based on a search for the optimal solution with respect to the regional goals of all stakeholders. Therefore, at this time the two options are merely tools to *explore the feasible value space*. They were aimed at providing information which could motivate the stakeholders within the steering group to continue the process of developing new forms of land-use, and to start demonstration and development projects. The results may be seen as an exploration of opportunities, and the conclusions as indicative. They should not be interpreted as recommendations or blueprints for land-use planning.

Options

Because only two options for replacing old activities by new ones could be worked out, the project team decided that one of them should come as close as possible to the present land-use, and to the actual impact of the implementation of new activities. The other option should be radical in the way new land-use would restore the original landscape of the region.

With regard to the way old activities are replaced by new, the two options can be described as follows:

Option 1: Nature and agriculture, combined with water catchment
The first option is cautious, with respect to replacing activities. It assumes the following principles:

- The major forms of land-use within the system are respected. Old, monofunctional activities are only replaced by new multiple-land-use activities with the same major product. Thus grassland remains 'grassland' but it acquires a multiple use: grass production is combined with the production and management of nature and ecological structures; forestry remains 'forestry', but becomes multiple by managing nature and offering more facilities for recreation.
- In the new land-use activity 'grassland', natural values are reinforced; in the case of grasslands near the brooks, the new grassland varieties 'wet species rich grassland' and 'hay land' are used. Outside these areas 'pastures' are used, and in the very dry parts 'species-rich dry grassland'.
- As far as possible, existing policy plans for changes in regional land-use have been taken into consideration. Therefore, in parts of the valleys the old activities are replaced by the new activities 'nature management' and 'water collection'.

Option 2: Recreation and agriculture, combined with water catchment
The second option is more radical. The original landscape qualities of the rural area around Winterswijk, where still present, are restored or improved. The agricultural activities which in the previous century led to a small-scale demarcation between low grassland and arable farming in the higher sections are respected. With regard to the replacement of old activities by new ones, the model is based on the following principles:

- There are considerable shifts in the major activities, but existing woods and nature areas are maintained, where they do not interfere with the new water-collection activities.
- Wet parts of the area become 'wet species-rich grassland'. The portions that are less wet will be used as 'pastures' and 'crop growing' is concentrated on the drier soils. Part of the multiple-land-use activities are aimed at supplying agricultural products which appeal to recreationists. These activities take place on dry soils, and 'forestry' on the very dry soils.
- As far as possible, existing policy plans for regional land-use changes are taken into consideration. Therefore, in parts of the valleys the old activities are replaced by the new activities 'nature management' and

'water collection'.

Common points of departure
The criteria for entering the multiple-land-use activities in the model are given above. In the case of land-use activities which do not depend on the resource characteristics, it was necessary to make assumptions about the extent to which they depend on other land-use activities. In dairy farming, the number of cows was related to the area of grassland, so as to respect the environmental standards for phosphate and nitrogen emissions. In the case of pig farming - an activity that is highly dependent on imported fodders - the number of pigs was set at about two-thirds of the current number, i.e., the expected result of the autonomous development due to production restriction policies (decision by the project team; Vreke, 1997). And finally, all possible forms of recycling were assumed to be carried out, so that all waste products from the area are used for the production of clean water, fodder, fertiliser, or energy carriers.

In both options one new recreation centre has been assumed. The possible consequences of the altered land-use for the recreational sector have been estimated in terms of the number of recreationists attracted by the landscape qualities in the two options (Veeneklaas & Langers, 1997). The employment and net value added are based on the estimated overnight stays (Veeneklaas & Langers, 1997).

XIII.6.2 Benefits of the two options for multiple land-use

Socioeconomic
The socioeconomic consequences of the two options are represented in Figures XIII.3a & b respectively. Both options result in a higher level of employment than might be expected in the case of autonomous development (Veeneklaas & Langers, 1997; Figure XIII.3a). Figure XIII.3b shows that the total net value added in the options is slightly smaller than in the case of autonomous development. However, the net value added was arrived at in a situation where the fodder import is halved, which means that the potential of meat production is limited. A fodder import comparable to the import in autonomous development can easily compensate for these small differences.

Both options are accompanied by an employment shift from land-use related employment to the services sector (Figure XIII.3a). This effect is greater in the second option than in the first. The net value added also reveals a shift - again, as expected - from agriculture to the services (recreation) and other sectors (conservation, water catchment, and energy production). In the present situation, because of the absence of benefits provided by biomass production or water

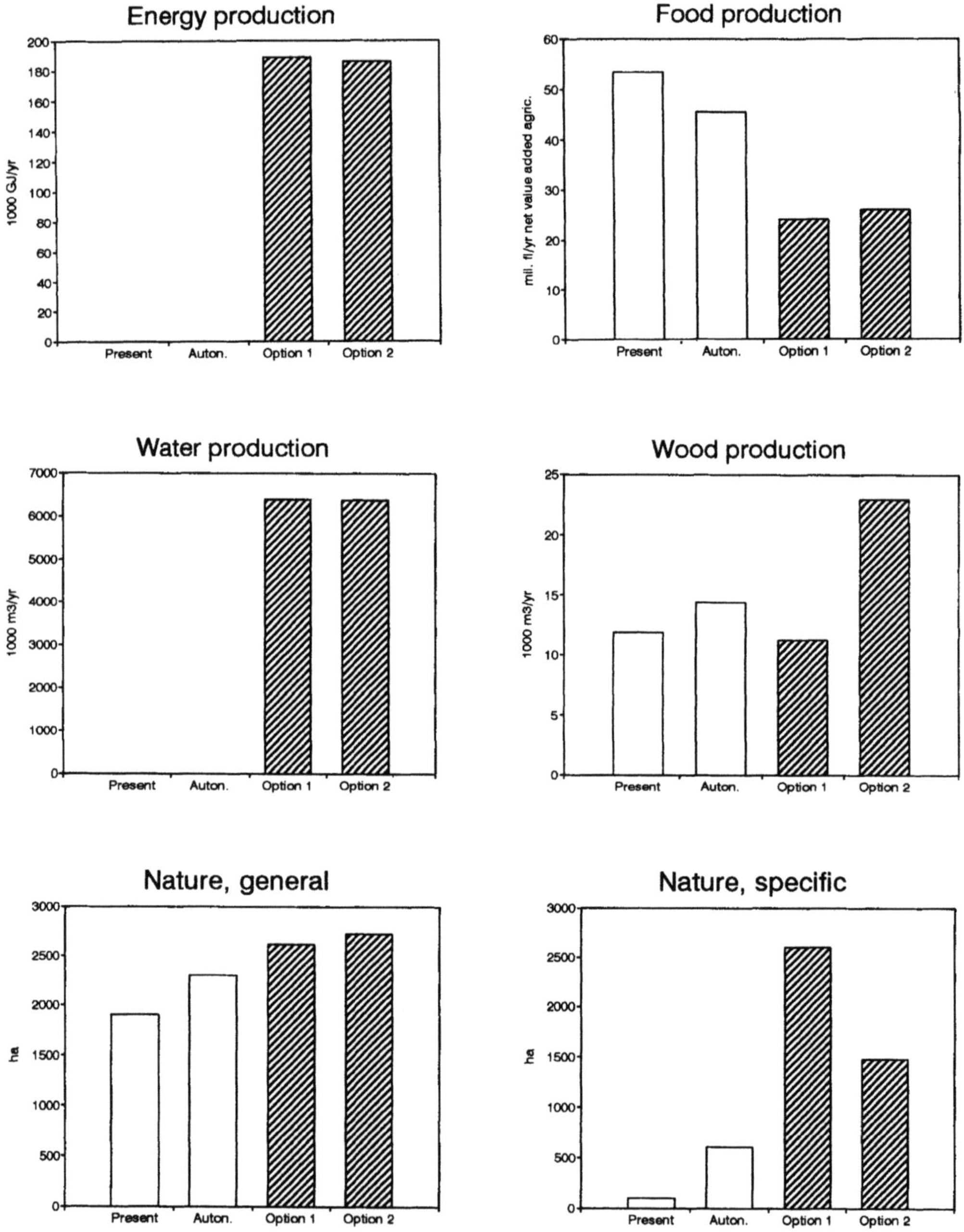

Figure XIII.5: The products in the two options

Products

The products supplied in the options are represented in Figure XIII.5. There is hardly any difference in energy production, water-collection, or nature conservation. The amount of energy supplied is compatible with the needs of 2000-3000 households. The amount of extra drinking water that can be collected is about the same as is currently being pumped out of deep, sandy layers.

The production of wood and 'specific' nature differs considerably in the two options. In the first one, wood production is the same as in the autonomous development, while in the second, it is greater. The area with specific nature is larger in both options than in the case of autonomous development, but the difference is greater in the first option.

Landscape

The scenic consequences of the first option are represented in Figure XIII.6. When this figure is compared with Figure XII.2 (Chapter XII), which represents current land cover, it reveals an increase in the number of landscape varieties. The quantitative consequences of this option for land-use are shown in Table XIII.3. While the total amount of grassland is larger, the amount of grassland for cows is considerably smaller, which means that the number of cows is also smaller (Table XIII.4). The increased grassland area is situated in a large reserve of poor grassland, both wet and dry. The amount of arable land - including that used to cultivate green maize - is much smaller. Due to the switch to a land-use plan that broadens the crop base, the reduction in the area of green maize cultivation is greater than for 'arable land' as a whole. The area of woods is somewhat smaller, because of the new basins for water catchment near the brooks. While the total area of open nature appears smaller, this impression is deceiving, since the row does not include the important category of 'species-rich grassland'.

The scenic consequences of the second option are presented in Figure XIII.7. Compared with Figure XII.2, which represents the current land cover, there are drastic changes. Large areas of arable land appear. Agricultural activities aimed at attracting tourists have a considerable impact on the landscape. Table XIII.3 shows a considerable decrease in the grassland area. The remaining grasslands are used almost exclusively as pasture and hay land. The total arable land area has been expanded, but, again, the area reserved for green maize area is smaller. There are fewer woods than in the first option, and the number of cows is even smaller. As in the first option, the open area consists exclusively of grassland.

Table XIII.3: Land cover in the two options (in ha)

Land cover	Present situation	Autonomous development	Option 1: Nature & agriculture	Option 2: Recreation & agriculture
Grassland	7300	6700	8100	5400
- Pasture	7300	6700	6500	5100
- Meadow	0	0	400	300
- Poor grass	0	0	1200	16
Arable land	3400	3000	2300	4000
- Maize	2000	1700	400	600
Forests	1900	2300	1500	1800
Specialised agriculture	0	0	0	800
Open nature	100	600	500	500
Buildings	1000	1000	1000	1000

Table XIII.4: Livestock in the two options

	Present situation	Autonomous development	Option 1: Nature & agriculture	Option 2: Recreation & agriculture
Dairy cows (x1000)	17	19	10	9
Pig places (x1000)	66	40	40	40
Fodder import (x1000 ton/year)	102	99	59	41

Comparison

Comparing the results of the two options for new land-use with both the present situation and autonomous development results in roughly the same changes as far as socioeconomic consequences, products, and environmental impact are concerned. This is noteworthy in view of the fact that the second option is designed to bring about greater changes in land-use structure. At the same time, it is reassuring. The results are robust in the sense that they are insensitive to the extent and the precise location of various new land-use activities.

Both options generate more employment, most of it in the services sector. The socioeconomic prospect rely heavily on developing the recreational quality of the area, and also depend on the possibility of maintaining high livestock

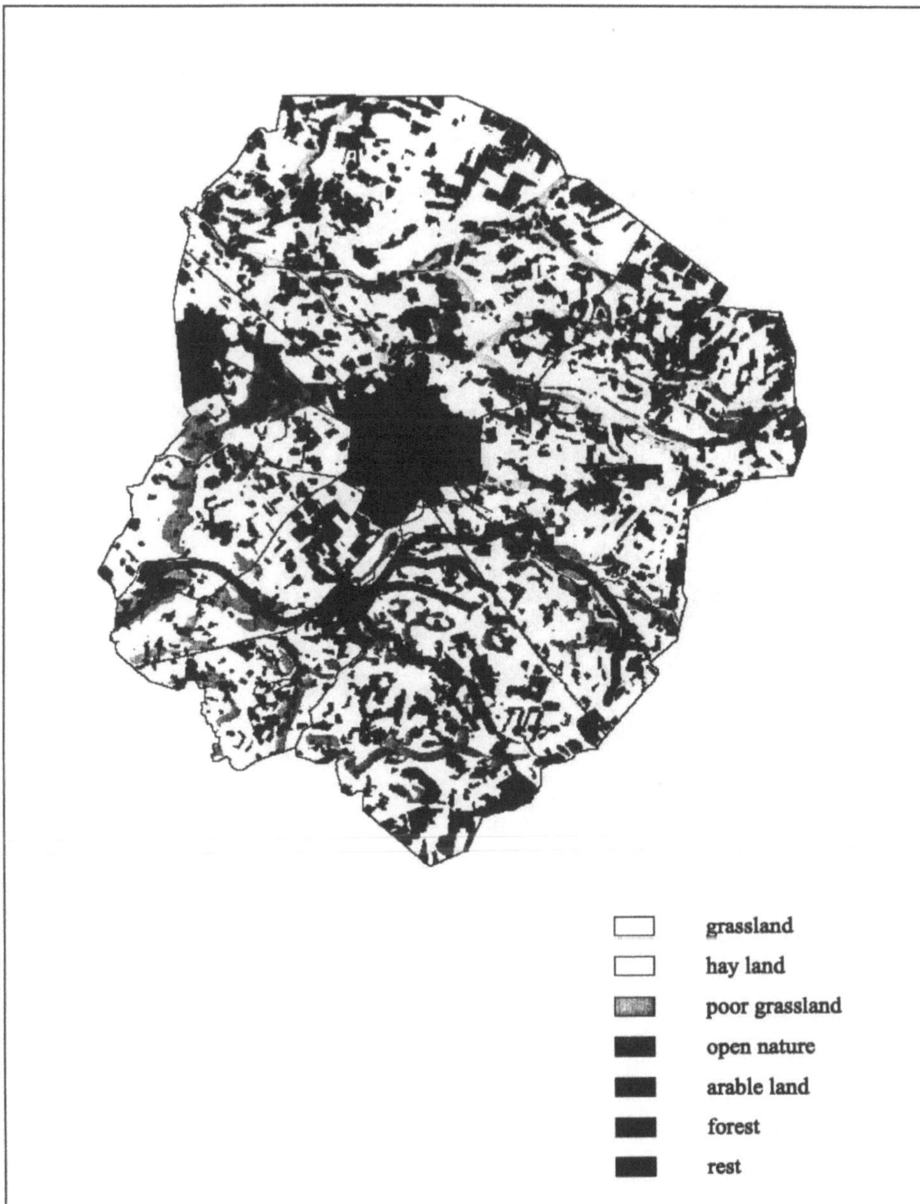

Figure XIII.6: The landscape in option 1: nature and agriculture, combined with water catchment

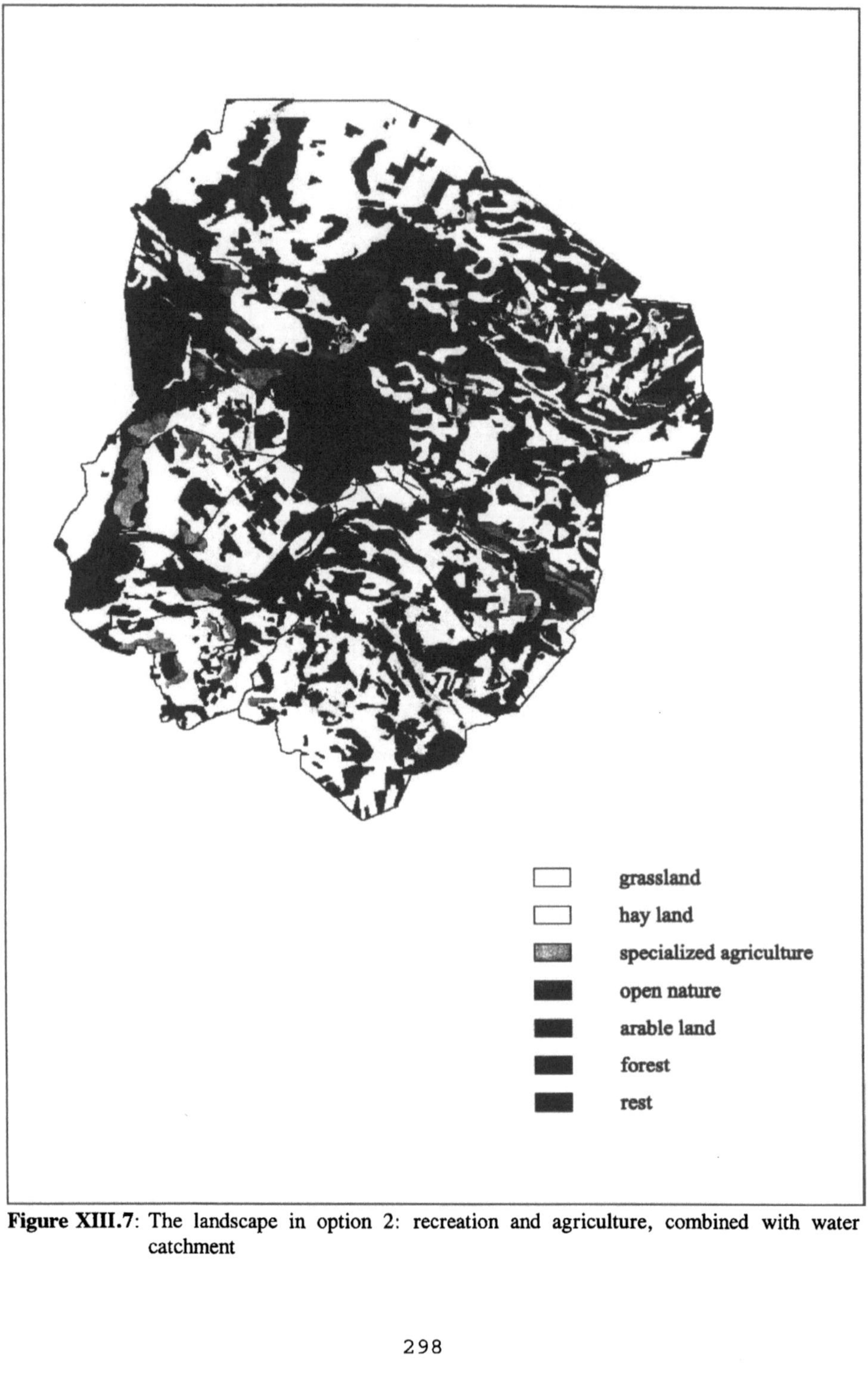

Figure XIII.7: The landscape in option 2: recreation and agriculture, combined with water catchment

production levels in an environmentally sustainable manner. This can succeed, if there is a consistently high level of fodder imports. Should these imports decline - for environmental reasons - then the productivity of the grassland and 'arable land' should be higher, but should not result in a decrease in the production of other products: this is admittedly quite a tall order for new technology.

There are, however, a number of differences between the options, especially in the amount of wood produced, the possibilities for nature and recreation, and the energy consumed by the livestock sector. The first option, aimed at the development of region-specific nature, offers fewer prospects for extending the recreational sector, because the changes in agriculture are insufficient to enhance the attractiveness for overnight recreationists. The second option, aimed at agricultural changes that attract recreation, produces more wood, consumes less energy, and results in a less unique nature.

XIII.6.3 Proposal and follow-up

The quantified regional benefits of the assumed new multiple land-use were, together with information on the multiple-land-use enterprises from the Working Group Technology, presented to the steering group of the project (De Kuijer *et al.*, 1997a). Furthermore, a proposal was drawn up for demonstration and development projects designed to operationalize the new land-use activities and the organizational structures. There were recommendations to monitor the development of new land-use and enterprises, to continue to inform the stakeholders about the potential benefits of this development, and to monitor the actual developments (Beeren *et al.*, 1997; Vos & De Graaf, 1997; Zwart, 1997). On the basis of the information put forward, the steering group decided to launch a follow-up project and to start designing the demonstration and research projects, as the first step towards operationalizing multiple land-use.

XIII.7 Discussion: problems encountered by applying the proposed procedure

The major question to be answered in Part Three was Can the phases proposed in Chapter VI be implemented in the way described, and is the proposed procedure effective and efficient in a context where gathering support and stimulating developments are the overall aims, and where the decision making process places constraints on the available time and budget for information-gathering? Does it still have the quality of information-gathering that it is supposed to have? The effectiveness of the procedure will be discussed first, followed by an examination of the actual application of the procedure within the constraints of external decision making.

Effectiveness

The study was performed within the context of a technology development programme. One of its aims was to show stakeholders the prospects of new forms of land-use based on new techniques. It was possible to provide the information desired within time and budget constraints, and this played an important role in decision making (De Kuijer, pers. comm.). At the conclusion of this study, the stakeholders indicated a readiness to take the next step and to invest in the actual technology development projects (De Kuijer & Van Wingerden, 1998). Thus one of the study's aims was achieved. However, information was provided on both regional benefits and new land-use enterprises, so this cannot be regarded exclusively as a benefit of the studies presented here.

Application

The working strategy of the project was such that the exploration of opportunities for new land-use was based on information from separate working groups. Consequently the basic information used could not be tested for qualities such as completeness. The completeness problem was partially compensated for by the fact that all decisions based on the information were discussed within the project team, and that the stop/go moments in the procedure were decided upon by the steering group in which the stakeholders were adequately represented. But it was not always possible to properly judge the quality of the information or, more importantly, to estimate the effects of more detailed information.

In order to *explore opportunities*, a model was designed in which the current land-use activities were replaced by new ones. The two options for replacement, which may been seen as directions of development, were chosen by the project team, not by the steering group. Because the direction of development to be explored is an important decision for the people involved, this choice should have been made by the steering group.

The technology working groups described the new activities on the basis of certain assumptions about what is 'realistic' on the one hand, and 'challenging' on the other. The accuracy of the options depended heavily on these assumptions. While these issues were discussed by the steering group, the impression remains that the technology working groups based their descriptions of new activities more on 'realism' than on 'challenge'. While this is a reasonable strategy, when the aim is to gain the support of stakeholders, it may be insufficient to generate an impulse aimed at developing new technology.

XIII.8 Concluding remarks: improvements to the proposed procedure

The conclusion of this analysis is that the procedure proposed in Chapter VI is capable of meeting the requirement to provide adequate information. Obviously, the results will depend on the quality of both the information-gathering process and the information itself. Although uncertainties remain with regard to the quality of the information, the quality of the information-gathering process was such that the stakeholders decided to continue the process of developing new land-use activities by designing demonstration and development projects (De Kuijer & Van Wingerden, 1998).

The applications discussed in Chapter XII & XIII show that the developed procedure can be embedded in a cyclical decision making process and that information gathered by others can be incorporated.

The use of the framework for selecting *valued system characteristics* shows that it may not be suitable for use by the people involved, but rather should be used by researchers to systematize the issues mentioned in interviews. If this reveals that certain types of characteristics are missing, then the researchers can consult the people involved again, to ascertain whether these omissions were deliberate, or important issues have been missed.

The decisions taken by the project team and the steering group tended to simplify the exploration of development opportunities by choosing a limited number of processes to be examined, and by ignoring sociocultural resource characteristics. Although this is understandable, it should be acknowledged that it is also a danger. Simplification could be taken to such lengths that the essence of the proposed procedure is lost, i.e., not all the different values at stake or available resources are taken into consideration in exploring the opportunities. In this way, one may lose sight of the main reason for the exercise: to find new and promising directions of development that may be sustainable.

XIII.8 Concluding remarks; improvements to the proposed procedure

[illegible]

PART FOUR

Summary, Perspectives, and References

CHAPTER XIV

Summary and Perspectives

CONTENTS

XIV.1 Introduction

In Part One of this work, the concept of sustainable development was elaborated, and a set of basic concepts developed, together with a methodology for scientific and applied research. On that basis, a procedure was inferred for a scientifically sound search for sustainable development. In particular, attention has been paid to methodologically appropriate gathering of the information needed for steering developments, resulting in a procedure for information-gathering. In Part Two, the proposed procedure was analyzed in order to improve it by applying it in four separate cases. These cases were projects in which the authors participated as a research team. In the first two projects, the methodology and the conclusions of the projects were re-analyzed by applying the newly proposed procedure, so that the strength and weakness of the latter could be established. In the second two projects, the application of the proposed procedure was simulated in order to analyze its efficiency. This analysis showed that modifications to the procedure were necessary. In Part Three, the application of the information-gathering procedure was presented as part of an overall decision making process. The theoretical and methodological framework, the proposed procedure, and the modifications needed formed the basis of our work on this decision making process. This part exemplifies that the conceptual framework and the proposed procedure provide a means for gathering appropriate and scientifically sound information on the opportunities for sustainable development, a means that was not yet available (Bingham *et al.*, 1995).

The concepts developed and the procedure proposed are shown in Box XIV.1. The concepts presented in italics in the text are defined in the box. Box XIV.2 gives an overview of the subjects discussed in the various chapters of this volume. The chapter ends with a discussion of the research performed.

XIV.2 Summary

XIV.2.1 Part One: Theory

The departure point of Part One is that the necessity to strive for sustainable development, i.e., attaining a broad spectrum of economic, social, cultural, and ecological goals while preventing negative economic, social, cultural, and ecological side-effects, is extremely pressing. Many people throughout the world have poor socioeconomic prospects and are endangered by environmental hazards, while the human, financial, and biophysical resources of the Earth are neither unlimited nor equally distributed (WCED, 1987; WRI, 1996). It is argued

Context:
the economic, social, cultural, and ecological surroundings of a *SEnS*

Development Opportunities (*DO*):
feasible opportunities for the development of a *SEnS* in the direction of acceptable *Value States*.

DO: see Development Opportunities

Functional Unit:
all the units of a *SEnS* that belong to the same combination of resources at the same time

Desired Value Space:
the complete set of *Value States* of the *SEnS* which are desired by the decision makers. The *Desired Value Space* may lie outside the *FVS*

Feasible Value Space (*FVS*):
the complete set of *Value States* of the *SEnS* which can be achieved at a given moment by steering at system level

FVS: see Feasible Value Space

People Directly Involved:
all persons or systems that value, use or provide - positively or negatively - the input, output, or structure elements of a *SEnS*, or the relation between input, output, and structure elements

People Indirectly Involved:
all persons or systems that are affected by the state or the functioning of the *SEnS*, but have no direct relation with the input, output, or structures of the system

Phase Space:
the complete set of states that a system can have

Resources:
the financial, sociocultural, and biophysical properties of a *SEnS* which can be used at the subsystem level to obtain or retain the *RSC* of that *SEnS*

Resource State:
the state of a *SEnS*, defined by the values of the variables that measure the *RSC*

Resource System Characteristics (*RSC*):
the set of properties of the *SEnS* that define the resources of the system, i.e., the set of input, output, and structure elements, or the relation between the input, output, and structure elements that are used or can be used to achieve a *Value State* of the *SEnS*

RSC: see Resource System Characteristics

SEnS: see Socio-Environmental System

Socio-Environmental System (*SEnS*):
a system in which economic, social, cultural, and ecological processes and structures are explicitly regarded as interconnected

---

Box XIV.1: Definition of concepts in the present volume developed.

Steering:
any deliberate action that is carried out in order to influence the course of developments

Sustainable Development:

1. any development of a *SEnS* with a high potential for continuity, because it attains economic, social, cultural, and ecological goals.
2. any development of a *SEnS* on which representatives of the people involved have reached agreement, based on a discussion about the opportunities for attaining economic, social, cultural, and ecological goals.

Use space:
the set of possible ways in whcih a certain *resource state* of a *SEnS* can be used to obtain or retain the *Valued System Characteristics* of that *SEnS*

Valuable System Characteristics:
the set of properties that the *SEnS* has or may acquire, i.e., the set of present or future input, output, and structure elements, and their mutual relations, that affect people or other systems

Value Space:
the complete set of *Value States* of a *SEnS*

Value State:
the state of a *SEnS*, defined by the values of the variables that measure the *VSC*

Valued System Characteristics (*VSC*):
the set of wanted or desired properties the *SEnS* has or may acquire, designated by the people involved or their representatives, i.e., the designated set of present or future input, output, and structure elements, and their mutual relations

VSC: see Valued System Characteristics

Box XIV.1: continued

here that sustainable development requires both a value-oriented approach, in which the present needs and wants of the people involved are addressed, and an approach oriented towards 'resource use', in order to take into account present as well as future needs. Consequently, the information needed for steering in the direction of a sustainable development is actually information on opportunities: which values can be realized and how can they be realized. It is concluded that the aim of information-gathering must be a shift from 'awakening awareness' to 'exploring opportunities', and that decision makers must be given information on opportunities for development that is scientifically sound, bite-sized, comprehensible, and immediately applicable to the decisions they are in a position to make (Bingham *et al.*, 1995; Underwood, 1995; Risser, 1996).

Obviously, the availability of decisive information on sustainable development will not automatically lead to new policies and thus to sustainable development. Policy makers should be provided with comprehensive information

General procedure	Phases of proposed procedure	Part One: Theory					Part Two: Analysis of procedure				Part Three: Application		
		II	III	IV	V	VI	VII	VIII	IX	X	XI	XII	XIII
							R	R					
								S	S	S			
												L	L
		II	III	IV	V	VI	VII	VIII	IX	X	XI	XII	XIII
I Initiative	1 Start-up	d	a	a	a	F	d	F	d	d	F	F	F
II Specific-ation	2 Delimiting socio-environmental system	d	F	a	a	F	d	d	d	d		F	F
	3 Selecting valued system characteristics	d		F	a	F	d	a	F	d		F	F
	4 Measuring the resource state	d			F	F	d	a		F		F	F
	5 Exploring the opportunities for development	d			F	F	d	F		a		F	F
III Operationa-lization		a											
IV Implementa-tion		a											
V Evaluation		a											

Box XIV.2: Overview of the contents of the chapters of this volume. F: focus is on the subject; d: subject is discussed; a: minor attention is paid to the subject; R: research application; S: supporting parties; L: real life application; See text for further explanation.

on the effects of steering, as well as feasible strategies and instruments for steering developments (Ter Keurs & Meelis, 1986). This volume is concerned exclusively with gathering information during the development of strategy and the preparation of plans. It focuses on a procedure for defining the object to be steered as a system, selecting the needs and wants at stake, measuring the availability of resources and resource use, and establishing a method for exploring development opportunities. This is designed to provide the information from which the opportunities for development and the measures to be taken can be derived.

Chapter II, which discussed the major issues in the search for *sustainable development*, was aimed at designing an overall strategy. Most approaches aimed at sustainable development focus on one set of goals, while underplaying others. They do not address the question of how the proposed measures are to be embedded in a decision making process, and appear to regard environmental problems as a technical issue (e.g., Kozlowski & Hill, 1993; Goodland & Daly, 1996; Robèrt *et al.*, 1997). The outlines of a more detailed strategy for sustainable development were worked out, based on the concept of sustainable development as a political issue. As such, it must be debated in relation to socioeconomic, cultural, and environmental processes and structures within a single system: the '*socio-environmental system*' (*SEnS*; Figure XIV.1). This debate is intended to bring about agreement on that development among the people involved or their representatives.

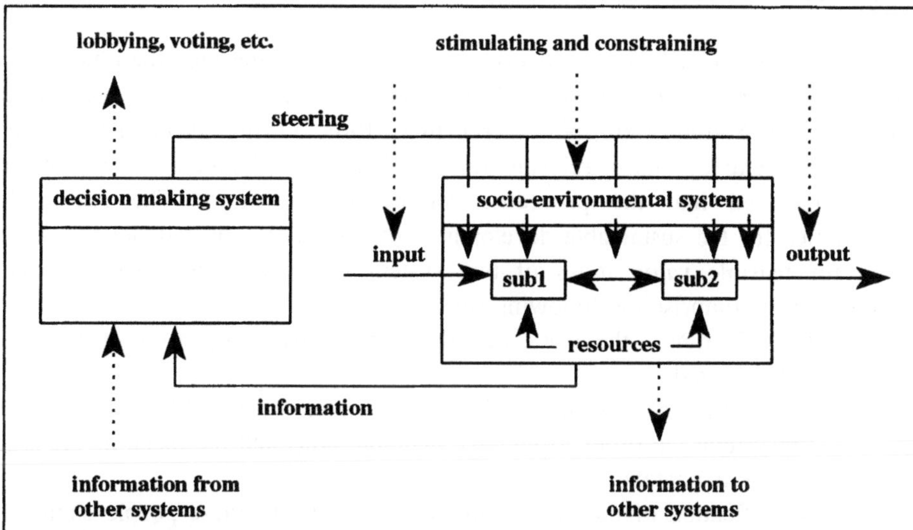

Figure XIV.1: Relationships between a decision-making system, a socio-environmental system, levels of scale, and the possibilities for steering. sub: subsystem.

To follow such a strategy, information will be needed on:

- the system that is to be developed in a sustainable way
- the needs and wants related to characteristics of the system
- the availability of resources that make it possible to satisfy those needs and wants
- the opportunities to modify the available resources and their use.

The analysis led to two definitions of sustainable development. One is formal, but not operational: *sustainable development* is a development of a *SEnS* with a high potential for continuity, because it attains certain economic, social, cultural, and ecological goals. The other definition is procedural, but does not guarantee continuity: *sustainable development* is a development of a *SEnS* on which representatives of the people involved have reached agreement, based on a discussion about the opportunities for attaining certain economic, social, cultural, and ecological goals.

At the end of the chapter it was concluded that at least two main areas require further study: the supply of information and the management of the debate and negotiations on decision making. It was suggested that a formal procedure for decision making may be drawn up, on the basis of the literature on consensus-building and experiences with 'environmental impact assessments' (e.g., Susskind & Cruikshank, 1987; Richard, 1993; Wood, 1995); that procedure is not elaborated in this volume.

In Chapter III, the problems and possibilities of defining the object to be steered as a *SEnS* were investigated. On the basis of general and hierarchical system theory (Simon, 1962; Von Bertalanffy, 1968; O'Neill *et al.* 1986), those aspects of system definition relevant for sustainable development were identified and incorporated into the procedure for defining a *SEnS*. The search for opportunities for the sustainable development of such a system was defined as the search for feasible *value states* of that system that match the goals set. This definition must comprise the following five parts:

1. Identification of the *SEnS*, i.e., the first tentative designation of the object to be steered.
2. Description of the hierarchical organization of the *SEnS* as a set of subsystems, in order to identify steerable units and the place of the *SEnS* within its context.
3. Identification of the people involved. Ideally, all these people should be represented in the decision making; at the very least, their needs and wants should be kept in mind by responsible decision makers.
4. Delimitation of the system in space and time. The spatial extent of the system, the time horizon of the developments, the detail level of the information, and the context of the system must be clearly defined.
5. An empirical description of the system in both structural and functional terms, i.e., the valued characteristics, the resources, and the relationship between them.

Defining the system serves not only to clarify the debate on the development of a *SEnS*, but also to facilitate the collection of adequate information on development opportunities on the basis of verbal or mathematical models. However, it has been demonstrated that even this first step toward a sustainable development of a *SEnS* entails a great many choices, and a specific sequence of questions to be answered.

Table XIV.1: A typology of the characteristics of a *socio-environmental system* related to types of needs and wants of the people involved, now and in the future, both at system level and higher-scale level (see Chapter IV for explanation).

<table>
<tr><td colspan="3">Present values for:</td><td colspan="2">the present</td><td colspan="2">the future</td></tr>
<tr><td rowspan="2"></td><td colspan="2">derived from:</td><td>processing</td><td>use</td><td>reserves for future use</td><td>capacities for future processing</td></tr>
<tr><td></td><td>type of system charac-teristic</td><td>processes</td><td>structures</td><td>structures for future use</td><td>structures for future processing</td></tr>
<tr><td rowspan="2">- direct users</td><td>direct users</td><td>separate</td><td></td><td></td><td></td><td></td></tr>
<tr><td>direct users' activities</td><td>combined</td><td></td><td></td><td></td><td></td></tr>
<tr><td rowspan="2">- higher-scale com-munities</td><td>higher-scale activities</td><td>combined</td><td></td><td></td><td></td><td></td></tr>
<tr><td>higher-scale com-munities</td><td>separate</td><td></td><td></td><td></td><td></td></tr>
</table>

In Chapter IV a framework was developed for the selection of the valued characteristics of a *SEnS*. It does not list values, but rather provides a typology of the actual and potential properties of the system that may be valued by the people involved, now and in the future, either directly or via a higher-scale community (Table XIV.1). The framework was based on a hierarchical description of a *SEnS* in time and space (Chapter III). The framework was used to analyze four existing valuation approaches, which focus on economic items (Cobb *et al.*, 1995), ecological considerations (R.S. de Groot, 1992), human basic needs (Doyal & Gough, 1991), and human wants (Henderson, 1991). By combining approaches it was possible to cover most types of characteristics, but the value of combinations of characteristics for the future was not recognised in any of the approaches.

Based on this analysis, it was concluded that the framework is useful as a means of verifying existing valuations, and that it can also be used to select the *valued system characteristics*.

The search for valued characteristics starts with step four, as defined in Chapter III:

4. Define the time and scale axes of the framework on the basis of the system defined.

And it continues with operationalizing step 5 in the definition of the system:

5. List the needs and wants or issues of the people involved
6. Enter the needs and wants in the cells in the framework proposed to identify the *valuable system characteristics*.
7 Check to see whether characteristics can also be placed somewhere else, referring to another type of needs and wants.

Although not discussed, this valuation is followed by two additional steps:

8. Operationalize the content of the cells by reduction via one or more techniques, such as partial lumping, using indicators, or prioritizing and selecting valuable characteristics.
9. Select the variables to be measured in order to assess developments.

The variables that measure the characteristics may be seen as the agreed dimensions of the *value space* of the system. In addition to representing the relevant goal variables, the *valued system characteristics* also form the basis for measuring the *resource state* of a *SEnS*.

Chapter V analyzes the possibilities to measure this *resource state*. It deals with the constraints on the development of a *SEnS* which arise as a result of the limited availability of resources and the limited possibilities to make use of the resources which are available. The *resource state* and the *use space* of a *SEnS* were defined (Box XIV.1; Figure XIV.2), and a method was developed to provide information on the *resource state* and the *use space* for the exploration of development opportunities. It was argued that to measure a *resource state*, both the interrelation between resources, and the relationship between resources and human activities must be described. The type of information required from the point of view of a sustainable development was compared with the information gained by means of conventional methods in such different areas of investigations as resource management (Barbier, 1989; WRI, 1996; Costanza *et al.*, 1997), land evaluation (FAO, 1976; Fresco *et al.*, 1995), and integrated approaches and concepts (Opschoor & Weterings, 1994; Cohen, 1995; Rotmans & De Vries, 1997). The conclusion was that it is still necessary to operationalize the measurement of interrelated resources. A method for measurement was proposed that consists of the

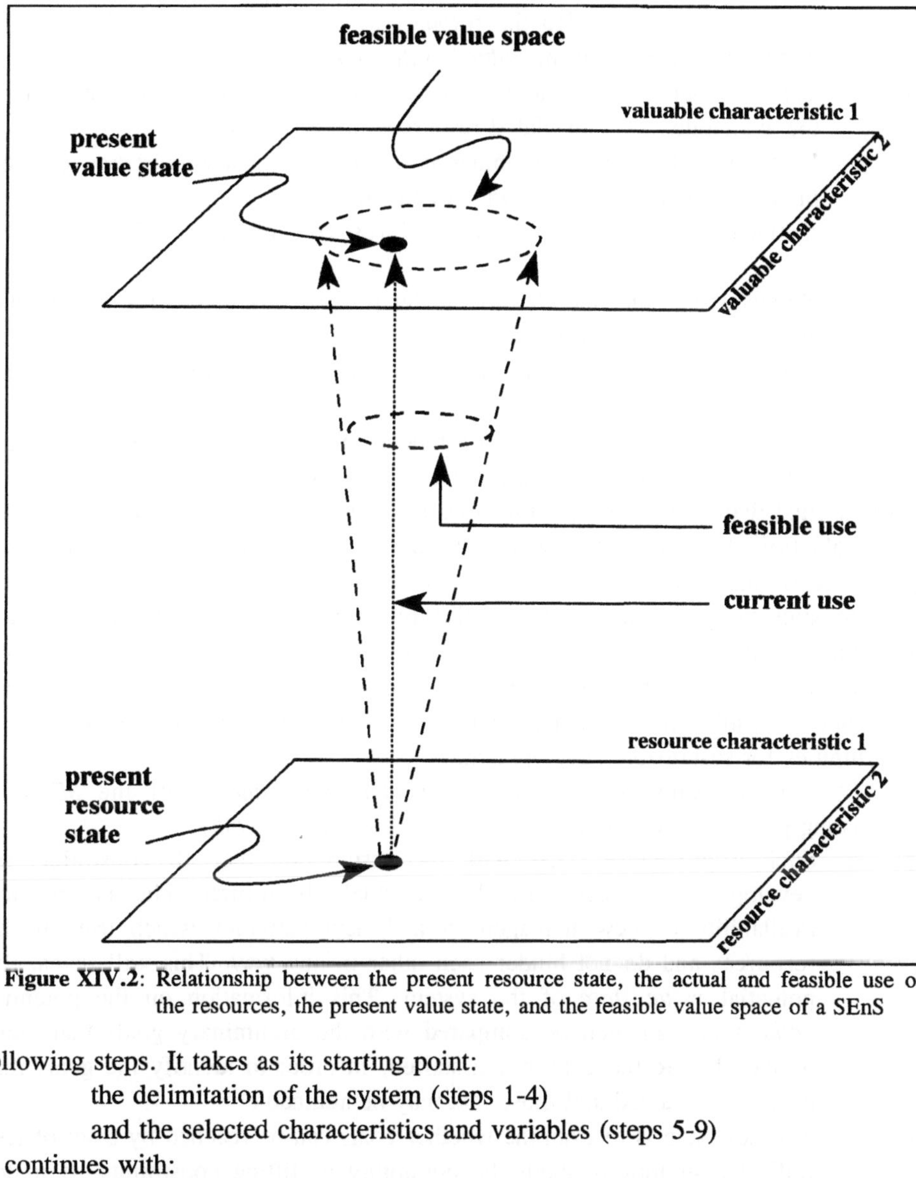

Figure XIV.2: Relationship between the present resource state, the actual and feasible use of the resources, the present value state, and the feasible value space of a SEnS

following steps. It takes as its starting point:

- the delimitation of the system (steps 1-4)
- and the selected characteristics and variables (steps 5-9)

It continues with:

10. the selection of the relevant processes and structures on the basis of the valued characteristics
11. on the basis of these processes and structures, resources are defined
12. then the requirements that define the *resources* are established and translated into system characteristics, in order to determine whether or not a specific part of the system represents a *resource*. These

characteristics are called the *resource system characteristics*, in order to distinguish them from the *valued system characteristics*.

13. The distribution of combinations of *resource characteristics* determines the availability of interrelated resources: the *resource state*.

14. Units with the same combination of *resource characteristics* can be presented as the *functional units* of the *SEnS*.

It was concluded that the following types of *resource characteristics* should be chosen:

- characteristics that facilitate or hinder processes or structures but are not altered by human steering or activities
- characteristics that facilitate or hinder processes or structures and can be steered
- characteristics that facilitate or hinder processes or structures, and are side-effects of human management or activities.

It was argued that if such an operationalization is based on recent developments in land evaluation, that evaluation should be extended to include other sectors than agriculture or nature alone. Furthermore, it should incorporate not only resource characteristics that relate a resource to different activities, but also characteristics that relate a human activity to different resources. And finally, it should make possible to explore modifications to resources as well as modifications to activities, and should result in a flexible and transparent instrument for exploring opportunities for sustainable development.

For an exploration of the *development opportunities* of the *SEnS*, a backcasting procedure was recommended (Schoonenboom, 1995).

15. Such a procedure starts with the assumption that all constraints on development that can be lifted have actually been lifted. This assumes the availability of new management and new activities which use all the resources and do not hinder each other in any way. This will reveal the potential *value state* of the system. This information on the potential *value state* can then be compared with the preliminary goals that were defined before the exploration started, in order to identify the goals that cannot be attained and those that may be attainable.

16. The next step is to examine whether goals can be reached by a set of less radical assumptions about the possibility of lifting constraints. If under these less radical assumptions a *value state* is reached that is likewise acceptable, a set of even less radical assumptions can be adopted, and so forth.

These steps will provide information about the feasible and desired *value space*, as well as about the least radical modifications by which it can be attained, i.e., modifications to the *use state* as well as the *resource state*. This method entails as an exploration of the *feasible value space*. This will show which developments are feasible and have the potential to reach a desired state of the system, thus identifying the opportunities for development.

In Chapter VI, the elements from the preceding chapters were integrated into the outlines of an information-gathering procedure embedded in a decision making process for steering developments. This procedure was divided into a logical and pragmatic sequence of phases which appears in many planning procedures (e.g., Glasbergen, 1995). Each phase focuses on a set of related decisions and is constructed in such a way that it results in a verified proposal. The five information-gathering phases and the relationship between them were discussed in detail (Figure XIV.3):

1. *Start-up*
2. Defining the *socioenvironmental system*
3. Selecting the *valued system characteristics*
4. Measuring the *resource state*
5. Exploring the *development opportunities*

These phases divide the sixteen steps mentioned above into five separate sets of related steps.

The focus in Chapter VI was on the coherence of the information-gathering within and between the phases. Coherence between information-gathering, management, and participation is crucial in preventing inconsistencies in decision making. The application of a procedure as proposed here is intended to ensure this coherence. When this procedure is embedded in a real-life decision making, its linearity may be lost, but that does not necessarily mean that coherence is also lost. It can be retained by applying the proposed procedure iterative, through the incorporation of restart moments. In any case, this will ensure that decisions made during the process which conflict with previous discussion are documented.

It was emphasized that the identification of the people involved plays an important role, since they are a crucial source of information about the needs and wants to be satisfied by the system and the possibilities to do so.

XIV.2.2 Part Two: Analysis of proposed methods and procedures

In Part One, the elaboration of the concepts of sustainable development resulted in the proposal of a conceptually and methodologically founded procedure for information-gathering on opportunities for sustainable development. Part Two dealt with the questions Can this developed methodology and proposed procedure

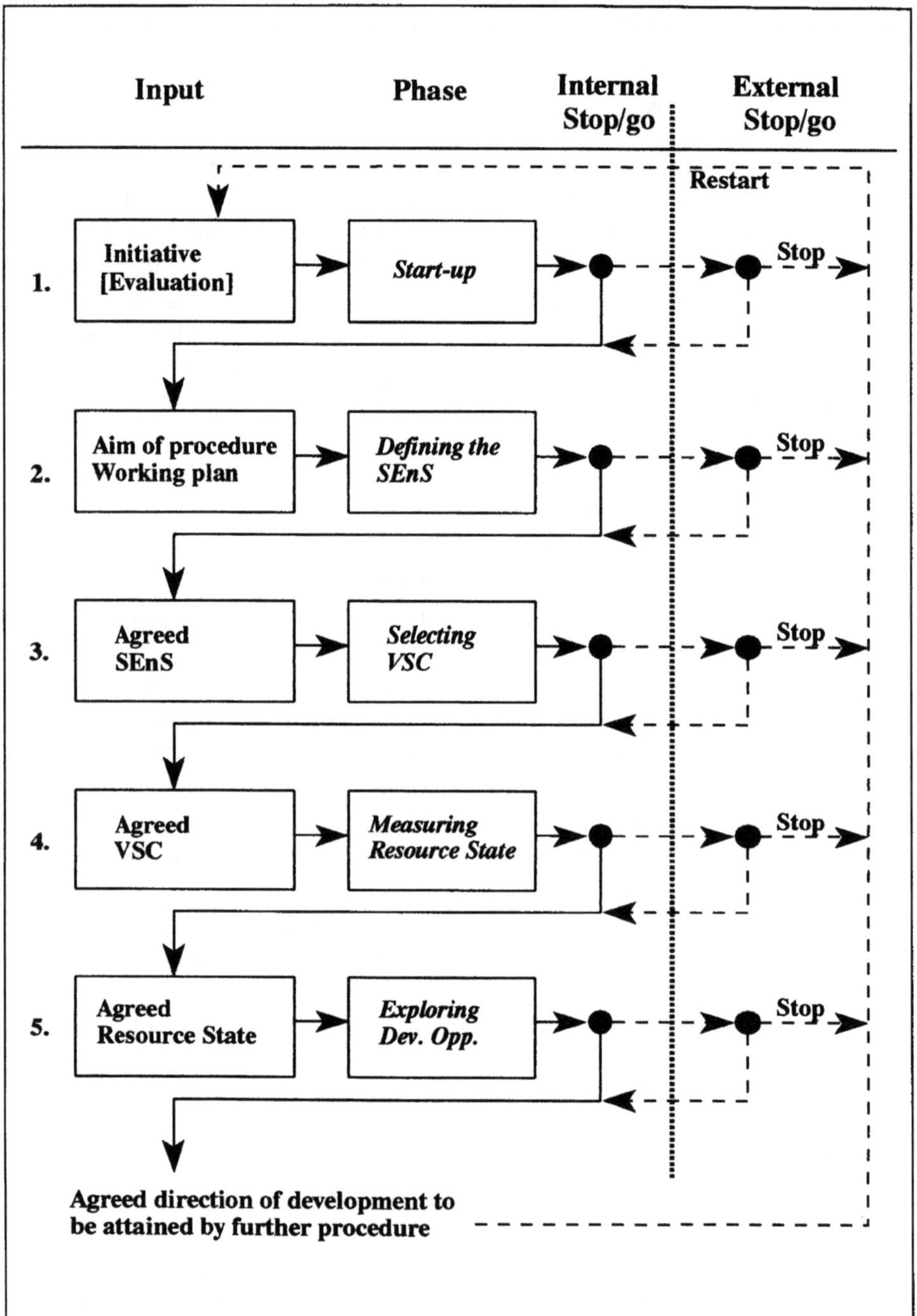

Figure XIV.3: Phases in the proposed procedure for information-gathering for sustainable development

be applied to support scientific research, parties involved in the debate on sustainable development, and decision making in planning processes? And how should the methods and procedure be modified to make it more suitable for application? Four case studies were analyzed; these exemplify parts of the proposed procedure and permit an initial examination of its strengths and weaknesses. Thus, the analysis focused on the quality of the proposed procedure in order to improve it. The case studies were divided into two groups.

The first two studies dealt with the complex of social, economic, and ecological problems and developments characteristic of Dutch agriculture (Chapters VII & VIII; Van der Weijden *et al.*, 1984a & b; Van der Wal *et al.*, 1985; Bakker, 1986; De Graaf, 1990; De Graaf & Tamminga, 1990). A method was followed in which the proposed procedure is used to examine the choices which - according to the proposed procedure - should have been made in the studies under discussion. These choices, and their consequences, are compared with those actually made in the studies.

The first case (Chapter VII) concerned the studies carried out as part of the project 'Dutch Integrated Agriculture' (Van der Weijden *et al.*, 1984a & b; Van der Wal *et al.*, 1985; Bakker, 1986). The analysis of the project using the proposed procedure for sustainable development revealed a number of strong and weak points in both the project and the proposed procedure.

- In the *Start-up* of the project a double aim was formulated while the proposed procedure implicitly stipulates that only one aim would be chosen in the *start-up* phase. It was concluded that while the double aim does not render the execution of the proposed procedure impossible, it does increase the risk that feasible solutions will be missed. When a double aim is formulated, it becomes even more important to be extremely explicit about the type of modifications to be assumed. No arguments were found to exclude the possibility of adopting more than one aim.
- It was concluded that the proposed procedure should be less strict with regard to system definitions. What in some respects can be considered as an 'too open' system may also meet the definition of a *SEnS*. However, if the *SEnS* is too open, care should be taken in the selection of boundaries, in order to keep the subsequent phases manageable. More attention should be given to what is seen as indirect involvement.
- The analysis showed that practical research can lead to a set of characteristics that includes items corresponding to all the types distinguished in the framework for selecting *valued system characteristics*. In that sense, the framework could be regarded as

superfluous. On the other hand, it was useful for purposes of verification on the level of goal variables, to make sure that all *valued characteristics* are taken into account when describing the *resource state* and its use.

- The analyses of the exploration of the *resource state* for an Integrated Agriculture showed that there are major financial and temporal constraints involved, which would have frustrated the application of the proposed procedure. But it also showed that the availability of a procedure for a systematic exploration could have led to the identification of new options which were within the budget of the project.

Chapter VIII analyzes the case 'Reduction in Production in the Dairy Sector in the Netherlands', a project aimed at exploring and - if possible - reconciling social, economic, and ecological visions on reduction of production in Dutch dairy farming (De Graaf, 1990; De Graaf & Tamminga, 1990). The emphasis was on the consistency of the steps taken in the execution of that project.

The project shows that even when each phase of the proposed procedure for gathering-information for sustainable development is executed correctly, the procedure as a whole may not supply the information aimed at in the *start-up* phase. The consistency of the procedure was violated by external decision making, in this case by decisions of the steering committee. Therefore, the evaluation of the project did not provide a great deal of information on the strength or weakness of the separate steps of the procedure. The discussion of the project shows that the proposed procedure as a whole can be used as a model for evaluating existing research. Moreover, it pointed up the importance of the setting in which the procedure is executed. In this respect, the following remarks were made:

- Whoever is executing the procedure should verify that external decisions pertaining to the procedure are in accordance with the original aims of the procedure. If this is not the case, the procedure should either be suspended, or restarted, i.e., the *start-up* should be executed again.
- Representatives of the people involved - in this case, the members of the parties involved in de debate - should participate in the external decision making. This will ensure that unilateral decisions do not alter the original aims of the procedure.
- Governmental organizations may take decisions that they assume to be 'politically realistic', but that interfere with an open exploration of development opportunities. This argues in favour of an independent process manager, to supervise the composition of the decision making group and see to it that there is an equitable distribution of positions within the decision making process.

In the next two chapters (Chapters IX & X), the proposed procedure was simulated in order to collect information on its effectiveness and efficiency. The application of the procedure was simulated assuming that little financial resources were available. Thus its usefulness was tested as an information-gathering instrument for parties involved in a debate about sustainable development that have few means.

The first case (Chapter IX) dealt with the provision of information for the debate on the development of a region. The object of study was Rijnmond, a region at the mouth of the river Rhine consisting of the Rotterdam harbour and the surrounding urban and rural areas (Dil *et al.*, 1993). The analysis was focused on the effectiveness and efficiency of the framework for the selection of *valued system characteristics* (Table XIV.1). It showed that systematic searching can provide a set of *valuable system characteristics* of such a region which can be used in the debate on the development of the region. A comparison was made between this set of valuable characteristics and a list of valued characteristics described by a group of organizations involved in drawing up a development plan for Rijnmond.

The study shows several ways to improve the guidelines for the use of the framework.

- The axes of the basic framework had to be divided, in order to make the framework suitable for the specific application at hand. It was concluded that this division is necessary in any application.
- The framework does not *a priori* include the specific direct users, activities, or aspects to be considered when selecting *valued characteristics*. According to the procedure, direct users, activities, and aspects to be considered are decided upon during the start-up or the system definition. In this study they were drawn from the development plan. However, cultural aspects were missing from the development plan and were not automatically included in the framework. In the same way, 'farming' was added to the list of relevant activities, since it appeared on the map of the region. If the discussion about the definition of the system had not been based on a map of the area, the need to add agriculture might not have been recognized. Thus supplementary information may be needed to complete the lists of direct users, activities, aspects, and higher-scale communities. It is advisable to use universal value lists from the literature to check the lists proposed in existing plans or in the start-up of a procedure for sustainable development.
- A characteristic could have been sought for each cell of the framework; this was not done, because it was assumed that there was no time or money to examine in full whether these characteristics reflect the real needs and wants of the people involved. But if it had been carried out, it

could easily have generated a huge list of potential valuable characteristics to be checked. It is preferable to use identified characteristics to search for new ones and the people who value them. The best way to controle the process of searching for *valuable characteristics* would appear to be by cooperating with the people involved (or their representatives) while using the framework iteratively as a search-and-select instrument. The power of the framework lies in the fact that it makes it possible to ask the right questions when the people involved are consulted on what they regard as the *valuable characteristics* of the *SEnS* to be developed.

The second case (Chapter X) examined the opportunities for a sustainable development of a dune and bulb-growing area (the 'Duin & Bollenstreek' in South Holland), by measuring its *resource state* and comparing the results with existing plans (Provincie Zuid-Holland, 1993; Projectgroep 'Toekomst Duin- en Bollenstreek', 1995).

The study produced the following results which are relevant for the modification of the developed procedure:

- It is possible to provide useful information about development opportunities, even with limited financial support and constrained availability of information.
- The selection of *resource characteristics* proved to be difficult. A formalized selection scheme was not available to answer such questions as: Which characteristics are essential for locating activities? And which essential characteristics are relevant for the level of scale of the simulation carried out here? In the end, the *resource system characteristics* selected had to be regarded as based on 'expert judgement'. Thus a formalized selection scheme should be added to the procedure.
- Although the *functional units* could fruitfully be used for exploring development opportunities, a table of *functional units* was not feasible, because it would have been too long. Therefore, the guideline that the *resource state* must be presented as a table of *functional units*, should be dropped. Using the data on the *resource state* of the *SEnS*, various figures and tables were presented, each showing a part of the information included in the *resource state*. The procedure could be further improved by adding recommendations on ways to present 'partial information' derived from the *resource state*.

XIV.2.3 Part Three: Application of the proposed procedure

This third part comprises the presentation of a example of the application of the proposed procedure in a real-life decision making situation. This project was based on the theoretical and methodological background as the procedure proposed and improvements of Part Two. Although the application was not aimed at regional planning or design, it came close to such a situation, since it was embedded in decision making on the development of new land-use activities. This application may be seen as a logical 'intermediate step' between the simulated applications in the previous chapters and a real-life planning application.

Chapter XI introduces two recent studies dealing with the opportunities for multiple land-use within the framework of the project 'Sustainable Land-Use' of the interdepartmental programme for Sustainable Technological Development (Aarts *et al.*, 1996; Musters & De Graaf, 1996; De Graaf & Musters, 1997; Jansen, 1997; De Kuijer *et al.*, 1997; De Kuijer & Van Wingerden, 1998). It deals with the a search for the regional opportunities of new land-use activities in the rural area around Winterswijk, a sandy-soil region of the Netherlands (Musters & De Graaf, 1996). The project presented here is aimed at options for land-use, not as blueprints for realization. It is not a land-use planning project, but rather a development-stimulation project that shows the opportunities of new land-use activities. Several organizations participated in the collection and processing of data. Chapter XII discusses the execution of the procedure to explore the potential to combine land-use processes. The decision making at the end of that study led to the start of a second study designed to explore the potential benefits of new land-use activities (Chapter XIII). Both chapters describe the part of the project that deals with the information-gathering on development opportunities on the regional level of scale.

The conclusion of the studies was that the proposed procedure is capable of fulfilling its aim. Obviously, the results will depend on the quality of both the information-gathering process and the information itself. Although uncertainties remain with respect to the quality of the information, the quality of the information-gathering process was such that the stakeholders decided to continue the process of developing new land-use activities by designing demonstration and development projects. With regard to the application, it was concluded that:

- The developed procedure can be embedded in a cyclical decision making process.
- The use of the framework for selecting *valued system characteristics* shows that it is not feasible to simply ask the people involved to fill in the blanks in the framework. Rather it should be used by researchers to

systematize the issues mentioned during interviews. If this shows that certain types of characteristics are missing, the researchers can again consult the people involved, to establish whether these omissions are deliberate or important issues have indeed been omitted.

- The decisions taken by the project team and the steering group tended to simplify the exploration of *development opportunities*, by choosing a limited number of processes to be taken into consideration, and by ignoring sociocultural resource characteristics. Although this is understandable from the point of view of process management, it should be acknowledged as a potential danger. Simplification could be carried so far that the essence of the procedure is lost. In this way, one may lose sight of the aims of the explorations, namely, to find new and promising directions of development that could be sustainable.

XIV.3 Perspectives

XIV.3.1 Improvements to the proposed procedure

We conclude that the proposed procedure for information-gathering can be fruitfully used for purposes of *support of research*, *support of parties*, and *support of planning* aimed at sustainable development. However, this does not mean that work on the procedure has been completed. As will be clear from the discussion of the analysis and the application, there is room for improvement, and not all parts of the procedure have been worked out in detail. For example, the way the *resource system characteristics* are deduced from the *valued system characteristics* is open to various interpretations. This may be regarded as a weakness, but it also allows the procedure to operate as a flexible model that can be modified to suit specific applications. It has been shown that the procedure can be adjusted in several ways, and that this causes few problems. However, it must be acknowledged that the developed procedure has not yet been fully applied to actual regional planning, neither in a situation characterized by real conflicts. And it is obvious from the case studies that the idea that socio-culturale resources may also determine *development opportunities*, is not worked out sufficiently. A further examination of the suitability of the procedure is necessary. Recently we started a study focusing on the suitability of the procedure for systematizing the values of biodiversity and measuring its resource and use space on various scales.

An important shortcoming of the developed procedure is the fact that little attention is given to the actual design of variants of 'directions of change' or 'options for development'. In Chapter VI it was stated that a well-chosen, comprehensive set of options is needed. Any proposed option should be based on a consistent set of assumptions about attitudes, value priorities, the relationship between values, autonomous developments in society, technology, uncertainties, risks, etc. (WRR, 1995; Lélé & Norgaard, 1996; Roe, 1996). The design of 'alternatives' is a poorly developed part of the research dealing with opportunities (W.T. de Groot, 1992), although a great many 'scenario' studies have been published (e.g., Schoute *et al.*, 1994). The concepts developed in this volume make it possible to formulate the problem of designing options more precisely: How do we design models that calculate potential *values states* of the *SEnS* as a function of both *use state* and *resource state* in such a way that these calculated states give a clear picture of the *feasible value space* of the *SEnS*, so that the acceptable or desirable parts of that *feasible value space* can be discussed and selected? Recommendations for solving this problem can be deduced from the procedure. For example, the models should obviously be consistent with the aims defined in the *start-up*, i.e., they should take into account the free space and the constraints of the decisions to be made. Furthermore, the use of backcasting to explore the development opportunities, as discussed in Chapter V, may be a point of departure for further searching into the conceptual and methodological aspects of this problem.

The procedure proposed here does not address the actual problems of decision making and participation, such as how to handle 'non-issues', i.e., issues that do not play a role in the political agenda of the moment, or how to deal with the possibility that the most powerful groups will use their power to obstruct the participation of their opponents. This is in fact what happened during the discussions on the 'Development Plan for Rijnmond' (Dil *et al.*, 1993). Industrial, business and government organizations dominated the discussions to such an extent that the environmental groups ultimately withdrew from the planning process. Their demands that more attention should be paid to several serious issues were not recognized (Van de Klundert & Glasbergen, 1995). The procedure is not designed to handle problems of this kind, although it may enable participants and non-participants to assess which issues have not been addressed during a decision making process.

XIV.3.2 Some fundamental issues

When designing and applying the proposed procedure, we also came across a number of fundamental issues that need to be discussed here.

The scope of the present study was limited in the sense that the procedure was designed for national, regional, and local levels of scale. On a number of occasions, we have noted that the procedure is not directly applicable at higher levels of scale. The higher the level of scale, the more uncertainties and lack of appropriate methods may be expected, for example, for selecting valuable characteristics, measuring the resource state, and exploring development opportunities. Further conceptual and methodological studies on the possibilities to develop effective and efficient procedures are needed. This is especially true of the global level of scale, where it is even hard to reach consensus about the environmental problems at stake (Van der Meij, *et al.*, 1995; Anonymous, 1997), although results have been reached in certain fields, such as global warming and biodiversity loss (WCED, 1987; UNEP, 1992; Masood, 1997). At high levels of scale, a problem-oriented rather than an opportunity-oriented approach may be more appropriate, as is illustrated in the discussions on reducing CFK and CO_2 emissions (Masood, 1997). The quest for a sustainable development of the Earth may depend to a large extent on the willingness of countries to take into account the major environmental problem at a global level when making decisions on their own development.

The information-gathering procedure proposed here is closely related to the rational planning of regional development in Western Europe in general, and the Netherlands in particular (e.g., Faludi & Van der Valk, 1994). Is the procedure applicable in less democratic or less developed societies? Is it applicable in societies in which some of the people involved have excessive power to influence decision making? As long as it is possible - and relevant - to study or plan developments, while taking into considerations economic, social, cultural, and ecological aspects, the procedure can be applied. If, however, it is not possible to take all aspects into consideration, for example, due to lack of co-operation on the part of the authorities, then the procedure cannot be seen as directed towards sustainable development and should not be employed. But, even where an approach aimed at sustainability has been adopted, the social or cultural circumstances may have unexpected effects on the applicability of a procedure developed in the tradition of Western rationalism (Lélé & Norgaard, 1996). Therefore, the wisdom of applying the procedure in a society that differs greatly from that of the Netherlands should be closely examined.

We have not dealt here in any detail with strategies for steering, plans for development, the implementation of plans, or the monitoring of effects. These aspects of planning and decision making are as important as the availability of comprehensive, decisive information. Leaving them out could result in what W.T. de Groot (1992) calls 'naive rationalism', i.e., the belief that following the right procedures will ensure the implementation of rational instruments and plans. The use of an information-gathering procedure does not in itself guarantee that the information will be used in decision making; in fact decision making may even be hindered by information. Furthermore, it may be difficult, if not impossible, to steer developments both democratically and rationally in a desired direction. W.T. de Groot (1992) mentions three ways to overcome the problem of 'naive rationalism', the first of which is explicitly addressed in the procedure proposed here, i.e., by striving to increase the participation of non-scientists, especially implementers and representatives of the people involved, thus creating and reinforcing the commitment to the information gathered. The second approach - which devotes explicit attention to the implementation context and the definition of its constraints - is described in the procedure as one of the two aims of the *start-up* phase: an explicit debate on the freedom and constraints of the decisions to be made. The third approach is to make implementation an explicit field of investigation; a first step in this direction has been taken by describing and analyzing the application of the procedure. In any case, the risk of 'naive rationalism' should not be employed as an argument against designing a scientifically sound, state-of-the-art procedure for sustainable development. Scientists should not shirk their responsibility to design the best possible instruments for decision making, simply because the use of these instruments is uncertain.

It was argued that one of the foundations of a debate on sustainable development is that all values at stake, all resources available, and all possibilities to use those resources should be taken into consideration. But can such a requirement of completeness be upheld in real-life situations? Is it not in fundamental conflict with the need to make decisions, whether the information is sufficient or not? Sometimes immediate decisions are needed, as in the case of disasters. But even leaving aside emergency decisions, choices have to be made in order to keep a process going. For example, in Part Three it was found that the decision makers tended to simplify the exploration of development opportunities by choosing a limited number of land-use processes for combining into the new, multiple-land-use activities. It was concluded that this endangers the exploration of opportunities; but how exactly does it endanger that exploration? How much can one simplify the search for sustainable development before the essence of the idea of sustainable development is lost? We cannot answer this question. What can be said on this issue is that by applying a procedure that is designed to incorporate all values, resources, and resource uses, one at least knows in what respects the information on which decision making is based is yet incomplete.

CHAPTER XV

References

Aarts, H.F.M., J.T.J. Beeren, H.J. de Graaf, O.C.H. de Kuijer, R. Kwak, A.B.W.M. van der Pas, J.W.B. Tiggeloven (1996). Illustratieproces Duurzaam Landgebruik: Duurzaam landgebruik Winterswijk. Eindrapportage fase A. DTO-werkdocument VD-DOC.39, Programma DTO, Delft.

Aarts H.F.M. & O.C.H. de Kuijer [red.] (1997a). Illustratieproces Duurzaam Landgebruik. Duurzaam Landgebruik. Van wensen en mogelijkheden naar voorbeeldsystemen. DTO-werkdocument VD-4. Programma DTO, Delft.

Aarts H.F.M. & O.C.H. de Kuijer [red.] (1997b). Illustratieproces Duurzaam Landgebruik; Duurzaam Landgebruik; van voorbeeldsystemen naar systeemonderzoek. DTO-werkdocument VD-5. Programma DTO, Delft.

Adger, W.N., K. Brown, R. Cervigni & D. Moran (1995). Total economic value of forests in Mexico. Ambio 24 (5): 286-296.

Aldred, J. (1994). Existence value, welfare and altruism. Environmental Values 3: 381-402.

Alfaro, R., J. Bouma, L.O. Fresco, D.M. Jansen, S.B. Kroonenberg, A.C.J. van Leeuwen, R.A. Schipper, R.J. Sevenhuysen, J.J. Stoorvogel & V. Watson (1995). Sustainable land use planning in Costa Rica: A methodological case study on farm and regional level. In: Fresco, L.O., L. Stroosnijder, J. Bouma & H. van Keulen [eds] (1995). The Future of the Land. Mobilising and Integrating Knowledge for Land Use Options. John Willey & Sons Ltd., Chichester.

Allen, T.F.H. & T.B. Starr (1982). Hierarchy. Perspectives for Ecological Complexity. The University of Chicago Press, Chicago.

Anonymous (1985). Interimwet beperking varkens- en pluimveehouderijen. Staatsblad 1985 1: 1-5.

Anonymous (1994). Britain's very own Brundtland strategy. Nature 367: 395-396.

Anonymous (1997). Environmental Scares: Plenty of Gloom. The Economist December 1997.

Archibugi, F. & P. Nijkamp [eds] (1989). Economy and Ecology: Towards Sustainable Development. Kluwer Academic Publishers, Dordrecht.

Asselt, M.B.A. van & J. Rotmans (1997). Uncertainties in perspectives. In: Rotmans, J. & B. de Vries [eds] (1997). Perspectives on Global Change. The TARGETS Approach. Cambridge University Press, Cambridge.

Atkinson, S. & T. Tietenberg (1991). Market failure in incentive-based regulation: The case of emission trading. Journal of Environmental Economics and Management, vol.21, no. 1: 17-32.

Bailey, K.D. (1990). Social Entropy Theory. State University of New York Press, Albany.

Bakker, J.J., B. van Dessel & F.J. van Zadelhoff (1989). Natuurwaardekaart 1988. Natuurgebieden, bossen en natte gronden in Nederland. Ministery of Agriculture, Nature conservation and Fisheries, Den Haag.

Bakker, Th.M. (1979). Landbouw en voeselvoorziening onder autarkische omstandigheden. Interne Nota nr. 236, Landbouw-Economisch Instituut, The Hague.

Bakker, Th.M. (1984a). Horizonten van zelfvoorziening. Mogelijkheden voor de voedselvoorziening van Nederland in autarkische omstandigheden. Publicatie 1.17, Landbouw-Economisch Instituut, The Hague.

Bakker, Th.M. (1984b). A model for the agricultural sector in The Netherlands. Agricultural Economics Research Institute (LEI), The Hague.

Bakker, Th.M. (1984c). RPD-onderzoek Scenario's produktiebeheersing. Verslagen van de interviews. Interne notitie van het project Productiebeheersing in de melkveehouderij, Landbouw-Economisch Instituut, The Hague.

Bakker, Th.M. (1985a). Eten van eigen Bodem. Een modelstudie. Proefschriften uit het LEI, no.1, Landbouw-Economisch Instituut, The Hague.
Bakker, Th.M. (1985b). Geïntegreerde landbouw: Een markeringspunt tussen monoloog en dialoog. Bedrijfsontwikkeling 16 (3): 109-110.
Bakker, Th.M. (1986). Geënsceneerde landbouw. Bouwen aan en spelen met een model van de Nederlandse landbouw. Landbouw-Economisch Instituut, The Hague.
Balstad Miller, R. (1994). Interactions and collaboration in global change across the social and natural sciences. Ambio 23 (1): 19-24.
Barbier, E.B. (1989). Economics, Natural-Resource Scarcity and Development. Conventional and Alternative Views. Earthscan Publications Limited, London.
Baumol, W.J. & W.E. Oates (1988). The Theory of Environmental Policy. Cambridge University Press, Cambridge.
Bauwens, A.L.G.M. (1983). Werkgelegenheid en werkgelegenheidsbeleid in land- en tuinbouw. Landbouw-Economisch Instituut, The Hague.
Beck, U. (1992). Risk Society. Towards a New Modernity. Sage Publications, London.
Becker, C.D. & E. Ostrom (1995). Human ecology and resource sustainability: The importance of institutional diversity. Annu. Rev. Ecol. Syst. 26: 113-133.
Beckerman, W. (1994). 'Sustainable development'. Is it a useful concept? Environmental Values 3: 191-209.
Beeren, J.T.J., H.F.M. Aarts, H.J. de Graaf, D. Jansen, O.C.H. de Kuijer, R. Kwak, A.B.W.M. van der Pas, J.W.B. Tiggeloven (1997). Illustratieproces Duurzaam Landgebruik. Projectvoorstel Praktijkontwikkelingsproject, Duurzaam Landgebruik Winterswijk. DTO-werkdocument VD-DOC.030. Programma DTO, Delft.
Bennett, R.J. & R.J. Chorley (1978). Environmental Systems. Philosophy, Analysis and Control. Methuen & Co LTD, London.
Bergh, J.C.J.M. van den (1991). Dynamic Models for Sustainable Development. Thesis, Tinbergen Institute, Amsterdam.
Berke, P.R. (1995). Towards planning for sustainable development. Landscape and Urban Planning 32: 73-75.
Bertalanffy, L. von (1968). General System Theory. Foundations, Development Applications. George Braziller, New York.
Bingham, G., R. Bishop, M. Brody, D. Bromley, E.(T.) Clark, W. Cooper, R. Costanza, T. Hale, G. Hayden, S. Kellert, R. Norgaard, B. Norton, J. Payne, C. Russell & G. Suter (1995). Issues in ecosystem valuation: Improving information for decision making. Ecological Economics 14: 73-90.
Biologische Station Zwillbrock (1995). Plan van aanpak aardkundige waarden in het WCL-gebied Winterswijk. Eigen uitgave Biologisch Station Zwillbrock, Vreden.
Bockstael, N., R. Costanza, I. Strand, W. Boynton, K. Bell & L. Wainger (1995). Ecological economic modelling and the valuation of ecosystems. Ecological Economics 14: 143-159.
Booth, D.E. (1994). Ethics and the limits of environmental economics. Ecological Economics 9: 241-252.
Bosatlas (1988). De grote Bosatlas. 50e editie. Wolters-Noordhoff Atlasproducties, Groningen.
Brink, B.J.E. ten, R. van den Berg & N.D. van Egmond (1994). Environmental utilisation space and the Environmental Outlook. Milieu 9 (5): 229-235.
Brinkman, R. (1995). Recent developments in land use planning, with special references to FAO. In: Fresco, L.O., L. Stroosnijder, J. Bouma & H. van Keulen [eds] (1995). The Future of the Land. Mobilising and Integrating Knowledge for land Use Options. John Willey & Sons

Ltd., Chichester.
Brown, L.R., C. Flavik, H.F. French, L. Startke, J.N. Abramovitz, C. Bright, S. Dunn, G. Gardner, A. Platt McGinn, J. Mitchell, M. Renner & J. Tuxill (1998). State of the World 1998. A Worldwatch Institute Report on progress Toward a Sustainable Society. W.W. Norton & Company, New York.
Bruijn, H. de & E. ten Heuvelhof (1994). Management of Environmental Policy Networks. In: Dutch Committee for Long-Term Environmental Policy [ed.] (1994). The Environment: Towards a Sustainable Future. Kluwer Academic Publishers, Dordrecht.
Budiansky, S. (1995). Chaos in Eden. New Scientist 14 october 1995: 32-35.
Cairns, J. & T.V. Crawford [eds] (1991). Integrated Environmental Management. Lewis Publishers Inc.,
Carson, R. (1962). Silent Spring. Houghton Mifflin, Boston.
CE (1982). Het CE-scenario "een realistisch alternatief". Centrum voor Energiebesparing, Delft.
Chadwick, G. (1971). A Systems View of Planning. Towards a Theory of the Urban and Regional Planning Process. Pergamon Press, Oxford.
Christensen, N.L., A.M. Bartuska, J.H. Brown, S. Carpenter, C. D'Antonio, R. Francis, J.F. Franklin, J.A. MacMahon, R.F. Noss, D.J. Parsons, C.H. Peterson, M.G. Turner & R.G. Woodmansee (1996). The report of the ecological society of America committee on the scientific basis for ecosystem management. Ecological Applications 6 (3): 665-691.
Clark, C.W. (1973). The economics of overexploitation. Science 181: 630-634.
Clayton, A.M.H. & N.J. Radcliffe (1996). Sustainability. A Systems Approach. Earthscan Publications Ltd, London.
Cobb, C., T. Halstead & J. Rowe (1995). The Genuine Progress Indicator. Summary of Data and Methodology. Redefining Progress, San Francisco.
Cocossis, H. & A. Parpairis (1992). Tourism and the environment - some observation on the concept of carrying capacity. In: Briassoulis, H. & J. van der Straaten [eds] (1992). Tourism and the Environment. Regional, Economic and Policy Issues. Kluwer Academic Publishers, Dordrecht.
Cohen, J.E. (1995). How Many People Can the Earth Support? W.W. Norton & Company, New York.
Cohen, J.E. (1997). Population, economics, environment and culture: An introduction to human carrying capacity. Journal of Applied Ecology 34: 1325-1333.
Common, M. & C. Perrings (1992). Toward an ecological economics of sustainability. Ecological Economics 6: 7-34.
Common, M.S., R.K. Blamey & T.W. Norton (1993). Sustainability and environmental valuation. Environmental Values 2: 299-334.
Conacher, A.J. (1994). The integration of land-use planning and management with environmental impact assessment: some Australian and Canadian Perspectives. Impact Assessment 12: 347-372.
Costanza, R. [ed.] (1991). Ecological Economics: The Science and Management of Sustainability. Columbia University Press, New York.
Costanza, R. (1996). Ecological economics: Reintegrating the study of humans and nature. Ecological Applications 6 (4): 978-990.
Costanza, R., d'Arge, R., de Groot, R., Farber, S., Grasso, M., Hannon, B., Limburg, K., Naeem, S., O'Neill, R.V., Paruelo, J., Raskin, R.G., Sutton, P. & van den Belt, M. (1997). The value of the world's ecosystem services and natural capital. Nature 387: 253-260.

Costanza, R. & B.C. Patten (1995). Defining and predicting sustainability. Ecological Economics 15: 193-196.

Crowards, T. (1997). Nonuse values and the environment: Economic and ethical motivations. Environmental Values 6 (2): 143-167.

Crul, M. & F. Schelleman (1995). Long-term environmental planning and the implementation of cleaner technologies. UNEP Industry and Environment April-September 1995: 15-20.

Daly, H.E. & J.B. Cobb (1989). For the Common Good. Beacon Press, Boston.

Denters, T., R. Ruesink & B. Vreeken (1994). Van muurbloem tot straatmadelief: Wilde planten in en om Amsterdam. Stichting Uitgeverij Koninklijke Nederlandse Natuurhistorische Vereniging, Utrecht.

Diepen, C.A. van, H. van Keulen, J. Wolf & J.A.A. Berkhout (1991). Land evaluation: from intuition to quantification. Advances in Soil Science 15: 139-204.

Dieperink, C. (1995). Between salt and salmon: Network management in the Rhine catchment area. In: Glasbergen, P. [ed.] (1995). Managing Environmental Disputes. Network Management as an Alternative. Kluwer Academic Publishers, Dordrecht.

Dietz, F.J. (1994). Natuurlijke hulpbronnen in de neoklassieke theorie. In: Dietz, F.J., W.A. Hafkamp & J. van der Straaten [red.] (1994). Basisboek milieu-economie. Boom, Amsterdam.

Dietz, F.J. & J. van der Straaten (1994). A note on the distributional effects of environmental policy goal setting. Milieu 9 (5): 253-256.

Dil. E., R.Adams, A.Doe, R. Kouprie, D. Weijsters [red.] (1993). Plan van aanpak en beleidsconvenant: ROM-project Rijnmond: Stuurgroep ROM-Rijnmond, Rotterdam.

Dodds, S. (1997). Towards a 'science of sustainability': Improving the way ecological economics understands human well-being. Ecological Economics 23: 95-111.

Doorn, J. van & F. van Vught (1978). Forecasting. Methoden en technieken voor toekomstonderzoek. Van Gorcum, Assen.

Dovers, S.R. & J.W. Handmer (1993). Contradictions in sustainability. Environmental Conservation 20 (3): 217-222.

Doyal, L. & I. Gough (1991). A Theory of Human Need. Macmillan Education Ltd, Basingstoke.

Driessen, P. & W. Vermeulen (1995). Network management in perspective. Concluding remarks on network management as an innovative form of environmental management. In: Glasbergen, P. [ed.], 1995. Managing Environmental Disputes. Network Management as an Alternative. Kluwer Academic Publishers, Dordrecht.

Droste, B. von & P. Dogsé (1991). Sustainable development: The role of investment. In: Goodland, R., H. Daly, S. El Serafy & B. von Droste [eds] (1991). Environmentally Sustainable Economic Development: Building on Brundtland. UNESCO, Paris.

Dutch Committee for Long-Term Environmental Policy [ed.] (1994). The Environment: Towards a Sustainable Future. Environment and Policy 1, Kluwer Academic Publishers, Dordrecht.

Dutch Parliament (1990). Actieplan Gebiedsgericht Milieubeleid. Tweede kamer 1990-1991, 21896, nrs. 1-2, The Hague.

Edel, B., J. Boom, e.a. (1980). Geïntegreerde landbouw - naar een landbouw die past bij hedendaagse doelstellingen van maatschappij en milieu. Vereniging Milieudefensie, Amsterdam.

Ester, P. & T. Mandemaker (1993). Socialization of environmental policy objectives: Tools for environmental marketing. In: Dutch Committee for Long-Term Environmental Policy [ed.] (1993). The Environment: Towards a Sustainable Future. Kluwer Academic Publishers, Dordrecht.

EU (1982). De toestand van de landbouw in de gemeenschap. Verslag 1981 van de Europese Commissie, Europese Unie, Brussel/Luxemburg.
EU (1983). Gemeenschappelijk landbouwbeleid: Voorstellen van de Commissie, Commissie van de Europese Gemeenschappen, Document COM (83) 500 def, Brussel.
EU (1985). Perspectieven voor het gemeenschappelijk landbouwbeleid. Groenboek van de commissie. Publicatie van de Dienst Landbouwvoorlichting- Directoraat-Generaal voor Informatie Commisie van de Europese gemeenschap, Brussel.
Faludi, A. (1986). Critical Rationalism and Planning Methodology. Pion Limited, London.
Faludi, A. (1987). A Decision-centred View of Environmental Planning. Pergamon Press, Oxford.
Faludi, A. & A. van der Valk (1994). Rule and Order Dutch Planning Doctrine in the Twentieth Century. The GeoJournal Library, Volume 28, Kluwer Academic Publishers, Dordrecht.
FAO (1976). A framework for land evaluation. Soils Bulletin 32, Food and Agriculture Organisation of the United Nations, Rome.
FAO (1983). Guidelines: Land evaluation for rainfed agriculture. FAO soils Bullitin 52, Food and Agriculture Organisation of the United Nations, Rome.
FAO (1984). Guidelines: Land evaluation for rainfed agriculture. FAO Soils Bulletins 52, Food and Agriculture Organisation of the United Nations, Rome.
FAO (1993). Agro-ecological assessments for national planning: the example of Kenya. FAO Soils Bulletin 67, Food and Agriculture Organisation of the United Nations, Rome.
Fearnside, P.M. (1997). Human carrying capacity estimation in Brazilian Amazonia as a basis for sustainable development. Environmental Conservation 24 (3): 271-282.
Folke, C., C.S. Holling & C. Perrings (1996). Biological diversity, ecosystems, and the human scale. Ecological Applications 6 (4): 1018-1024.
Frazier, J.G. (1997). Sustainable development: Modern elixir or sack dress? Environmental Conservation 24 (2): 182-193.
Freeman III, A. Myrick (1993). The Measurement of Environmental and Resource Values. Theory and Methods. Resources for the Future, Washington DC.
Fresco, L., H. Huizing, H. van Keulen, H. Luning & R. Schipper (1990). Land evaluation and farming systems analysis for land use planning. FAO Guidelines: Working Document, Food and Agriculture Organisation of the United Nations, Rome.
Fresco, L.O. (1995). Imaginable futures: A contribution to thinking about land use planning. In: Fresco, L.O., L. Stroosnijder, J. Bouma & H. van Keulen [eds] (1995). The Future of the Land. Mobilising and Integrating Knowledge for Land Use Options. John Willey & Sons Ltd., Chichester.
Fresco, L.O., L. Stroosnijder, J. Bouma & H. van Keulen [ed.] (1995). The Future of the Land. Mobilising and Integrating Knowledge for land Use Options. John Willey & Sons Ltd., Chichester.
Friend, J. & A Hickling (1987). Planning under Pressure. The strategic choice approach. Pergamon Press Oxford.
Funtowicz, S.O. & J.R. Ravetz (1990). Uncertainty and Quality in Science for Policy. Kluwer Academic Publishers, Dordrecht.
Funtowicz, S.O. & J.R. Ravetz (1991). A new scientific methodology for global environmental issues. In: Costanza, R. [ed.] (1991). Ecological Economics. The Science and Management of Sustainability. Columbia University Press, New York.
Gaston, K. [ed.] (1996). Biodiversity. A Biology of Numbers and Differences. Blackwell Science, Oxford.
Giampietro, M. (1994a). Using hierarchy theory to explore the concept of sustainable

development. Futures 26 (6): 616-625.
Giampietro, M. (1994b). Sustainability and technological development in agriculture. BioScience 44 (10): 677-689.
Giampietro, M. (1997). Socioeconomic constraints to farming with biodiversity. Agriculture, Ecosystems and Environment 62: 145-167.
Glasbergen, P. (1993). Naar een nieuwe vorm van sturing bij omgevingsvraagstukken. Landinrichting 33 8: 11-16.
Glasbergen, P. [ed.] (1995). Managing Environmental Disputes. Network Management as an Alternative. Kluwer Academic Publishers, Dordrecht.
Golley, F.B. (1993). A History of the Ecosystem Concept in Ecology. More Than the Sum of Parts. Yale University Press, New Haven.
Goodland, R. (1991). The case that the world has reached its limits: More precisely that current throughput growth in the global economy cannot be sustained. In: Goodland, R., H. Daly, S. El Serafy & B. von Droste [eds] (1991). Environmentally Sustainable Economic Development: Building on Brundtland. UNESCO, Paris.
Goodland, R. & H. Daly (1996). Environmental sustainability: Universal and non-negotiable. Ecological Applications 6 (4): 1002-1017.
Goodland, R., H. Daly, S. El Serafy & B. von Droste [eds] (1991). Environmentally sustainable economic development: Building on Brundtland. UNESCO, Paris.
Goodland, R.J.A., H.E. Daly & S. El Serafy (1993). The urgent need for rapid transition to global environmental sustainability. Environmental Conservation 20 (4): 297-309.
Graaf, H.J. de, M.J. Linders & H. van der Wal (1980). Toegankelijkheid & toepasbaarheid van het projectonderzoek geïntegreerde landbouw. Afdeling Biologie en Samenleving van de Subfaculteit Biologie, Rijksuniversiteit Utrecht.
Graaf, H.J. de (1990). Landbouwmodeluitkomsten en natuur- en milieu-effecten. Werkdocument w13. Milieubiologie, Rijksuniversiteit Leiden.
Graaf, H.J. de & C.J.M. Musters [red.] (1997). Illustratieproces Duurzaam Landgebruik. Duurzaam Landgebruik. Perspectieven voor het landelijk gebied van Winterswijk. DTO-werkdocument VD-2. Programma DTO, Delft.
Graaf, H.J. de, H. van der Wal & W.J. ter Keurs (1987). Scenario's productiobeheersing melkveehouderij: Voorspellen of verkennen met het LEI-Model. Notitie N. 15, Milieubiologie, Rijksuniversiteit Leiden.
Graaf, H.J. de & G. Tamminga (1990). Productiebeheersing in de melkveehouderij: Verkenningen van de gevolgen voor landbouw, natuur en milieu. Onderzoekverslag 70, Landbouw-Economisch Instituut en Milieubiologie Rijksuniversiteit Leiden, The Hague.
Gregory, R. & P. Slovic (1997). A construnctive approach to environmental valuation. Ecological Economics 21: 175-181.
Grizzle, R.E. (1994). Environmentalism should include human ecological needs. BioScience 44 (4): 263-268.
Grontmij/Biologisch Station Zwillbrock (1994). Gebiedsvisie natuur, bos en landschap Winterswijk. Grondmij Advies & Techniek, Zeist / Biologisch Station Zwillbrock, Vreden.
Groot, R.S. de (1992). Functions of Nature. Evaluation of Nature in Environmental Planning, Management and Decision Making. Wolters-Noordhoff, Groningen.
Groot, W.T. de (1992). Environmental Science Theory. Concepts and Methods in a One-World, Problem-oriented Paradigm. Studies in Environmental Science 52. Elsevier, Amsterdam.
Haila, Y. & R. Levins (1992). Humanity and Nature. Ecology, Science and Society. Pluto Press, London.

Hampicke, U. (1994). Ethics and economics of conservation. Biological Conservation 67: 219-231.

Hanegraaf, M.C., Meij, T. van der, P. Vos & H.J. de Graaf (1992a). Natuur en milieu in de landbouw: Relaties voor beleid en onderzoek. Deel 3. Zware metalen en arseen. Milieubiologie, Rijksuniversiteit Leiden.

Hanegraaf, M.C., Meij, T. van der, P. Vos & H.J. de Graaf (1992b). Natuur en milieu in de landbouw: Relaties voor beleid en onderzoek. Deel 4. Bestrijdingsmiddelen. Milieubiologie, Rijksuniversiteit Leiden.

Harris, C.C. & G. Brown (1992). Gain, loss and personal responsibility: The role of motivation in resource valuation decision-making. Ecological Economics 5: 73-92.

Haughton, G. & C. Hunter (1994). Sustainable cities. Kingsley Publications, London.

Hawksworth, D.L. [ed.] (1995). Biodiversity. Measurement and estimation. Chapman & Hall, London.

Heijman, W.J.M. (1991). Depletable Resources and the Economy. Wageningen economic studies 21, Agriculture University, Wageningen.

Heijungs, R. [ed.] (1992). Environmental Life Cycle Assessment of Products. Backgrounds. Centrum voor Milieukunde, Leiden.

Henderson, H. (1991). Paradigms in Progress. Life beyond Economics. Knowledge Systems Inc., Indiapolis.

Hetsen, H. (1993). Spatial conditions for a sustainable agriculture: rehgional differentiation in the Netherlands. Landscape and Urban Planning 27: 131-139.

Heywood, V.H. [ed.] (1995). Global Biodiversity Assessment. UNEP, Cambridge University Press, Cambridge.

Hidding, M.C. (1993). In search of new concepts of sustainable development of rural areas in the Netherlands. Landscape and Urban Planning 27: 259-264.

Hoanh, C.T., H. van Keulen & T.P. Tuong (1997). A model assessing water management strategies for integrated land-use planning in the Mekong Delta of Vietnam. In: Teng *et al.* [eds] (1997). Applications if Systems Approaches at the Farm and Regional Levels. Kluwer Academic Publishers, Dordrecht.

Hofkes, M. (1996). Objectieve duurzaamheid bestaat niet. Economisch Statische Berichten, 12-6-1996: 531-532.

Holling, C.S. (1986). The resilience of terrestrial ecosystems: Local surprise and global change. In: William C. Clark and R.E. Munn [eds] (1986). Sustainable Development of the Biosphere. Cambridge University Press, Cambridge.

Hourcade, J.C., J.M. Salles & D. Théry (1992). Ecological economics and scientific controversies. Lessons from some recent policy making in the EEC. Ecological Economics 6: 211-233.

Hueting, R. (1990). The Brundtland report: A matter of conflicting goals. Ecological Economics 2: 109-117.

Hueting, R. & P. Bosch (1991). Note on the correction of national income for environmental losses. In: Kuik, O. & H. Verbruggen [eds] (1991). In search of Indicators of Sustainable Development. Kluwer Academic Publishers, Dordrecht.

Hueting, R. & L. Reijnders (1996). Duurzaamheid is een objectief begrip. Economisch Statische Berichten, 8-5-96: 425-427.

Huizing, H., M.C. Bronsveld, S. Chandrapatya, M. Lathan, M. Omakupt, S. Panichapong, S. Patinavin, B. Saengwan & A. Sajjapengse (1995). Knowledge transfer to farmers and the use of information systems for land use planning in Thailand. In: Fresco, L.O., L.

Stroosnijder, J. Bouma & H. van Keulen [eds] (1995). The Future of the Land. Mobilising and Integrating Knowledge for Land Use Options. John Willey & Sons Ltd., Chichester.
Ierland, E. C. van (1993). Macroeconomic Analysis of Environmental Policy. Ph.D. thesis, University of Amsterdam, The Netherlands.
Jansen D. (1997). Intern work document. Winand Staring Centre for Integrated Land, Soil and Water Research (SC-DLO), Wageningen.
Jennings, M.D. & J.P. Renagold (1991). A Theoretical Basis for Managing Environmental Sensitive Areas. Environmental Conservation 18: 211-218.
John, K.H., R.G. Walsh & R.L. Johnson (1994). An integrated model of human-wildlife interdependence. Ecological Economics 11: 65-75.
Johnson, C.W. (1995). Planning and designing for the multiple use role of habitats in urban/suburban landscapes in the Great Basin. Landscape and Urban Planning 32: 219-225.
Jongh, P.E. de, W.J. ter Keurs & S. Algra (1978). De 'Landbouwverkenningen' en de 'Struktuurviesie Landbouw'. Wel grotere problemen voor de boeren, maar nog steeds geen oplossingen. Bèta 2: 48-55.
Jurgens, C.R. (1993). Strategic planning for sustainable rural development. Landscape and Urban Planning 27: 253-258.
Kessler, J.J. (1994). Usefulness of the human carying capacity concept in assessing ecological sustainability of land-use in semi-arid regions. Agriculture, Ecosystems and Environment 48: 273-284.
Kessler, J.J. (1997a). An introduction to: Strategic Environmental Analysis (SEA). SNV Netherlands development organisation, The Hague.
Kessler, J.J. (1997b). Strategic Environmental Analysis (SEA). A framework for planning of environmental care in development policies and interventions. reader with theoretical background ans application guidelines. AIDEnvironment, Amsterdam.
Keurs, W.J. ter & E. Meelis (1986). Monitoring the biotic aspects of our environment as a policy instrument. Environmental Monitoring and Assessment 7: 161-168.
Klundert, B. Van de, & P. Glasbergen (1995). The role of mediation in the process of integrated planning: Environmnetal planning in the area of the world's biggist harbor. In: Glasbergen. P. [ed.] (1995). Managing Environmental Disputes. Network Management as an Alternative. Kluwer Academic Publishers, Dordrecht.
Kolasa, J. & C.D. Rollo (1991). Introduction: The heterogeneity of heterogeneity: A glossary. In: Kolasa, J. & S.T.A. Pickett [eds] (1991). Ecological Heterogeneity. Springer-Verlag, New York.
Kormondy, E.J. (1969). Concepts of Ecology. Prentice-Hall, New Jersey.
Kozlowski, J. (1993). Towards 'ecological reorientation' of professional planning. In: Kozlowski, J. & G. Hill [eds] (1993). Towards Planning for Sustainable Development. A Guide for the Ultimate Environmental Threshold (UET) Method. Avebury, Aldershot.
Kozlowski, J. & G. Hill [eds] (1993). Towards Planning for Sustainable Development. A Guide for the Ultimate Environmental Threshold (UET) Method. Avebury, Aldershot.
Krijger, A. (1986). Twee jaar superheffing in Nederland. Economisch Statisische Berichten, ESB 3-12-86: 1189-1195.
Kuijer, Oskar de (1995). Op weg naar een duurzame voeding in 2040. Programma Voeden. Programmabureau Interdepartementaal Onderzoekprogramma Duurzame Technologische Ontwikkeling, Delft.
Kuijer, O.C.H. de, H.F.M. Aarts, J.T.J. Beeren, H.J. de Graaf, D. Jansen. R. Kwak, A.B.W.M. van der Pas & J.W.B. Tiggeloven (1997a). Illustratieproces Duurzaam Landgebruik;

Resultaten project Duurzaam Landgebruik, Resultaten fase B/C. DTO-werkdocument VD-7. Programma DTO, Delft.
Kuijer, Oskar de, Jaques Beeren & Leo Klep (1997b). Meervoudig duurzaam landgebruik: Een wenkend perspectief voor 2020, Brochure Interdepartementaal onderzoekprogrmama Duurzame technologische ontwikkeling, The Hague.
Kuijer, Oskar de & Paulijn de Bruijne (1996). Land in zicht. Duurzame Landgebruik in 2040. Programmabureau Interdepartementaal Onderzoekprogramma Duurzame Technologische Ontwikkeling, Delft.
Kuijer, O.C.H. de & W.S. van Wingerden (1998). Duurzaam Meervoudig Landgebruik Winterswijk. Programmaplan ontwerpfase vastgesteld door Stuurgroep Duurzaam Landgebruik 20 februari 1998. DTO-document, Programma DTO, Delft.
Kuik, O. & H. Verbruggen [eds] (1991). In Search of Indicators of Sustainable Development. Kluwer Academic Publishers, Dordrecht.
Landbouworganisaties Winterswijk (1995). Landbouwvisie mei 1995. Projectbureau Buitengebied Winterswijk, Winterswijk.
Latesteijn, H.C. van, D. Scheele & I.J. Schoonenboom (1994). Paradigms of sustainability and perceptions of environmental utilisation space. Milieu 9 (5): 244-252.
LEI & CBS (1996), Tuinbouwcijfers, Landbouw-Economisch Instituut (LEI-DLO) & Centraal bureau voor de Statistiek (CBS), 's-Gravenhage & Heerlen.
LEI & MIBI (1984). Scenario's productiebeheersing melkveehouderij. Landbouw-Economisch Instituut en Milieubiologie Rijksuniversiteit Leiden, The Hague.
Lélé, S. & R.B. Norgaard (1996). Sustainability and the scientist's burden. Conservation Biology 10 (2): 354-365.
Lemons, J. & D.A. Brown [ed.] (1995). Sustainable Development. Science, Ethics, and Public Policy. Kluwer Academic Publishers, Dordrecht.
Leroy, P. (1994). De ontwikkeling van het milieubeleid en van de milieubeleidstheorie. In: Glasbergen, P. (red.) (1994). Milieubeleid. Een beleidswetenschappelijke inleiding. VUGA Uitgeverij B.V., The Hague.
LNV (1992). Structuurschema Groene Ruimte. het landelijke gebied de moeite waard. Ontwerp-planologische kernbeslissing. Ministry of Agriculture, Nature Management and Fisheries, The Hague.
Lockwood, M. (1997). Integrated value theory for natural areas. Ecological Economics 20: 83-93.
LV (1984). Voorlichtingsbrief Superheffing. J3191, Minister van Landbouw en Visserij, The Hague.
Lörzing & Keijsers (1996). Recreatie in WCL-Winterswijk. Recreatieve zonering & mobiliteitsaspecten. Concept 4 april 1996, Lörzing & Keijsers.
Ludwig, Donald., Ray Hilborn & Carl Walters (1993). Uncertainty, Resource Exploitation, and Conservation: Lessons from History. Science Vol. (260): 17 & 36.
Maarel, E. van der & P.L. Dauvelier (1978). Naar een Globaal Ecologisch Model voor de ruimtelijke ontwikkeling van Nederland. Ministerie van Volkshuisvesting en Ruimtelijke Ordening. Studierapport Rijks Planologische Dienst, The Hague.
Mansvelt, J.D. van & J.A. Mulder (1993). European features for sustainable development: A contribution to the dialogue. Landscape and Urban Planning 27: 67-90.
Margerum, R.D. & S.M. Born (1995). Integrated Environmental Management: Moving from Theory to Practice, Journal of environmental Planning and Management 38 (3): 371-391.
Masood, Ehsan. (1997). Battling for science takes its toll on UN climatre panel stalwarts. Nature 390: 649-650.

Max-Neef, M. (1992). Development and human needs. In: Ekins, P. & M. Max-Neef [eds] (1992). Real-Life Economics. Understanding Wealth Creation. Routledge, London.
May, R.M. [ed.] (1981). Theoretical Ecology. Principles and applications. [2e ed.] Blackwell Scientific Publications, Oxford.
May, R.M., 1994. The economics of extinction. Nature 372: 42-43.
Mazzotti, F.J. & C.S Morgenstern (1997). A scientific framework for managing urban natural areas. Landscape and Urban Planning 38: 171-191.
McLeod, S.R. (1997). Is the concept of carrying capacity useful in variable environments? OIKOS 79: 529-542.
Meadows, D.H., D.L. Meadows & J. Randers (1991). Beyond the Limits. Confronting Global Collapse; Envision a Sustainable Future. Earthscan Publications Ltd., London.
Meadows, D.L., D,H. Meadows, J. Randers & W. Behrens (1972). The Limits to Growth: A Report for the Club of Rome's Project on the Predicament of Man. Universe Books, New York.
Meester, G (1985). De groene zorgen van Europa. Drie studies over het EG-landbouwbeleid. Proefschriften uit het LEI no.2. Landbouw-Economisch Instituut, The Hague.
Meester, G & D. Strijker (1985). Het Europese landbouwbeleid voorbij de scheidslijn van zelfvoorsiening. Voorstudies en achtergronden V46, Wetenschappelijke Raad voor het Regeringsbeleid, Staatsuitgeverij, The Hague.
Meij, T. van der, M.C. Hanegraaf, P. Vos & H.J. de Graaf (1992a). Natuur en milieu in de landbouw: Relaties voor beleid en onderzoek. Deel 1. Algemene inleiding en belangrijkste resultaten. Uitgave Milieubiologie, Rijksuniversiteit Leiden.
Meij, T. van der, M.C. Hanegraaf, P. Vos & H.J. de Graaf (1992b). Natuur en milieu in de landbouw: Relaties voor beleid en onderzoek. Deel 2. Natuurwaarden. Uitgave Milieubiologie, Rijksuniversiteit Leiden.
Meij, T. van der, J.H.W. Hendriks, C.J.M. Musters, H.J. de Graaf, m.m.v. M.A.W. Noordervliet (1995). Ontwikkelingen in de Natuur. Visies op de Levende Natuur in de Wereld en Scenario's voor Behoud daarvan. Wetenschappelijke Raad voor het Regeringsbeleid, Voorstudies en achtergronden V87, Sdu Uitgeverij, The Hague.
Meij, T. van der, H.J. de Graaf & A. Steekelenburg (1997a). Economische en ecologische doelen in Rijnmond nog niet op hetzelfde spoor. ROM Magazine jaargang 15 nr. 3 maart 1997: 12-15.
Meij, T. van der, C.J.M. Musters, G.J. Oudshoorn & H.J. de Graaf (1997b). MilieuGebruiksRuimte van de Duin en Bollenstreek. Rapport 9701. Milieubiologie Rijksuniversiteit leiden.
Melchers, M. & R. Daalders [red.] (1996). Sijsjes en drijfsijsjes: De vogels van Amsterdam. Schuyt & co, Haarlem.
Melchers, M. & G. Timmermans (1991). Haring in het IJ. De verborgen dierenwereld van Amsterdam. Stadsuitgeverij Amsterdam.
Midden, C.J.H. (1995). Direct participation in macro-issues: A multiple group approach. An analysis and critique of the Dutch national debate on energy policy, fairness, competence, and beyond. In: Renn, O., T. Webler & P. Wiedemann (1995). Fairness and Competence in Citizen Participation. Evaluating Models for Environmental Discourse. Kluwer Academic Publishers, Dordrecht.
Mierlo, J.G.A. van (1988). Pressiegroepen in de Nederlandse politiek. SMO-Boek, Stichting Maatschappij en Onderneming, The Hague.

Milieu Overleg Duin- en Bollenstreek (1994). De strijd om de ruimte, natuur en landschap in de knel? Reactie op de provinciale Nota Keuzevarianten en Keuzebepalingen voor de Duin- en Bollenstreek en Leidse Regio. Milieu Obverleg Duin- en Bollenstreek, Lisse.

Moffatt, I. (1996). An evaluation of Environmental Space as the basis for sustainable Europe. Int. J. Sustain. Dev. World Ecol. 3: 49-69.

Mohr, E. (1994). Environmental norms, society, and economics. Ecological Economics 9: 229-239.

Munasinghe, M. (1993). Environmental economics and biodiversity management in developing countries. Ambio 22 (2-3): 126-135.

Munda, G., P. Nijkamp & P. Rietveld (1994). Qualitative multicriteria evaluation for environmental management. Ecological Economics 10: 97-112.

Musters, C.J.M., M.J. Dix & H.J. de Graaf (1995). Het bepalen van de milieugebruiksruimte van oppervlaktewater. De geschiktheid van verschillende instrumenten daarvoor. Rapport 95-02, Milieubiologie, Rijksuniversiteit leiden.

Musters, C.J.M. & H.J. de Graaf (1992). Duurzame ontwikkeling: Een kwestie van wensen en mogelijkheden. Methode voor het integreren van doelstellingen. Milieubiologie, Leiden.

Musters, C.M.J. & H.J. de Graaf (1996). Illustratieproces Duurzaam Landgebruik: Werkplan gebiedsperspectief, Concept Phase B en C Concept-werkdocument VD-DOC.038, Programma DTO, Delft.

Musters, C.J.M., H.J. de Graaf & W.J. ter Keurs (1994a). Een methodiek om milieugebruiksruimte te bepalen. Ministerie van Volkshuisvesting, Ruimtelijke Ordening en Milieubeheer, Publicatiereeks Stoffen, Veiligheid, Straling nr. 1994/21.

Musters, C.J.M., H.J. de Graaf & W.J. ter Keurs (1997). Vooronderzoek Operationalisatie Biodiversiteit: Uitgangspunten, definities, werkwijze. Projectvoorstel 9706. Milieubiologie, Rijksuniversiteit Leiden.

Musters, C.J.M., H.J. de Graaf, M.A.W. Noordervliet & W.J. ter Keurs, (1994b). Measuring environmental utilisation space: Can it be done? Milieu 9 (5): 213-220.

Myers, N. (1993). Population, environment, and development. Environmental Conservation 20 (3): 205-216.

Myers, N. (1996). Economists and ecologists. BioScience 46 (9): 717-719.

Nieuwehuize, J. (1981). De landbouw - een verhaal apart. Voedingsbond FNV (Confederation of Duth Trade Unions), Utrecht.

Norton, B.G. (1991). Toward Unity among Environmentalists. Oxford University Press, New York.

Norton, B.G. (1994). Economists' preference and the preferences of economists. Environmental Values 3: 311-332.

Norton, B.G. & B. Hannon (1997). Environmental values: A place-based theory. Environmental Ethics 19 (3): 227-245.

Norton, B.G. & R.E. Ulanowicz (1992). Scale and biodiversity policy: A Hierarchical Approach. Ambio 21: 244-249.

Ofomata, G.E.K. & J.E. Umeuduji (1994). Topographic constraints to urban land uses in Enugu, Nigeria. Landscape and Urban Planning 28: 129-141.

Ojima, D.S., K.A. Galvin & B.L. Turner II (1994). The global impact of land-use change. BioScience 44 (5): 300-304.

O'Neill, R.V., D.L. DeAngelis, J.B. Waide & T.F.H. Allen (1986). A Hierarchical Concept of Ecosystems. Monographs in Population Biology 23. Princeton University Press, Princeton.

Oosterbaan, J. & H. Overbeek (1992). Ecologische Structuur Zuid-Holland. Fase 1 'Streefbeeld

ESZH'. Rijkswaterstaat Directie Zuid-Holland & Provincie Zuid-Holland Dienst Ruimte en Groen, Den Haag.

Opschoor, J.B. & S.W.F. van der Ploeg (1990). Duurzaamheid en kwaliteit: Hoofddoelstellingen van milieubeleid. In: Commissie Lange Termijn Milieubeleid [red.] (1990). Het Milieu: denkbeelden voor de 21ste eeuw. Kerckebosch BV, Zeist.

Opschoor, H. & J. van der Straaten (1993). Sustainable development: An institutional approach. Ecological Economics 7: 203-222.

Opschoor, J.B. & R. Weterings (1994). Environmental utilisation space: An introduction. Milieu 9 (5): 198-205.

Oskam, A.J., D.D. van der Stelt-Scheele, J. Peerlings, *et al.* (1987). De superheffing- is er een alternatief. Landbouw Universiteit Wageningen.

Parikh, J.K. & J.P. Painuly (1994). Population, consumption patterns and climate change. A socioeconomic perspective from the south. Ambio 23 (7): 434-437.

Parker, K. (1993). Economics, sustainable growth, and community. Environmental Values 2: 233-245.

Pearce, D. [ed.] (1993). Blueprint 3: Measuring Sustainable Development. Earthscan Publications Limited, London.

Perrings, C. [ed.] (1995). The economic value of biodiversity. In: Heywood, V.H. [ed.] (1995). Global Biodiversity Assessment. Cambridge University Press.

Peterson, M.J. & T.R. Peterson (1993). A rhetorical critique of 'nonmarket' economic valuations for natural resources. Environmental Values 2: 47-65.

Pezzey, J. (1992). Sustainability: An interdisciplinary guide. Environmental Values 1: 321-362.

Platt, J., 1969. Theorems on boundaries in hierarchical systems. In: Whyte, L.L., A.G. Wilson & D. Wilson [eds] (1969). Hierarchical Structures. American Elsevier Publishing Company, Inc., New York.

Ponting, C. (1991). A Green History of the World. Sinclair-Stevenson, London.

Projectgroep 'Toekomst Duin- en Bollenstreek' (1995). Pact van Teylingen. Werken aan het Pact (concept) Op weg naar Corso 2, fase 3 project 'Toekomst Duin- en Bollenstreek'. Projectbureau Toekomst Duin- en Bollenstreek, Lisse, 26 juni.

Proops, J.L.R., M. Faber, R. Manstetten & F. Jöst (1996). Achieving a sustainable world. Ecological Economics 17: 133-135.

Provincie Gelderland (1996a). Streekplan Gelderland. Provinciale Staten van gelderland, Arnhem.

Provincie Gelderland (1996b). Gelders Milieuplan voor de jaren 1996-2000. Provinciale Staten van gelderland, Arnhem.

Provincie Gelderland (1996c). Waterhuishoudingsplan Gelderland voor de jaren 1996-2000. Provinciale Staten van Gelderland, Arnhem.

Provinciale VVV Zuid-Holland (s.a.). Zuid-Holland Noord 1: 50000. Kaart voor vakantie en vrije tijd. Provinciale VVV Zuid-Holland, Delft.

Provincie Zuid-Holland (1987). Streekplan Zuid-Holland West, beschrijving en toelichting. Gedeputeerde Staten van Zuid-Holland, Provincie Zuid-Holland, Den Haag.

Provincie Zuid-Holland (1990). Milieubeleidsplan, Gedeputeerde Staten van Zuid-Holland, Den Haag.

Provincie Zuid-Holland (1992). Een beeld van het Zuidhollandse landschap. Samenvatting. Provincie Zuid-Holland, Den Haag.

Provincie Zuid-Holland (1993). Keuzevarianten en keuzebepalingen voor de Duin- en Bollenstreek en Leidse regio. Gedeputeerde Staten van Zuid-Holland, Provincie Zuid-Holland/Dienst Ruimte en Groen, Den Haag.

Provincie Zuid-Holland (1995a). Milieubeleidsplan 1995-1999. Provinciale Staten van Zuid-Holland, Den Haag.

Provincie Zuid-Holland (1995b). Regioprofiel Duin- en Bollenstreek. Provincie Zuid-Holland, Den Haag.

PWZH (1980). Waterstaatkundige kaart 1:25000. Provinciale Waterstaat Zuid-Holland, Den Haag.

Rabbinge, R. & M.K. van Ittersum (1995). Tension between aggregation levels. In: Fresco, L.O., L. Stroosnijder, J. Bouma & H. van Keulen [eds] (1995). The Future of the Land. Mobilising and Integrating Knowledge for Land Use Options. John Willey & Sons Ltd., Chichester.

Rebele, F. (1994). Urban ecology and special features of urban ecosystems. Global Ecology and Biogeography Letters 4: 173-187.

Redclift, M. (1993). Sustainable development: Needs, Values, Rights. Environmental Values 2: 3-20.

Rees, W.E. (1990). The Ecology of sustainable development. The Ecologist 20 (1): 18-23.

Renn, O., T. Webler & P. Wiedemann [eds] (1995). Fairness and Competence in Citizen Participation. Evaluating Models for Environmental Discourse. Kluwer Academic Publishers, Dordrecht.

Rennings, K. & H. Wiggering (1997). Steps towards indicators of sustainable development: Linking economic and ecological concepts. Ecological Economics 20: 25-36.

Richard, D. (1993). Regulatory negotiation: An assessment of its application. Environmental Impact Assessment Review 13: 189-198.

Riebsame, W.E., W.J. Parton, K.A. Galvin, I.C. Burke, L. Bohren, R. Young & E. Knop (1994). Integrated modelling of land use and cover change. BioScience 44 (5): 350-356.

Risser, P.G. (1996). Decision-makers must lead in defining some environmental science. Ecological Applications 6 (1): 24-26.

RIVM (1991). Nationale Milieuverkenning 2. 1990-2010. Rijksinsituut voor Volksgezondheid en Milieuhygiene. Samson H.D. Tjeenk Willink bv, Alphen aan den Rijn.

Robèrt, K.H., H. Daly, P. Hawken & J. Holmberg (1997). A compass for sustainable development. Int. J. Sustain. Dev. World Ecol. 4: 79-92.

Robinson, J.G., 1993. The Limits to caring: Sustainable living and the Loss of biodiversity. Conservation Biology 7 (1): 20-28.

Roe, E.M. (1996). Sustainable development and cultural theory. Int. J. Sustain. Dev. World Ecol. 3: 1-14.

ROM-Rijnmond (1992a). Streefbeeld Mainport, Streefbeeld Ruimtelijke Inrichting. Doelstellingen Milieu. Projectbureau ROM-Rijnmond, Rotterdam.

ROM-Rijnmond (1992b). Startconvenant ROM-project. Projectbureau ROM-Rijnmond Rotterdam.

Rooy, P.T.J.C. van (1995). Op weg naar totaal waterbeheer (3): Planvorming. H2O 28 (22): 666-672.

Rooy, P.T.J.C. van, J.W. van Sluis & H.H. Tolkamp (1995). PRIMAVERA: Decision support system for integral water management. European Water Pollution 5 (1): 6-9.

Rosen, R. (1991). Life Itself: A Comprehensive Inquiry Into the Nature, Origin and Fabrication of Life. Columbia University Press, New York.

Rotmans, J. & B.J.M. de Vries [eds] (1997). Perspectives on Global Change. The TARGETS Approach. Cambridge University Press, Cambridge.

RPD (1983). Het project Productiebeheersing melkveehouderij. Memo: nr 40, Rijksplanologische

Dienst, The Hague.
RPD (1984a). Het project Productiebeheersing melkveehouderij. Memo: 34/12, Rijksplanologische Dienst, The Hague.
RPD (1984b). Het project Productiebeheersing melkveehouderij. Notulen van de eerste vergadering van de begeleidingscommissie. Rijksplanologische Dienst, The Hague.
RPD (1985). Het project Productiebeheersing melkveehouderij. Notulen van de zesde vergadering van de begeleidingscommissie. Rijksplanologische Dienst, The Hague.
RPD (1987). Het project Productiebeheersing melkveehouderij. Selectie van varianten. Memo 31-03.1987, Rijksplanologische Dienst, The Hague.
Runhaar, J & C.L.G. Groen (1993). Kwetsbaarheidskaarten Zuid-Holland: Voorstudie natuur. CML-report 95, Centrum voor Milieukunde, Leiden.
RUU (1978). Projectvoorstel Geïntegreerde Landbouw, Landbouwprojectgroep, Subfaculteit Biologie RUU, Rijksuniversiteit Utrecht.
Sagoff, M. (1988). The Economy of the Earth: Philosophy, Law, and Environment. Cambridge University press, Cambridge.
Schoonenboom, I.J. (1995). Overview and state of the art of scenario studies for the rural environment. In: Schoute, J.F.Th., P.A. Finke, F.R. Veeneklaas & H.P. Wolfert [eds] (1995). Scenario Studies for the Rural Environment. Kluwer Academic Publishers, Dordrecht.
Schoute, J.F.Th, P.A. Finke, F.R. Veeneklaas & H.P. Wolfert [eds] (1994). Scenario Studies for the Rural Environment. Kluwer Academic Publishers, Dordrecht.
Shogren, J.F., J.A. Herriges & R. Govindasamy (1993). Limits to environmental bonds. Ecological Economics 8: 109-133.
Simon, H.A. (1962). The architecture of complexity. Proceedings of the American Philosophical Society 106 (6): 467-482.
Sips, H., R. Brinkman, M. Lok & L. Oosterhuis (1994). Environmental utilisation space and Dutch environmental policy. Milieu 9 (5): 209-212.
SNM (1979). Door productiebeheersing naar een ander beleid, Stichting Natuur en Milieu, Utrecht.
SNM (1982). Naar een milieuvriendelijke landbouw (discusssie nota). Werkgroep Landbouw van de Stichting Natuur en Milieu, Utrecht.
STD (1991). Voorstel voor een Nationaal Onderzoek Programma Duurzame Technologische Ontwikkeling. Programma Milieu en Technologie, Wb91-26, The Hague.
STD (1992). Het onderzoek programma DTO op hoofdlijnen. Concept. Document wp\dto\dto.030, The Hague.
STD (1995). Duurzaam landgebruik, definitiestudie. AB-DLO, PE-LUW, MIBI-RUL, & Heidemij Advies B.V. DTO-werkdocument VD-0, Delft.
STD (1996). Duurzaam landgebruik, Map resultaten fase A en werkplannen fase B/C DTO, AB-DLO, Heidemij Advies B.V., MIBI-RUL & Stichting WCL. DTO-werkdocument VD-1, Delft.
STD (1997). DTO sleutel Voeden; spectrum van een duurzame voedselvoorziening. Interdepartementaal onderzoekprogrmama Duurzame technologische ontwikkeling, The Hague.
Steekelenburg, A. (1995). Inrichting Duin- en Bollenstreek vraagt om regisuer met toekomstvisie. Milieu Aktief 95/3: 5-8.
Steverson, B.K. (1995). Contextualism and Norton's convergence hypothesis. Environmental Ethics 17: 135-150.

Stinner, D.H., B.R. Stinner & E. Martsolf (1997). Biodiversity as an organizing principle in agroecosystem management: Case studies of holistic resource management practioners in the USA. Agriculture, Ecosystems and Environment 62: 199-213.

Stolwijk, H. (1996). Het subjectieve karakter van duurzaamheid. Economisch Statistische Berichten, 12-6-1996: 534-536.

Stoorvogel, J.J., R.A. Schipper & D.M. jansen (1995). USTED: a methodoly for a quantative analysis of land use scenarios. Netherlands Journal of Agricultural Science 43: 5-18.

Straaten, D. van [ed.] (1993). Methodology, evaluation and scope of Environmental Impact Assessment. First Report: Evaluation of the EIA-Process. Report No. 197. North Atlantic Treaty Organization, Brussel.

Susskind, L. & J. Cruikshank (1987). Breaking the Impasse. Consensual Approaches to Resolving Public Disputes. Basic Books, New York.

Sutherland, W.J. (1996). From Individual Behaviour to Population Ecology. Oxford University Press, Oxford.

Szakál, F. & Cs.B. Illés (1993). Problems and prospects of redirecting Hungarian agriculture towards sustainable development. Landscape and Urban Planning 27: 223-229.

Tamminga, G.F, & A.D. Verhoog. (1988). Dutch Regionalized Agricultural Model, (DRAM). Interne notities over de ontwikkeling van het DRAM. Landbouw-Economisch Instituut, The Hague.

TDN (1988). Topografische Kaart van Nederland 1:25000. Topografische Dienst Nederland, Emmen.

Tikunov, V.S. & A.M. Trofimov (1995). Complex ecological and economic systems: Problems of study. Intern. J. Environmental Studies 48: 135-147.

Titulaer, G.H.J. (1983). Boer Blijven, Plaats en toekomst van de Nederlandse landbouw. Nederlands Agrarisch Jongeren Kontakt, 's-Gravenhage.

Tjallingii, S.P. (1995). Ecopolis. Strategies for Ecologically Sound Urban Development. Backhuys Publishers, Leiden.

Tjallingii, S.P. (1996). Ecological Conditions, Strategies and Structures in Environmental Planning. Ph.D. thesis, Technische Universiteit Delft.

UNEP (1992). Convention on biological diversity. United Nations Environmental Programme, Nairobi.

Underwood, A.J. (1995). Ecological research and (and research into) environemtnal management. Ecological Applications 5 (1): 232-247.

Upreti, G. (1994). Environmental conservation and sustainable development require a new development approach. Environmental Conservation 21 (1): 18-29.

Vandermeer, J., M. van Noordwijk, J. Abderson, C. Ong & I. Perfecto (1998). Global change and multi-species agroecosystems: Concepts and issues. Agriculture, Ecosystems & Environment 67: 1-22.

Vatn, A. & D.W. Bromley (1994). Choices without prices without apologies. Journal of Environmental Economics and Management 26: 129-148.

Veen, P.J. & F.J. Jorna (1995). Winterswijk-Oost. Een verkenning van de mogelijkheden voor natuurontwikkeling en voor aangepast agrarisch beheer in een kleinschalig cultuurlandschap. Concept. Vereniging Natuurmonumenten, 's-Graveland.

Veeneklaas, F.R., H. van Keulen, S. Cissé, P. Gosseye & N. van Duivenbooden (1995). Competing for limited resources: Options for land use in the fifth region of Mali. In: Fresco, L.O., L. Stroosnijder, J. Bouma & H. van Keulen [eds] (1995). The Future of the Land. Mobilising and Integrating Knowledge for Land Use Options. John Willey & Sons

Ltd., Chichester.
Veeneklaas, F.R. & F. Langers (1997). Economische effecten veranderend landschap: Verblijfsrecreatie. Interne notitie DLO-Staring Centrum, Wageningen.
Veer, J. de (1983). Onderzoeksvoorstel "Scenario's produktiebeheering". Landbouw-Economisch Instituut, The Hague.
Verbruggen, H. & R. Dellink (1996). Duurzaam gebruik en duurzame ontwikkeling. Economisch Statistische Berichten, 12-6-1996: 532-533.
Vereijken, P. (1990). Proefbedrijf OBS. In: Vereijken, P. & F.G. Wijnands [red.] (1990). Geïntegreerde akkerbouw naar de praktijk. Strategie voor bedrijf en milieu. Publicatie nr.50, Proefstation voor de Akkerbouw en de Groenteteelt in de Vollegrond, Lelystad.
Vereijken, P. [red.] (1997a). Programmeringstudie Multifunctionele Landbouw. 1.Innovatieve ideeën en expertise binnen DLO. DLO-NL, Wageningen
Vereijken, P. (1997b). A methodical way of prototyping integrated and ecological arable farming systems (I/EAFS) in interaction with pilot farms. European Journal of Agronomy 7: 235-250.
Vereijken, P., C. Edwards, A. El Titi, A. Fougeroux & M. Way (1985). Study Group Management of Arable Farming Systems for Integrated Crop Protection. Bulletin IOBC/WPRS, International Organization for Biological Control of Noxious Animals and Plants, West Palaeartic Regional Section, ISBN 92-9067-001-0.
Vereijken, P.H. & P. de Jonge (1980-1983). Proefbedrijf Ontwikkeling Bedrijfs-Systemen: Verslagen over 1980, 1981 en 1983. Proefstation voor de Akkerbouw en de Groenteteelt in de Vollegrond, Lelystad.
Vos, P., & H.J. de Graaf [red.] (1997). Illustratieproces Duurzaam Landgebruik; Geïntegreerd Informatiesysteem voor de evaluatie en bijsturing van duurzaam landgebruik, basisontwerp van een monitoringsysteem voor het voorbeeldgebied Winterswijk. DTO-werkdocument VD-3. Programma DTO, Delft.
Vos, P., E. Meelis & W.J. ter Keurs, in prep. A framework for the design of ecological monitoring systems for nature conservation and environmental management. Environmental Biology, Leiden University.
Vos, P., A.B.M. Orleans, M.P.J.M. Janssen, E. Meelis & W.J. ter Keurs (1990). Natuur- en milieumeetneten voor het beleid. Hoofdrapport: Het ontwerpen van meetnetten. Milieubiologie & Instituut voor Theoretische Biologie, R.U. Leiden.
Vreke, J. (1997). Opzet economische evaluatie project 'Duurzaam Landgebruik'. Interne notitie DLO-Staring Centrum, Wageningen.
Vries, H.J.M. de (1989). Sustainable Resource Use. An Enquiry into Modelling and Planning. Thesis, Rijksuniversiteit Groningen.
Vries, H.J.M. de (1994). Environmental utilisation space in model worlds. Milieu 9 (5): 236-243.
VROM (1989). Kiezen of verliezen. Nationaal Milieubeleidsplan. Ministerie van Volkshuisvesting, Ruimtelijke Ordening en Milieubeheer, SDU uitgeverij, The Hague.
VROM (1998). Uitkomsten klimaatconferentie Kyoto en een eerste indicatie van de betekenis daarvan voor Nederland. Interne notitie Ministerie van Volkshuisvesting, Ruimtelijke Ordening en Milieubeheer, The Hague.
Wal, H. van der, H.J. de Graaf, N.A. van Brussel, W.J. van der Weijden & W.J. ter Keurs m.m.v. Th. Bakker (1985). Speelruimte voor een geïntegreerde landbouw: Verkenningen met behulp van een model. Werkdocumenten W8, Wetenschappelijke Raad voor het Regeringsbeleid, The Hague.
Wallace, R.R. & B.G. Norton (1992). Policy implications of Gaian theory. Ecological Economics

6: 103-118

Warner, M. (1996). Integration of Regional Land Use Planning with Environmental Impact Assessment: A practical land suitability assessment approach. Impact Assessment 1996: 155-189.

Warren, K.J. & J. Cheney (1993). Ecosystem ecology and metaphysical ecology: A case study. Environmental Ethics 15: 99-116.

WCED (1987). Our Common Future. World Commission on Environment and Development, Oxford University Press, Oxford.

WCL Winterswijk (1996). Gebiedsperspectief voor het waardevolle cultuurlandschap. Provincie Gelderland, Arnhem.

Weijden, W.J. van der, H. van der Wal, H.J. de Graaf, N.A. van Brussel & W.J. ter Keurs, m.m.v. Th. Bakker (1984a). Bouwstenen voor een geïntegreerde landbouw. Voorstudies en achtergronden V44, Wetenschappelijke Raad voor het Regeringsbeleid, Staatsuitgeverij, The Hague.

Weijden, W.J. van der, H. van der Wal, H.J. de Graaf, N.A. van Brussel, W.J. ter Keurs & Th. Bakker (1984b). Towards an Integrated Agriculture. Summary of the Preliminary Study V44, Netherlands Scientific Council for Government Policy, Werkdocument W4, The Hague.

Weijden, W.J. van der, H. van der Wal, H.J. de Graaf, N.A. van Brussel & W.J. ter Keurs, m.m.v. Th. Bakker (1985a). Bouwstenen voor een geïntegreerde landbouw. Bedrijfsontwikkeling 16 (3): 76-82.

Weijden, W.J. van der, H. van der Wal, H.J. de Graaf, N.A. van Brussel & W.J. ter Keurs (1985b). Geïntegreerde landbouw: Geen blauwdruk maar uitdaging. Bedrijfsontwikkeling 16 (3): 103-108.

Weijden, W.J. van der, H. van der Wal, H.J. de Graaf, N.A. van Brussel en W.J. ter Keurs (1985c). Geïntegreerde landbouw: Naast beperkingen ook verruimingen. Milieudefensie 1 (6).

Weinreich, J.A. & C.J.M. Musters (1989). Toestand van de natuur. Veranderingen in de Nederlandse natuur. SDU Uitgeverij, The Hague.

Weizsäcker, E.U. von & F. Schmidt-Beek (1994). Signs of hope for the 21st century? In: Dutch Committee for Long-Term Environmental Policy [ed.] (1994). The Environment: Towards a Sustainable Future. Environment and Policy 1, Kluwer Academic Publishers, Dordrecht.

Werner, R. (1993). Ecological and economic efficient and sustainable use of agricultural landscapes. Landscape and Urban Planning 27: 237-248.

Westra, L. (1997). Why Norton's approach is insufficient for environmental ethics. Environmental Ethics 19: 279-297.

Weterings, R. & J.B. Opschoor (1994). Environmental utilisation space and reference values for performance evaluation. Milieu 9 (5): 221-228.

Weterings, R.A.P.M. & J.B. Opschoor (1992). De milieugebruiksruimte als uitdaging voor technologie-ontwikkeling. Publicatie RMNO 74, Raad voor het Milieu- en Natuuronderzoek, Rijswijk.

Wood, C. (1995). Environmental Impact Assessment. A comparative review. Longman Scientific & Technical, Harlow.

Wossink, G.A.A. (1993). Analysis of future agricultural change. A farm economics approach applied to Dutch arable farming. Agricultural University Wageningen.

WRI (1994). World Resources 1994-95. World Resources Institute, Oxford University Press, Oxford.

WRI (1996). World Resources, 1996-97. World Resources Institute, Oxford University Press, New York.

WRR (1980). Beleidsgerichte toekomstverkenning. Deel 1: Een poging tot uitlokking. Rapporten aan de regering 19, Staatsuitgeverij, The Hague.

WRR (1983) Beleidsgerichte toekomstverkenning . Deel 2 Een verruiming van perspectief. Rapporten an de regering nr. 25, Wetenschappelijk Raad voor het Regeringsbeleid, Staatsuitgeverij, The Hague.

WRR (1992a). Environmental Policy: Strategy, Instruments and Enforcement. Summary of the 41 th Report, Reports to the Government, Netherlands Scientific Council for Government Policy, The Hague.

WRR (1992b). Grond voor keuzen. Vier perspectieven voor de landelijke gebieden in de Europese gemeenschap. Rapporten aan de Regering 42, Wetenschappelijke Raad voor het Regeringsbeleid, The Hague.

WRR (1994). Besluiten over grote projecten. Rapporten aan de Regering 46. Wetenschappelijk Raad voor het Regeringsbeleid, Sdu Uitgeverij, The Hague.

WRR (1995). Sustained risks: A lasting phenomemon. Report to the Government 44, Netherlands Scientific Council for Government Policy, the Hague.

Yunlong, C. & B. Smit (1994). Sustainability in agriculture. A general review. Agriculture, Ecosystems and Environment 49: 299-307.

Zandvoort Kappelhoff Advies (1994). Ruimte voor Economische Activiteit: Toerisme en Recreatie; een onderzoek naar de ruimtelijke problematiek van de toeristisch-recreatieve sector in Nederland. Zandvoort Kappelhoff Advies.

Zubrow, E.B.W., J.R. Schumm, S. Finn, G.A. Panetski & J. Van Ness (1995). The biological reserve: The future's last stand. Futures 27 (4): 437-446.

Zwart, K.B. (1997). Illustratieproces Duurzaam Landgebruik; Onderzoekprogramma multifunctioneel landgebruik, oplossen van knelpunten in voorbeeldsystemen duurzaam landgebruik. DTO-werkdocument VD-6. Programma DTO, Delft.